Sensing and Imaging for Defect Detection

Sensing and Imaging for Defect Detection

Guest Editors

Haitao Wang
Yongkai Zhu
Fei Fei

MDPI

Basel • Beijing • Wuhan • Barcelona • Belgrade • Novi Sad • Cluj • Manchester

Guest Editors

Haitao Wang
College of Automation Engineering
Nanjing University of Aeronautics and Astronautics
Nanjing
China

Yongkai Zhu
College of Automation Engineering
Nanjing University of Aeronautics and Astronautics
Nanjing
China

Fei Fei
College of Automation Engineering
Nanjing University of Aeronautics and Astronautics
Nanjing
China

Editorial Office

MDPI AG
Grosspeteranlage 5
4052 Basel, Switzerland

This is a reprint of the Special Issue, published open access by the journal *Sensors* (ISSN 1424-8220), freely accessible at: https://www.mdpi.com/journal/sensors/special_issues/V9X2BNWIW6.

For citation purposes, cite each article independently as indicated on the article page online and as indicated below:

Lastname, A.A.; Lastname, B.B. Article Title. *Journal Name* **Year**, *Volume Number*, Page Range.

ISBN 978-3-7258-6788-2 (Hbk)
ISBN 978-3-7258-6789-9 (PDF)
https://doi.org/10.3390/books978-3-7258-6789-9

Contents

About the Editors

Haitao Wang

Haitao Wang is a Prof. at the School of Automation, Nanjing University of Aeronautics and Astronautics. For a long time, he has been engaged in the research of non-destructive testing technologies such as sensor technology, electromagnetic, ultrasonic, laser-ultrasonic, phased array and image processing of digital ray defects. He is Director of the Metrology Society of China, Deputy Director of the Key Laboratory of Non-destructive Testing of the Ministry of Industry and Information Technology, Standing Director of Jiangsu Instrument and Meter Society, Standing Director of Jiangsu Graphics and Image Society, Director of Non-destructive Testing Instrument Professional Committee, Chair of the Air-Coupled Ultrasonic and Laser Ultrasonic Special Topic at the 19th World Congress on Nondestructive Testing (WCNDT2016). He served as an editorial board member of the international journal *Structural Durability & Health Monitoring*, as well as the domestic core journals *Computer Measurement & Control* and *Non-destructive Testing*. He also served as a Guest Editor of the international journal *Sensors*.

Yongkai Zhu

Yongkai Zhu is an Associate Professor in the College of Automation Engineering at Nanjing University of Aeronautics and Astronautics. His research interests are highly focused on advanced sensing technologies of aerospace, including detection solutions for aeroengine and detection techniques for aircraft assembly and sensing technologies for aircraft control systems. Furthermore, he is deeply engaged in structural health monitoring (SHM) for composite materials, the design of optical imaging systems, and the fabrication of micro-nano optical devices.

Fei Fei

Fei Fei is an Assistant Professor with the College of Automation Engineering, Nanjing University of Aeronautics and Astronautics. His research interests include smart sensors and energy harvesting devices.

sensors

MDPI

Communication

Wheel Defect Detection Using a Hybrid Deep Learning Approach

Khurram Shaikh [1], Imtiaz Hussain [2,*] and Bhawani Shankar Chowdhry [3]

[1] Department of Electronic Engineering, Mehran University of Engineering and Technology, Jamshoro 76062, Pakistan; khuram.shaikh@admin.muet.edu.pk
[2] Department of Electrical Engineering, National University of Sciences and Technology, Islamabad 44000, Pakistan
[3] NCRA-Condition Monitoring Lab, Mehran University of Engineering and Technology, Jamshoro 76062, Pakistan; bhawani.chowdhry@faculty.muet.edu.pk
* Correspondence: imtiaz.hussain@pnec.nust.edu.pk

Abstract: Defective wheels pose a significant challenge in railway transportation, impacting operational performance and safety. Excessive traction and braking forces give rise to deviations from the intended conical tread shape, resulting in amplified vibrations and noise. Moreover, these deviations contribute to the accelerated damage of track components. Detecting wheel defects at an early stage is crucial to ensure safe and comfortable operation, as well as to minimize maintenance costs. However, the presence of various vibrations, such as those induced by the track, traction motors, and other rolling stock subsystems, poses a significant challenge for onboard detection techniques. These vibrations create difficulties in accurately identifying wheel defects in real-time during operational activities, often resulting in false alarms. This research paper aims to address this issue by using a hybrid deep learning-based approach for the accurate detection of various types of wheel defects using accelerometer data. The proposed approach aims to enhance wheel defect detection accuracy while considering onboard techniques' cost-effectiveness and efficiency. A realistic simulation model of the railway wheelset is developed to generate a comprehensive dataset. To generate vibration data in various scenarios, the model is simulated for 20 s under different conditions, including one non-faulty scenario and six faulty scenarios. The simulations are conducted at different speeds and track conditions to capture a wide range of operating conditions. Within each simulation iteration, a total of 200,000 data points are generated, providing a comprehensive dataset for analysis and evaluation. The generated data are then utilized to train and evaluate a hybrid deep learning model, employing a multi-layer perceptron (MLP) as a feature extractor and multiple machine learning models (support vector machine, random forest, decision tree, and k-nearest neighbors) for performance comparison. The results demonstrate that the MLP-RF (multi-layer perceptron with random forest) model achieved an accuracy of 99%, while the MLP-DT (multi-layer perceptron with decision tree) model achieved an accuracy of 98%. These high accuracy values indicate the effectiveness of the models in accurately classifying and predicting the outcomes. The contributions of this research work include the development of a realistic simulation model, the evaluation of sensor layout effectiveness, and the application of deep learning techniques for improved wheel flat detections.

Keywords: wheel defects; deep learning; wheel flats; false flange; nonlinear dynamics; MLP

1. Introduction

The railway wheelset plays a vital role in railway transportation, ensuring the safe and comfortable operation of the vehicles. The conical shape of the wheel tread is designed to maintain optimal performance. However, excessive traction and braking forces at the wheel–rail interface can cause the wheel's exterior perimeter to deviate from its desired conical shape. This either results in a flat or more conical surface of the wheel tread.

This deviation leads to increased vibrations, noise, and a decrease in ride comfort and operational safety. Moreover, wheel defects contribute to the accelerated growth of cracks on the rail tracks, ultimately leading to premature failure of the rail system [1]. Statistical analysis of mechanical components of trains between 2004 and 2007 revealed that wheelset faults accounted for 44.7% of train accidents, making them the most significant cause [2]. Therefore, early detection of defective wheels is crucial. To achieve this, an automated approach capable of accurately distinguishing between healthy and damaged wheels needs to be developed. Consequently, there is a great interest among railway administrations and rolling stock operators in finding effective methods for early detection and identification of wheel flats [3].

Rolling stock inspection is normally performed at fixed intervals and carried out periodically in workshops. Periodic inspection is expensive due to the unavailability of rolling stock during maintenance time. Moreover, it is also inefficient, which usually results in over- or under-maintenance of the components [4]. In the last few decades, various techniques have been employed in the railway industry, and various monitoring approaches have been proposed to automatically inspect wheel conditions. Most of these techniques are based on the concept that the wheel–rail interaction forces increase in a defective wheel [5,6]. The most popular techniques which have gained plenty of attention from railway researchers are the onboard techniques and wayside measurement techniques [7].

The wayside inspection involves the installation of various sensors along the railway track at specific locations, allowing for the evaluation of all passing wheels [8,9]. These techniques utilize vibration, acoustic, image detection, and ultrasonic technologies to monitor the condition of the trains [10–12]. Hot axle box detectors (HABDs) are commonly employed in wayside monitoring systems to identify faulty overheating axle bearings by using thermal imaging. However, HABDs are unable to detect damage at early stages, as minor faults do not typically cause a noticeable increase in temperature. Additionally, HABDs are prone to false alarms triggered by environmental conditions [13]. Trackside acoustic array detectors (TAADS) utilize arrays of microphones to capture the audible noise produced by passing axle bearings. However, the operating frequency range of these microphones (normally 22–44 kHz) makes them susceptible to errors resulting from background noise [14]. To enhance the accuracy of detection, many modern wayside monitoring systems incorporate machine learning and deep learning algorithms [5,15]. For example, in [5], a multi-sensor data fusion approach combined with an unsupervised early damage detection methodology has been proposed, capable of automatically distinguishing between defective and healthy wheels, particularly considering small flat sizes. This methodology relies on the evaluation of acceleration and shear time histories recorded on the rails during the passage of traffic loads. In [15], an improved YOLOv3 framework is developed for rail wheel surface defect detection, achieving classification detection of four types of wheel tread defects with an average mean average precision (MAP) accuracy of 0.92. Various types of sensors, including strain gauges and accelerometers, are utilized to capture input signals for detecting wheel flats.

However, wayside monitoring systems are often costly due to the requirement for multiple sensors and high-end computing capabilities for comprehensive diagnosis and effective wheel condition monitoring. For instance, in the Swedish railway network, the wayside equipment for monitoring rolling stock consists of almost 200 wayside inspection devices [16]. Moreover, this method also requires detailed information about a target vehicle (e.g., number of axles and type of wagon) for accurate condition monitoring [17]. The high cost and maintenance issues associated with this method limit its widespread use. Additionally, the deployment and security of wayside equipment are important considerations that need to be addressed.

Due to the challenges associated with wayside monitoring techniques, several onboard techniques have been proposed in the literature [18–23]. One such technique, presented in [10], detects wheel flat defects by measuring the vertical acceleration on the axle box. This algorithm operates in the time domain and can identify the presence of wheel flats at an early stage and estimate their severity. The results demonstrate that the proposed wheel flat index is effective for detecting small flats and estimating their severity. In [18], multiple models and a fuzzy logic-based technique are introduced for detecting conicity in railway wheelsets. This method indirectly identifies the conicity condition by analyzing the lateral acceleration of the wheelset. Another onboard detection method proposed in [7], known as the angle domain synchronous averaging technique (ADSAT), utilizes vertical axle-box vibration acceleration (ABVA) to monitor the conditions of axle-box bearings. The results indicate that this method not only achieves better detection than traditional methods but also mitigates the influence of background noise. In [19], a lightweight 1D convolutional neural network (CNN) architecture, guided by Bayesian optimization, is presented for wheel flat (WF) detection using car body accelerations. Additionally, model-based onboard techniques are proposed in [20–23] to detect wheelset conditions under different environmental conditions, utilizing axle vibration and gyroscopic data for onboard diagnosis.

The advantage of on-board monitoring systems is that the wheel is monitored continuously and not only when the vehicle passes a trackside monitoring site. This allows for the timely detection of emerging wheel defects [10], thus, allowing immediate action to perform maintenance after the formation of a wheel flat, without requiring a visit to a trackside monitoring site [10,24]. Furthermore, if the onboard monitoring system provides positioning, the occurrence of a wheel defect can be linked to a position on the track. The track at this position can then be inspected and, in cases where a track defect is identified, appropriate maintenance actions can be issued to avoid further damage to the rail and other passing vehicles [24]. The early detection of wheel flats can also be correlated with braking occurrences, if the braking system is monitored, to develop strategies for reducing their frequency [25]. However, onboard detection methods require equipping all wheels with sensors for comprehensive diagnosis and effective wheel condition monitoring, which can be costly and pose maintenance challenges.

Onboard detection techniques face a substantial challenge due to the presence of diverse vibrations originating from multiple sources within the railway system. These vibrations, generated by the track, traction motors, and various rolling stock subsystems, introduce significant complexity and pose obstacles to the accurate identification of wheel defects in real-time during operational activities [7,26]. The intricate nature of these vibrations makes it difficult to distinguish between genuine wheel defects and false alarms, leading to potential disruptions and unnecessary maintenance interventions. Addressing the challenge of vibration interference requires innovative approaches that can effectively separate the signals related to wheel defects from the surrounding vibrations. By employing machine learning algorithms, it becomes possible to enhance the accuracy of onboard detection systems and mitigate the impact of background vibrations. These techniques aim to extract relevant features and patterns associated with wheel defects, enabling reliable and real-time identification even in the presence of complex vibration environments [27].

This paper presents a novel technique to address the challenges faced by onboard techniques. The proposed technique utilizes axle vibration data exclusively, thereby reducing the reliance on multiple sensors. Moreover, the proposed technique incorporates a hybrid deep learning approach, which plays a crucial role in mitigating the influence of background vibrations. The utilization of hybrid deep learning algorithms allows for enhanced detection performance by effectively distinguishing between genuine wheel defects and the vibrations induced by the track, traction motors, and other rolling stock subsystems. By harnessing the power of deep learning, the proposed technique aims to overcome the limitations and false alarms commonly encountered in onboard detection systems.

A significant novelty of this research is the automatic detection of defective wheels at a very early stage, representing a substantial improvement in the effectiveness of the proposed method and facilitating its full implementation in real-world applications. The research work's key contributions are as follows:

(a) Development of a realistic simulation model of the railway wheelset to generate a comprehensive dataset.
(b) Evaluation of the effectiveness of the proposed method considering a minimalist layout of sensors.
(c) Enhancement of wheel flat detection through the application of hybrid deep learning technique.

2. Methodology

The methodology section of this research paper outlines the approach used to develop a deep learning model for the detection of defects on railway wheelsets using vibration measurements. This section describes the data collection process, the deep learning architecture employed, the training procedure, and the model evaluation methodology. By leveraging the inherent capabilities of deep learning, the proposed methodology aims to automatically learn relevant features directly from the raw vibration measurements, eliminating the need for explicit feature extraction. The following subsections provide a detailed overview of the methodology, highlighting the steps taken to train and evaluate the deep learning model accurately.

2.1. Development of a Realistic Simulation Model

A nonlinear wheelset simulation model presented in [21–23,28] is used to develop the simulation model in Simulink to mimic the behavior of the actual wheelset dynamics. The model is modified to consider all the disturbances that are faced by an actual railway vehicle (e.g., irregularities in track in the lateral direction, variation in gauge and track geometry) to generate data that are close to real-time scenarios. Only the lateral and yaw dynamics of the system are considered, which are most affected by the wheel defects. The model is described in Equations (1)–(5).

$$\ddot{y} = \frac{2}{m_w}\left(\Psi\frac{F_a}{\gamma} - \frac{\dot{y}}{V_X}\frac{F_a}{\gamma}\right) \tag{1}$$

$$\ddot{\Psi} = \frac{1}{I_w}\left(\frac{2y_g y_t L_g}{r_0}\frac{F_a}{\gamma} - \frac{2yL_g\lambda_w}{r_0}\frac{F_a}{\gamma} - \frac{2\dot{\Psi}{L_g}^2}{V_X}\frac{F_a}{\gamma} - K_w\Psi\right) \tag{2}$$

$$\gamma = \sqrt{\left(\frac{L_g\dot{\Psi}}{V_X} + \frac{\lambda_w(y - y_t - y_g)}{r_0}\right)^2 + \left(\frac{\dot{y}}{V_X} - \Psi\right)^2} \tag{3}$$

$$\mu = u_0\left[(1-A)e^{(-B\gamma V_X)} + A\right] \tag{4}$$

$$F_a = \frac{2N\mu}{\pi}\left[\frac{k_A\varepsilon}{1+(k_A\varepsilon)^2} + \arctan(k_S\varepsilon)\right] \tag{5}$$

The simulations are in different conditions to develop a comprehensive dataset. The following scenarios are considered in the development of the dataset.

Variation in speed (V_X): In previous similar studies [18–24] speed was kept constant to detect the anomalies. However, in real-time scenarios, speed is not constant. Therefore, in this study, data are generated at variable speeds (25 km/h, 50 km/h, 75 km/h, 100 km/h, 120 km/h, and 150 km/h).

Variation in track conditions: Track condition also changes with time and location. Therefore, the data are generated by varying the track parameters (μ, k_A, ε, and k_S).

Variation in wheelset condition: Wheel profile plays an important role in the safe and reliable operation of railway vehicles. Due to wear and tear during the operation wheel profile is changed. The variation in the wheel profile is incorporated during the simulation by varying wheel conditions (λ_w).

Track disturbances: The creep forces generated at the wheel–rail interface are directly affected by the track irregularities, such as lateral variation and variation in gauge. Therefore, these disturbances are also considered in this study.

The data collection process is depicted in Figure 1. A simulation model of a nonlinear railway wheelset is developed in MATLAB/Simulink. Varying forward speed, varying wheel conditions, and varying track conditions are given as input to the model. Practically, a 3-axis accelerometer is placed on the left side of the axle (opposite side of the traction motor) to measure axle vibration. A preprocessing unit is used to remove any bias in the vibration data and to extract vibrations in lateral direction. Simulations are run for 20 s in each of the scenarios, including one non-faulty scenario and six faulty scenarios, to develop the dataset. Some of the results are given in Figure 2. It is quite evident from Figure 2 that the frequency and the amplitude of the lateral acceleration are directly affected by the changes in the speed as well as by the wheel profile variation. An accelerometer is placed on the axle of the wheelset to capture lateral vibrations during the simulation process. A total of 200,000 data points are generated in each scenario which is archived in CSV format for later use.

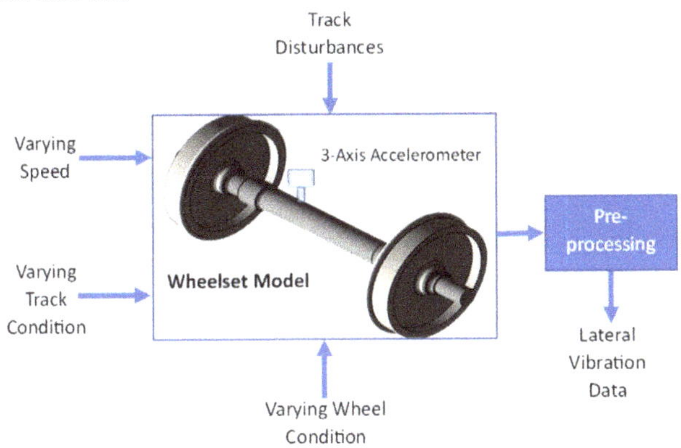

Figure 1. Data collection process.

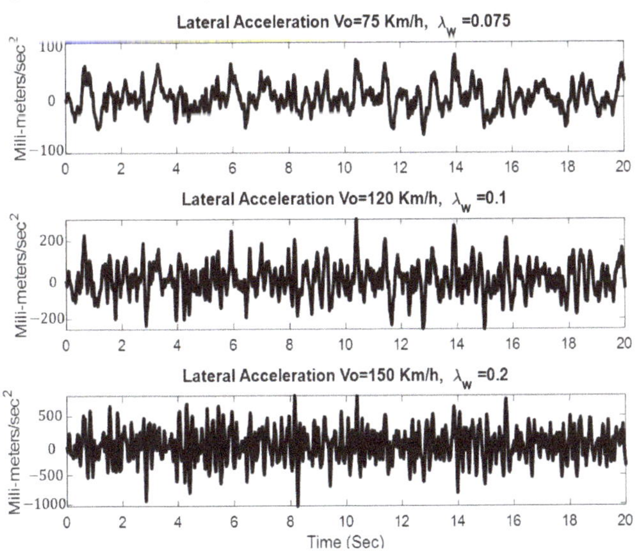

Figure 2. Lateral acceleration of the wheelset in different conditions.

2.2. Hybrid Detection Architecture

The overall block diagram for defect detection is shown in Figure 3. The defect detection architecture proposed in this research paper combines an MLP as a feature extractor with a machine learning model for the final classification of conicity values. The input features used in this architecture are derived from vibration data only. This is one of the main contributions of this work because in the presence of disturbances and variable parameters it is difficult to detect wheel defects using vibration data only.

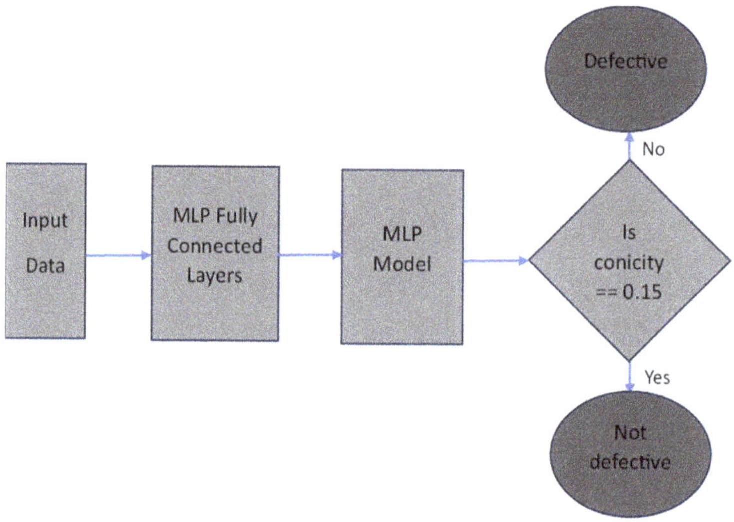

Figure 3. Model architecture.

To ensure effective feature extraction, the MLP undergoes a pre-training phase, as shown in Figure 4a. During pre-training, the model is initialized with appropriate weights and biases, enabling it to extract meaningful patterns from the input data. This initialization process sets a foundation for the MLP to learn and capture relevant features that are crucial for accurate defect detection in railway wheelsets.

2.3. MLP—Fully Connected Network

The model presented in this research paper is designed for defect detection in railway wheelsets using vibration data. The architecture, shown in Figure 5, consists of an MLP as a feature extractor and a subsequent machine-learning model for the final classification of conicity values. The feature extraction process begins with the MLP, which includes three parallel dense networks. Each network has multiple dense layers with decreasing numbers of units, incorporating the GELU activation function. Regularization is applied to the first dense layer of the first parallel network using a kernel regularizer.

During training, the model is initially pre-trained to initialize the weights and biases. This pre-training step helps the MLP extract meaningful patterns from the vibration data, facilitating effective feature extraction. Once the MLP is pre-trained, the last classification layer is removed. Instead, the layer where the outputs of all three branches are concatenated is used to extract the features learned by the MLP. This concatenated layer combines the features extracted from each parallel network with the input layer, providing a comprehensive representation of the input data. A training summary of the MLP is given in Table 1.

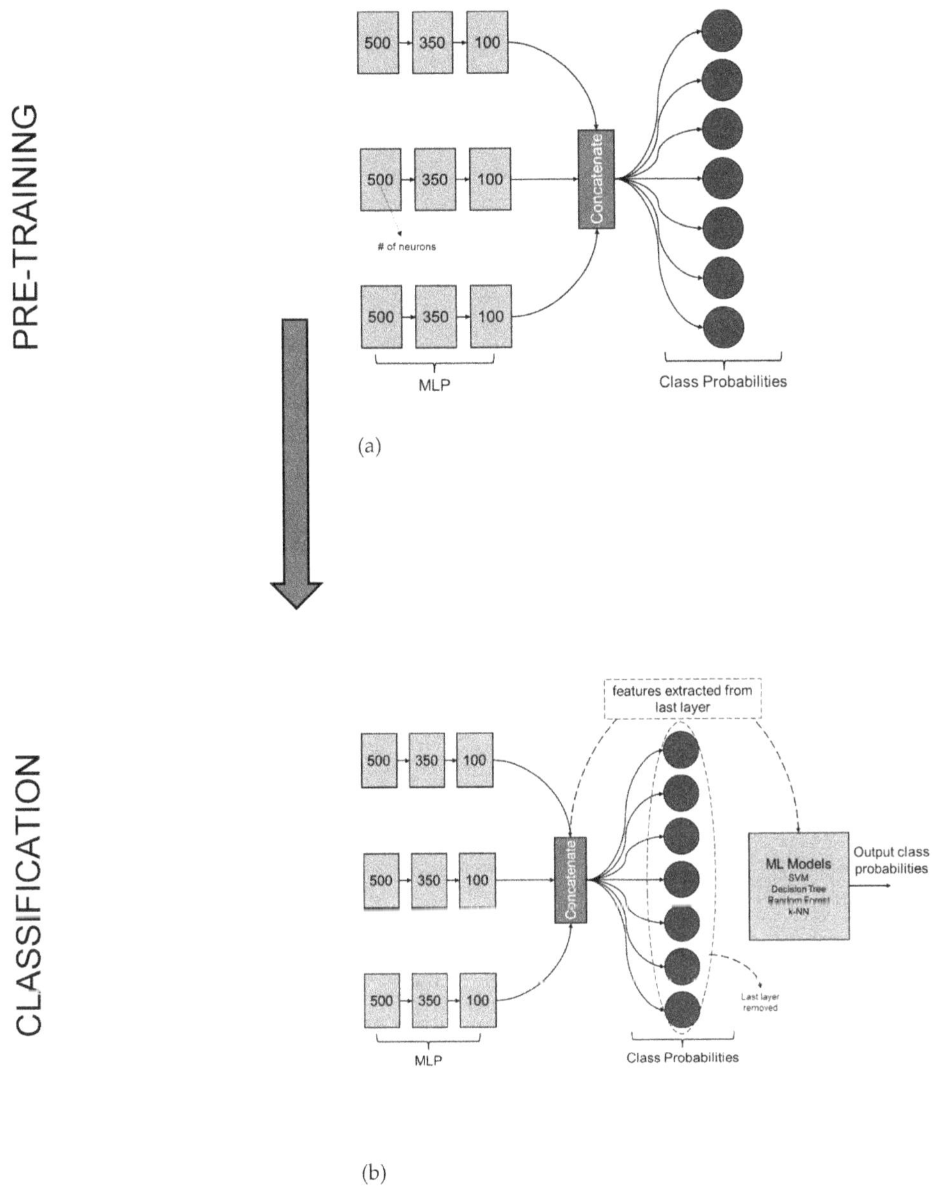

Figure 4. (**a**) Pre-training process. (**b**) MLP Network pruned with ML-based classifier. Once the MLP has been pre-trained and equipped with the ability to extract relevant features, these extracted features serve as inputs to the subsequent machine learning model, as shown in (**b**). The machine learning model leverages the extracted features to perform the final classification of conicity values. By utilizing the comprehensive and representative features obtained from the MLP, the machine learning model can make accurate predictions regarding the severity of defects in railway wheelsets.

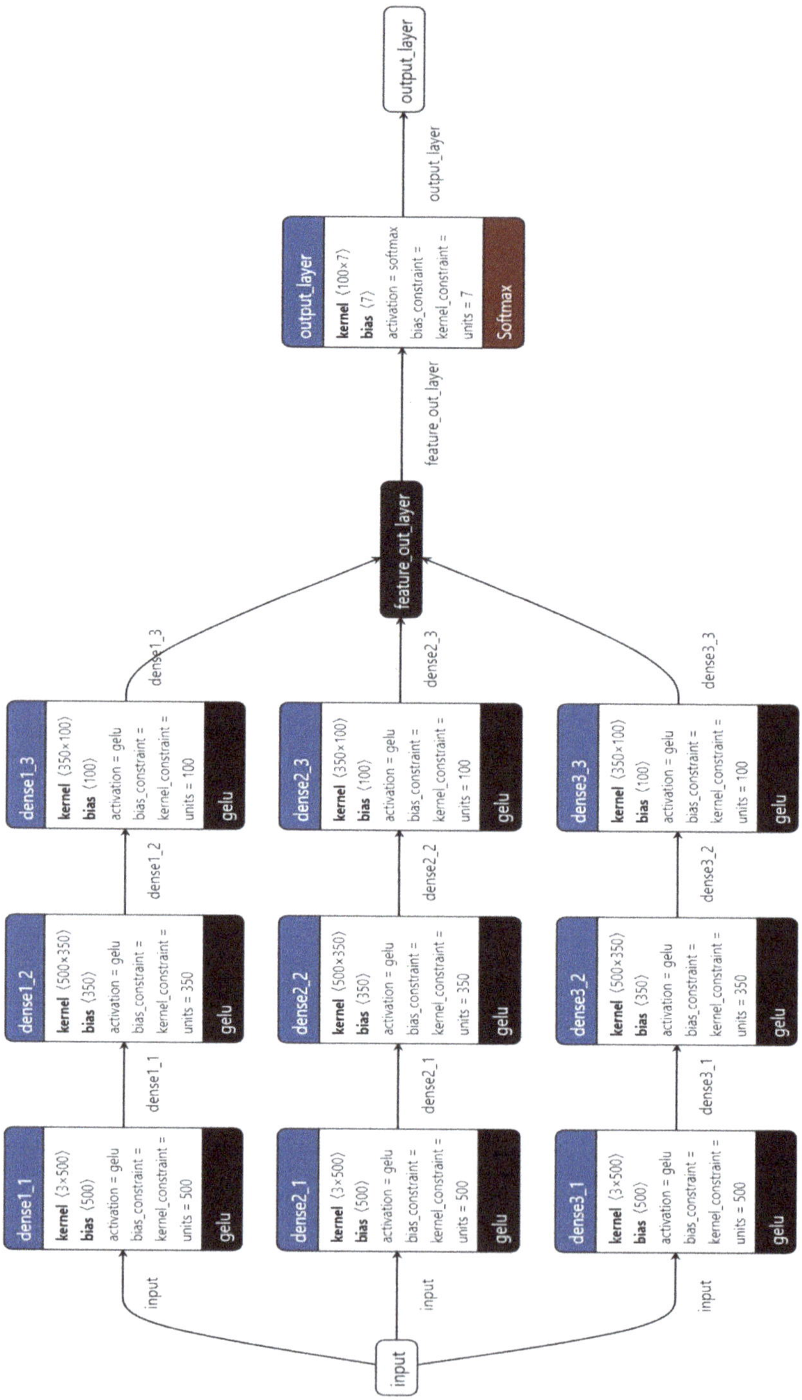

Figure 5. MLP networks' architecture with three parallel branches and concatenation layer.

Table 1. MLP training summary.

Total Parameters	**4471**
Trainable parameters	4471
Non-trainable parameters	0

These extracted features serve as the inputs to the subsequent machine learning model, which performs the final classification of conicity values. The machine learning model utilizes the extracted features to make accurate predictions regarding the severity of defects in railway wheelsets. By removing the last classification layer and extracting features from the concatenated layer, the model focuses on capturing and utilizing the learned representations of the input data. This approach enhances the effectiveness of defect detection by leveraging the discriminative power of the extracted features for classification purposes.

2.4. *Hyperparameters*

The details of the hyperparameter are given in Table 2. The model uses a Gaussian error linear unit (GELU) as an activation function with a categorical cross-entropy loss function and Adam optimizer.

Table 2. Hyperparameters.

Hyperparameters	
Loss function	Categorical cross entropy
Activation function	GELU
Optimization algorithm	Adam
# of training epochs	100
Batch size	100

2.4.1. GELU Activation Function

The GELU activation function outperforms ReLU in the presented architecture due to its smoothness, improved representation learning capabilities, and ability to alleviate the vanishing gradient problem. These advantages enable more stable training, enhanced feature extraction, and improved accuracy in defect detection for railway wheelsets. The GELU activation function is defined as follows in Equation (6):

$$GELU(x) = 0.5x \cdot \left(1 + \tanh\left(\sqrt{\frac{2}{\pi}} \cdot \left(x + 0.044715x^3\right)\right)\right) \tag{6}$$

2.4.2. Categorical Cross Entropy Loss Function

Categorical cross entropy is utilized as the loss function for the final classification model. This choice is motivated by the nature of the defect detection task, where the goal is to classify conicity values into different categories or severity levels. Categorical cross entropy is well-suited for multi-class classification problems, providing a measure of the dissimilarity between the predicted conicity values and the true labels. By minimizing this loss function, the model is encouraged to accurately classify the severity of defects in railway wheelsets.

$$Loss = -\sum_{i=1}^{\substack{output \\ size}} y_i \cdot log\hat{y}_i \tag{7}$$

2.4.3. Adaptive Moment Estimation (ADAM) Optimizer

The ADAM optimizer is a widely used optimization algorithm for training neural networks. It offers several advantages that contribute to its popularity. ADAM utilizes adaptive learning rates, adjusting the learning rate for each parameter individually based on the gradients' history. This adaptive nature allows for faster convergence and efficient parameter updates. The inclusion of momentum helps accelerate optimization by maintaining a running average of past gradients. ADAM also handles sparse gradients effectively, which is common in deep learning models. Additionally, it incorporates L2 regularization, preventing overfitting and improving generalization. Overall, these features make ADAM a versatile and effective optimizer for neural network training.

3. Results

3.1. Training Procedure

The entire training procedure for the proposed algorithm consists of pretraining an MLP for feature extraction and then feeding these feature vectors to ML model training. The steps are shown in Table 3.

Table 3. Training procedure.

Step	Step Name	Details
1	Data preparation	Prepare and arrange the entire dataset with the number of columns as the number of features and the output label as the last column. The number of rows is the number of datapoints.
2	Pre-training phase	An MLP network is trained as a classifier first with inputs as feature vectors and output as conicity labels. The model is trained for 50 epochs using the GELU activation function and ADAM optimizer.
3	Feature extraction	The trained model in Step 2 is then used where the last classification layer is removed, giving us the concatenated layer. The output of this layer is of size 3×100. This is a feature matrix from the pre-trained MLP network.
4	ML model training	Multiple ML models, such as SVM, DT, RF, and kNN, are then trained on the extracted feature matrix from Step 3.

3.2. Model Performance

The results are gathered by following a training procedure where the MLP network is first pre-trained. Then, using the trained model, the last classification layer is removed to extract features. These features are then fed to ML models one by one, and a comparison is drawn. Each model is analyzed for its accuracy, precision, recall, and F1 score. The performance metrics are summarized in Table 4.

Table 4. Performance comparison.

Model	Performance Metrics			
	Accuracy	Precision	Recall	F1 Score
MLP	88.6%	89.6%	88.6%	88.7%
MLP-RF	99.0%	99.2%	99.2%	99.1%
MLP-DT	98.9%	98.9%	98.0%	98.9%
MLP-KNN	95.0%	94.5%	94.5%	94.0%
MLP-SVM	83.1%	85.6%	83.1%	83.5%

RF = random forest, DT = decision tree; KNN = k-nearest neighbor, SVM = support vector machine.

Figure 6 shows the performance comparison of MLP, SVM, RF, DT, and k-N. The MLP-RF shows the highest detection accuracy, precision, and F1 Score.

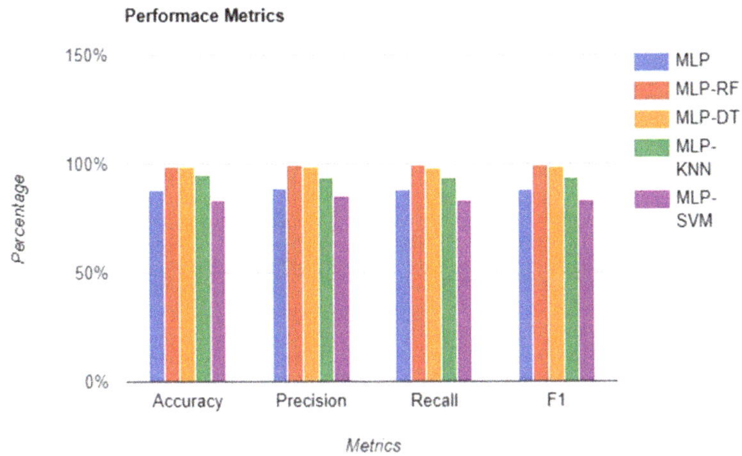

Figure 6. Bar chart of performance metrics of MLP, SVM, RF, DT, and k-N.

3.3. Results Interpretation

The performance results in the previous section show four types of performance metrics, namely accuracy, precision, recall, and F1 score. The results are summarized in Table 5.

Table 5. Summary of results.

Model	Results Interpretation
MLP	This model achieved an accuracy of 88.6%, which means it correctly classified 88.6% of the defects. The precision of 89.6% indicates that when it predicted a defect, it was correct 89.6% of the time. The recall of 88.6% indicates that it identified 88.6% of the actual defects. The F1 score, which considers both precision and recall, is 88.7%.
MLP-RF	This model achieved high accuracy of 99.0%, indicating that it performed exceptionally well in classifying defects. The precision and recall of 99.2% suggest it had a very low rate of false positives and false negatives. The F1 score of 99.1% reflects the overall effectiveness of the model in detecting defects.
MLP-DT	This model achieved an accuracy of 98.9%, indicating strong performance in defect detection. The precision of 98.9% suggests that it had a very low rate of false positives. However, the recall of 98.0% indicates it missed a small portion of actual defects. The F1 score of 98.9% reflects a good balance between precision and recall.
MLP-kNN	This model achieved an accuracy of 95.0%, indicating a relatively high performance in detecting defects. The precision and recall of 94.5% suggest a low rate of false positives and false negatives. The F1 score of 94.0% reflects a good overall performance, though slightly lower than the previous models.
MLP-SVM	This model achieved an accuracy of 83.1%, which indicates it had a moderate level of performance in detecting defects. The precision of 85.6% suggests it had a relatively low rate of false positives. However, the recall of 83.1% indicates it missed a significant portion of actual defects. The F1 score of 83.5% reflects the overall effectiveness of the model, considering both precision and recall.

3.4. Classification Results for Conicity Values

Figure 7 shows the error matrix (or confusion matrix) for predicted and ground truth results. Starting from MLP, the accuracy of valid conicity value (c = 0.15) is 82%, while for MLP-RF it is 99%, for MLP-DT it is 98.9%, for MLP-kNN it is 94.2%, and for MLP-SVM it is 79.3%. The confusion matrix is plotted for the unseen test data.

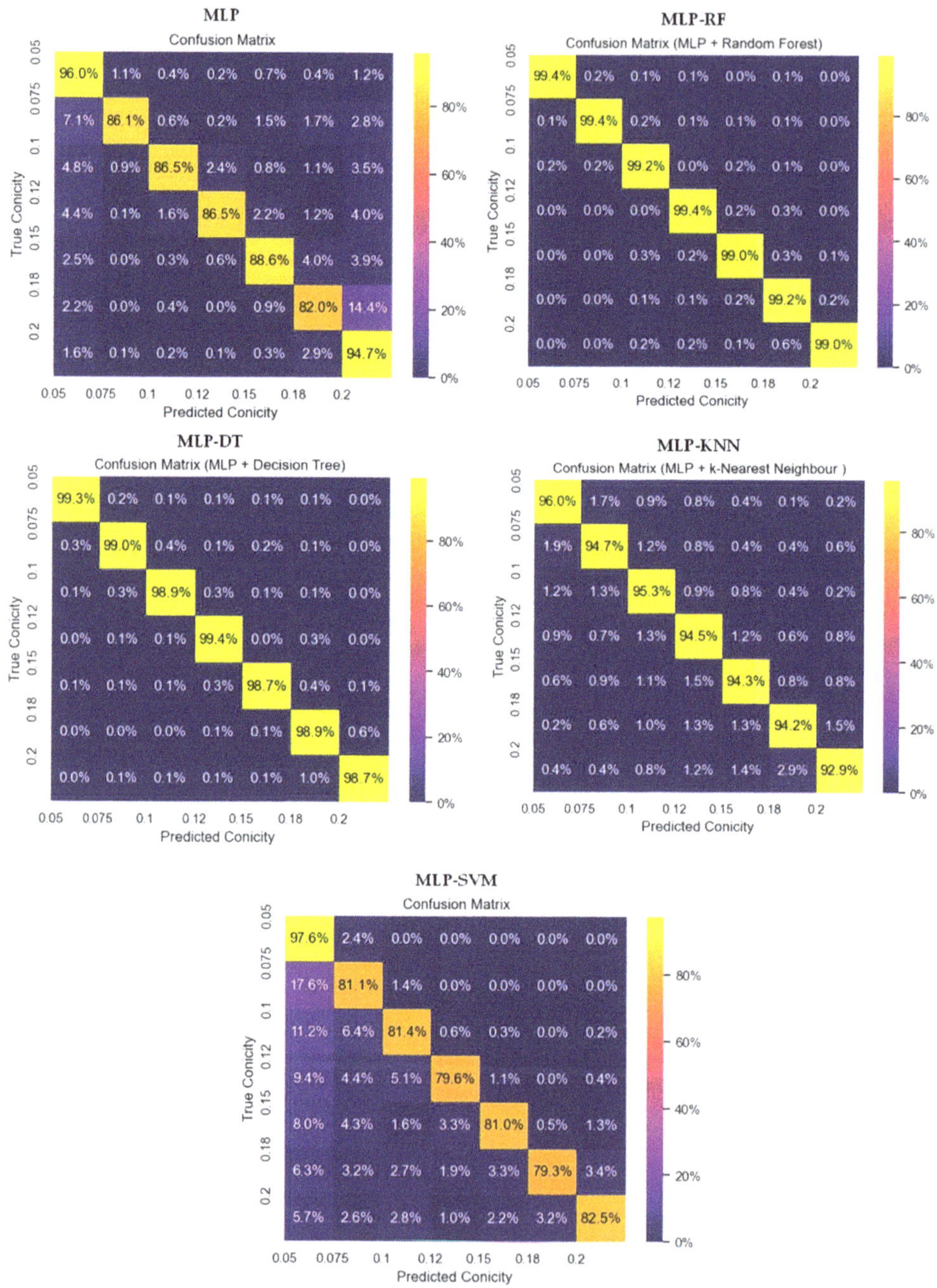

Figure 7. Classification results.

4. Conclusions

In conclusion, this research paper presents the implementation of a hybrid deep learning-based defect detection system. The system utilizes a multilayer perceptron (MLP) for feature extraction, which is pre-trained on collected data. Multiple machine learning models, including support vector machine (SVM), random forest (RF), decision tree (DT), and k-nearest neighbors (k-NN), are employed for classifying conicity target values. The performance of different models in detecting wheel defects was evaluated. The MLP model achieved an accuracy of 88.6% with balanced precision and recall. MLP-RF exhibited exceptional performance with high accuracy (99.0%), precision, recall, and F1 score. MLP-DT demonstrated strong performance with an accuracy of 98.9% and balanced precision and recall. MLP-kNN achieved a relatively high accuracy of 95.0% with good precision and recall. MLP-SVM showed moderate performance with an accuracy of 83.1%, relatively low recall, and moderate precision. Overall, the results indicate the effectiveness of MLP-RF, MLP-DT, and MLP-kNN in accurately detecting wheel defects, while MLP-SVM showed relatively lower performance.

These findings suggest the effectiveness of combining deep learning-based feature extraction with various machine learning models for accurate defect detection and classification. The achieved results highlight the potential of hybrid approaches in defect detection applications, showcasing the benefits of leveraging deep learning and traditional machine learning techniques in tandem. Future research can explore further enhancements to improve accuracy, explore different feature extraction methods, or investigate the applicability of the proposed system to other defect detection scenarios. One limitation of this study is the reliance on simulated data to train and evaluate the proposed model. While efforts have been made to create a realistic simulation model and generate comprehensive datasets, there may still be discrepancies between simulated data and real-world scenarios. It is important to validate the model's performance using real-world data to ensure its effectiveness in practical applications.

To advance the field of defect detection and classification using hybrid deep learning methodologies, several avenues for future research can be explored. These include fine-tuning the pre-trained MLP model specifically for defect detection, investigating alternative deep learning architectures, such as CNNs or RNNs, and evaluating the effectiveness of data augmentation techniques. The integration of transfer learning and real-time implementation in real-time settings should also be considered. Additionally, validation on diverse datasets, deployment in real-time environments, and integration with automated decision-making systems are promising areas for further investigation. To achieve practical implementation, the proposed model can be effectively deployed on an edge computing platform. This deployment enables the real-time detection of wheel defects, providing prompt and accurate identification of any issues. By leveraging the computational capabilities of the edge computing platform, the model can efficiently process the incoming data from sensors and rapidly analyze them for the presence of wheel defects. This approach ensures that potential defects are promptly detected, and appropriate actions can be taken to maintain operational safety and efficiency in railway transportation.

Author Contributions: Conceptualization, K.S. and I.H.; data curation, K.S. and I.H.; formal analysis, I.H.; funding acquisition, B.S.C.; investigation, K.S. and I.H.; methodology, K.S. and B.S.C.; project administration, I.H.; resources, K.S.; software, K.S. and I.H.; supervision, I.H.; validation, K.S. and B.S.C.; visualization, B.S.C.; writing—original draft, K.S.; writing—review and editing, I.H. and B.S.C. All authors have read and agreed to the published version of the manuscript.

Funding: This research received no external funding.

Institutional Review Board Statement: Not applicable.

Informed Consent Statement: Not applicable.

Data Availability Statement: All data, models, and codes generated or used during the study are available on request.

Acknowledgments: We greatly acknowledge the support provided by the National Centre of Robotics and Automation-Condition Monitoring Systems lab and the National University of Sciences and Technology.

Conflicts of Interest: The authors declare no conflict of interest.

Nomenclature

Symbol	Parameter	Value and/or Unit
F_a	Creep Force	N
μ	Friction coefficient	
k_A	Reduction factor around the adhesion	
r_0	Wheel radius	0.5 m
L_g	The half gauge of the track	0.75 m
λ_w	Wheel conicity	0.15 rad
ε	The gradient of the tangential stress in the adhesion area	
V_X	Vehicle's forward velocity	
Y	Lateral motion	meter
y_g	Variation in gauge	meter
y_t	Track disturbance in the lateral direction	meter
Ψ	Yaw angle	radians
k_S	Reduction factor in a slip	
I_w	Yaw moment of inertia of wheelset	700 Kgm^2
K_w	Yaw stiffness	5×10^6 N/rad
m_w	Wheel weight with induction motor	1250 Kg

References

1. Krummenacher, G.; Ong, C.S.; Koller, S.; Kobayashi, S.; Buhmann, J.M. Wheel Defect Detection with Machine Learning. *IEEE Trans. Intell. Transp. Syst.* **2018**, *19*, 1176–1187. [CrossRef]
2. Alemi, A.; Corman, F.; Lodewijks, G. Condition monitoring approaches for the detection of railway wheel defects. *Proc. Inst. Mech. Eng. Part F J. Rail Rapid Transit* **2017**, *231*, 961–981. [CrossRef]
3. Mohammadi, M.; Mosleh, A.; Vale, C.; Ribeiro, D.; Montenegro, P.; Meixedo, A. An Unsupervised Learning Approach for Wayside Train Wheel Flat Detection. *Sensors* **2023**, *23*, 1910. [CrossRef]
4. Shaikh, K.; Kalwer, I.H.; Mal, K.; Chowdhry, B.S. Signal-Based Indirect Wheel Profile Estimation Technique for Solid Axle Railway Wheelset. *Mehran Univ. Res. J. Eng. Technol.* **2021**, *40*, 443–449. [CrossRef]
5. Mosleh, A.; Meixedo, A.; Ribeiro, D.; Montenegro, P.; Calçada, R. Early wheel flat detection: An automatic data-driven wavelet-based approach for railways. *Veh. Syst. Dyn.* **2023**, *61*, 1644–1673.
6. Nielsen, J.C.O.; Johansson, A. Out-of-round railway wheels-a literature survey. *Proc. Inst. Mech. Eng. Part F J. Rail Rapid Transit* **2000**, *214*, 79–91.
7. Sun, Q.; Chen, C.; Kemp, A.H.; Brooks, P. An on-board detection framework for polygon wear of railway wheel based on vibration acceleration of axle-box. *Mech. Syst. Signal Process.* **2021**, *153*, 107540.
8. Alexandrou, G.; Kouroussis, G.; Verlinden, O. A comprehensive prediction model for vehicle/track/soil dynamic response due to wheel flats. *Proc. Inst. Mech. Eng. Part F J. Rail Rapid Transit* **2016**, *230*, 1088–1104.
9. Mosleh, A.; Costa, P.A.; Calçada, R. A new strategy to estimate static loads for the dynamic weighing in motion of railway vehicles. *Proc. Inst. Mech. Eng. Part F J. Rail Rapid Transit* **2019**, *234*, 183–200.
10. Bosso, N.; Gugliotta, A.; Zampieri, N. Wheel flat detection algorithm for onboard diagnostic. *Measurement* **2018**, *123*, 193–202.
11. Cavuto, A.; Martarelli, M.; Pandarese, G.; Revel, G.; Tomasini, E. Train wheel diagnostics by laser ultrasonics. *Measurement* **2016**, *80*, 99–107.
12. Amini, A.; Entezami, M.; Huang, Z.; Rowshandel, H.; Papaelias, M. Wayside detection of faults in railway axle bearings using time spectral kurtosis analysis on high-frequency acoustic emission signals. *Adv. Mech. Eng.* **2016**, *8*, 1687814016676000.
13. Chen, B.; Yan, Z.; Chen, W. Defect Detection for Wheel-Bearings with Time-Spectral Kurtosis and Entropy. *Entropy* **2014**, *16*, 607–626. [CrossRef]
14. Papaelias, M.; Amini, A.; Huang, Z.; Vallely, P.; Dias, D.C.; Kerkyras, S. Online condition monitoring of rolling stock wheels and axle bearings. *Proc. Inst. Mech. Eng. Part F J. Rail Rapid Transit* **2016**, *230*, 709–723.
15. Xing, Z.; Zhang, Z.; Yao, X.; Qin, Y.; Jia, L. Rail wheel tread defect detection using improved YOLOv3. *Measurement* **2022**, *203*, 111959.
16. Asplund, M. Wayside Condition Monitoring System for Railway Wheel Profiles: Applications and Performance Assessment. Ph.D. Dissertation, Luleå Tekniska Universitet, Luleå, Sweden, 2016.
17. Kilinc, O.; Vágner, J. Condition Monitoring of Metro Wheelsets Using Wayside Vibration and Acoustic Sensors. *Res. Sq.* **2022**, *1*, 1–23. [CrossRef]
18. Saba, E.; Kalwar, I.H.; Unar, M.A.; Memon, A.L.; Pirzada, N. Fuzzy Logic-Based Identification of Railway Wheelset Conicity Using Multiple Model Approach. *Sustainability* **2021**, *13*, 10249. [CrossRef]

19. Shi, D.; Ye, Y.; Gillwald, M.; Hecht, M. Designing a lightweight 1D convolutional neural network with Bayesian optimization for wheel flat detection using car body accelerations. *Int. J. Rail Transp.* **2021**, *9*, 311–341. [CrossRef]
20. Hussain, I. Multiple Model-Based Real-Time Estimation of Wheel-Rail Contact Conditions. Ph.D. Thesis, University of Salford, Salford, UK, 2012.
21. Hussain, I.; Mei, T.X.; Ritchings, R.T. Estimation of wheel-rail contact conditions and adhesion using the multiple model approach. *Veh. Syst. Dyn.* **2013**, *51*, 32–53.
22. Hussain, I.; Mei, T.X. Identification of the wheel-rail contact condition for traction and braking control. In Proceedings of the 22nd International Symposium on Dynamics of Vehicles on Roads and Tracks, Manchester, UK, 14–19 August 2011; pp. 14–19.
23. Saba, E.; Hussain, I.; Ali, M. Model-based indirect conicity estimation technique for solid axle railway wheelset. *Mehran Univ. Res. J. Eng. Technol.* **2021**, *40*, 753–760.
24. Baasch, B.; Heusel, J.; Roth, M.; Neumann, T. Train Wheel Condition Monitoring via Cepstral Analysis of Axle Box Accelerations. *Appl. Sci.* **2021**, *11*, 1432. [CrossRef]
25. Vale, C. Wheel Flats in the Dynamic Behavior of Ballasted and Slab Railway Tracks. *Appl. Sci.* **2021**, *11*, 7127. [CrossRef]
26. Li, C.; Luo, S.; Cole, C.; Spiryagin, M. An overview: Modern techniques for railway vehicle on-board health monitoring systems. *Veh. Syst. Dyn.* **2017**, *55*, 1045–1070.
27. Tsunashima, H. Condition Monitoring of Railway Tracks from Car-Body Vibration Using a Machine Learning Technique. Applied Sciences. *Appl. Sci.* **2019**, *9*, 2734. [CrossRef]
28. Mal, K.; Memon, T.D.; Kalwar, I.H.; Chowdhry, B.S. FPGA implementation of extended Kalman filter for parameters estimation of railway wheelset. *Comput. Mater. Contin.* **2023**, *74*, 3351–3370.

sensors

MDPI

Article

Anomaly Detection in Annular Metal Turning Surfaces Based on a Priori Information and a Multi-Scale Self-Referencing Template

Xinyu Suo †, Jie Zhang †, Jian Liu *, Dezhi Yang and Feitao Zhou

State Key Laboratory of Advanced Design and Manufacture for Vehicle Body, Hunan University, Changsha 410082, China; suoxinyu@hnu.edu.cn (X.S.); zhangjie@hnu.edu.cn (J.Z.); yangdezhi@hnu.edu.cn (D.Y.); feitaozhou@hnu.edu.cn (F.Z.)

* Correspondence: liujian@hnu.edu.cn

† These authors contributed equally to this work.

Abstract: To solve the problem of anomaly detection in annular metal turning surfaces, this paper develops an anomaly detection algorithm based on a priori information and a multi-scale self-referencing template by combining the imaging characteristics of annular workpieces. First, the annular metal turning surface is unfolded into a rectangular expanded image using bilinear interpolation to facilitate subsequent algorithm development. Second, the grayscale information from the positive samples is used to obtain the a priori information, and a multi-scale self-referencing template method is used to obtain its own multi-scale information. Then, the phase error and large-size anomaly interference problems of the self-referencing method are overcome by combining the a priori information with its own information, and an accurate response to anomalous regions of various sizes is realized. Finally, the segmentation completeness of the anomalous region is improved by utilizing the region growing method. The experimental results show that the proposed method achieves a mean *pixel AUROC* of 0.977, and the mean *M_IOU* of segmentation reaches 0.788. In terms of efficiency, this method is also much more efficient than the commonly used anomaly detection algorithms. The proposed method can achieve rapid and accurate detection of defects in annular metal turning surfaces and has good industrial application value.

Keywords: image anomaly detection; annular workpiece; a priori and self-information

1. Introduction

There are a lot of grooves, tapered surfaces, end faces and other annular turning surfaces in the manufacturing process of metal workpieces [1–3]. Surface anomaly detection in annular metal turning surfaces is an important process for the quality control of machining. In images, the annular metal turning surface is a circular structure, and its region of interest occupies only a small part of the whole image [4]. In grayscale images, the annular metal turning surface shows a uniform distribution of grayscale in the circumferential direction. Developing a general and efficient anomaly detection algorithm using the structure and grayscale characteristics of annular metal turning surfaces has important application value for automated anomaly detection in annular metal turning surfaces.

Two problems need to be overcome to achieve anomaly detection in annular metal turning surfaces: 1. large image resolution but small region of interest; 2. metal turning surface positive and negative samples are unbalanced, and negative samples are difficult to collect.

In an image containing a complete annular metal turning surface, the image size is determined by the outer diameter of the ring. The difference between the inner and outer diameter of the ring is usually not large, which results in the region of interest of the annulus occupying a small area in the image. Regarding the problem of high-resolution anomaly detection, in addition to the direct processing of the whole image [5,6], other common solutions are resizing [7] and cutting [8,9]. Zhao [10] et al. directly compressed the input image size, and experiments proved that such a strategy has little effect on the detection accuracy for large-size targets, but the recognition effect is poor for small-size targets, and the direct resize neutralizes or removes the pixels in small target regions. Some researchers have used a cutting method to cut large-size images into small-size images for processing and finally stitch the detection results together as the final output. The sliding window approach [11–13] has also been adopted by many researchers to transform the problem of detecting high-resolution images into a small sliding window detection problem. Coordinate transformation is also an effective way to extract the ring region of interest. Suo [4] and Deng [14] et al. directly unfolded the region of interest of annular workpieces to complete the inspection of the annular workpieces without reducing the resolution of the part being inspected. The conversion of the annular region into a rectangular region also facilitates subsequent algorithm development.

In terms of industrial anomaly detection needs, the collection of anomaly samples is more difficult due to the low product substandard rate. Meanwhile, anomalies have different manifestations, and it is difficult to cover all types of anomalies in the collection of anomaly samples, and unexpected anomalies are usually encountered in actual production. The difficulty of collecting negative samples leads to the difficulty of practical application of supervised target retrieval [15,16] or segmentation [17,18] methods. To address the problem of negative samples being difficult to collect, researchers have carried out extensive research work on positive-sample-based anomaly detection algorithms. Traditional anomaly detection algorithms can quickly achieve anomaly detection by manually designing the processing flow and using morphological processing [19,20] or template matching [21,22]. However, the traditional algorithm has high requirements for illumination, imaging angle and the state of the workpiece to be detected, and the selection of each parameter in the algorithm is complicated, and the algorithm is less robust. With the development of deep learning technology, the powerful feature extraction capability of deep networks has greatly enriched the technical methods of unsupervised anomaly detection. Positive sample anomaly detection methods using deep learning networks can be classified into two main categories [23]: image-reconstruction-based methods [24–26] and discriminative-embedding-based methods [27,28]. Image-reconstruction-based methods tend to learn the structural information of positive samples for image reconstruction and compare the reconstruction differences between the image to be detected and the abnormality-free map to achieve abnormality detection. Discriminative-embedding-based methods are more about learning the feature distribution of positive samples, and the determination of anomalous regions is achieved by calculating the distance between the features of the samples to be detected and the statistical positive sample features. Defard et al. [29] used VGG16 as a feature extraction layer, patch embedding for anomaly-free samples using multivariate Gaussian distribution to obtain a probabilistic representation of the normal class, and anomaly detection was achieved by calculating the Mahalanobis distance between the image to be detected and the normal class. Roth et al. [30] combined embeddings from ImageNet models with an outlier detection model to construct a maximally representative memory bank of nominal patch features. This method provides a further improvement in the speed of anomaly detection. The detection effect of deep-learning-based anomaly detection methods depends mainly on the learning effect of the network, and it is more difficult to improve the targeted effect of the trained network. Additionally, anomaly detection by deep learning methods generally results in too-large segmentation of small-sized anomalous regions.

Since the annular metal turning surface has a uniform grayscale in the circumferential direction, this grayscale characteristic is suitable for the self-referencing template method in the traditional anomaly detection method. The self-referencing method uses the periodic grayscale information of the image itself to construct an anomaly-free template, and the final anomaly detection result is obtained by the difference between the template and the image to be detected [31,32]. Ideally, the self-referencing template is anomaly free, but when there are large-size anomalous regions, the self-referencing template constructed using its own grayscale information contains a large amount of anomalous information, which causes serious misidentification problems [33]. Due to the use of grayscale differencing, the self-referencing template method may be incomplete in detecting anomalous regions with low contrast.

Combining the above analysis and the existing problems, this paper combines the circular structure of the annular metal surface and the characteristics of uniform imaging to unfold the annular region of interest, which not only maintains the pixel resolution of the region of interest but also reduces the amount of data processing and the background complexity of anomaly detection objects. Due to the lack of anomalous samples, this paper proposes to implement anomaly detection in annular metal surfaces based on a self-referencing algorithm. In this paper, we overcome the problem of the self-referencing template method failing to detect large-size anomalous regions by introducing a priori information and introduce the strategy of a multi-scale self-referencing template and region growth to improve the accuracy of low-contrast anomaly detection.

Our main contributions in this article are summarized as follows:

(1) A bilinear interpolation expansion method based on the principle of transformation between polar and Cartesian coordinate systems is invoked for the problem of extracting the region of interest of annular metal turning surfaces;
(2) For the problem of large-size anomaly detection, a general self-referencing template anomaly detection algorithm that fuses a priori information and its own grayscale information is proposed, and the anomaly segmentation effect is further improved by using a region growing method;
(3) A more systematic and comprehensive experimental analysis and demonstration of the selection basis and performance effect of the parameters in the algorithm of this paper are conducted, and the generality of the parameters is proved through experiments.

The remainder of this paper is organized as follows. The flow of the developed algorithm and the object to be tested are briefly introduced in Section 2. Section 3 details the implementation of the entire algorithm. The following Section 4 describes the experimental design and results. Finally, this work is concluded in Section 5.

2. Pipelines and Sample Acquisition

2.1. Pipeline

The process of manual anomaly detection is: first, find the region of interest; then, judge the rough anomaly region based on the a priori information in the mind; finally, finish the precise location of the anomaly region based on the information relating to texture or the grayscale of the image itself. Referring to the process of manual anomaly detection, this paper carries out the development of an anomaly detection algorithm for annular metal turning surfaces. The pipeline of the proposed method is illustrated in Figure 1, which is mainly divided into two parts: mean grayscale template construction and the defect detection process. Firstly, the annular region of interest (ROI) in positive samples I_g is unfolded into rectangular image f_g (corresponding to the manual detection of the search of interest).

In the mean grayscale template construction part, the grayscale value data of i columns in N normal expanded images are randomly extracted, and a total of $N \times i$ columns of waveform data are extracted, and the grayscale mean value of each row is calculated as the standard grayscale waveform to construct mean grayscale template MGT with the same size as the input image (corresponding to the a priori information of manual detection). In the defect detection process, for the expanded image to be detected f_d, the significant anomalous regions are screened using the mean grayscale template and the multi-scale self-referencing template (corresponding to the coarse anomaly detection for manual detection); the significant anomalous regions are restored using MGT data to obtain the restoration map R; again, using the multi-scale self-referencing template method to obtain the final anomaly difference score map $Score_{final}$, the high-response region in $Score_{final}$ is the anomaly region (corresponding to the precise location of the manually detected anomaly region); finally, the abnormal region is segmented by the region growth and morphology processing method, and the segmentation map $Result$ is obtained.

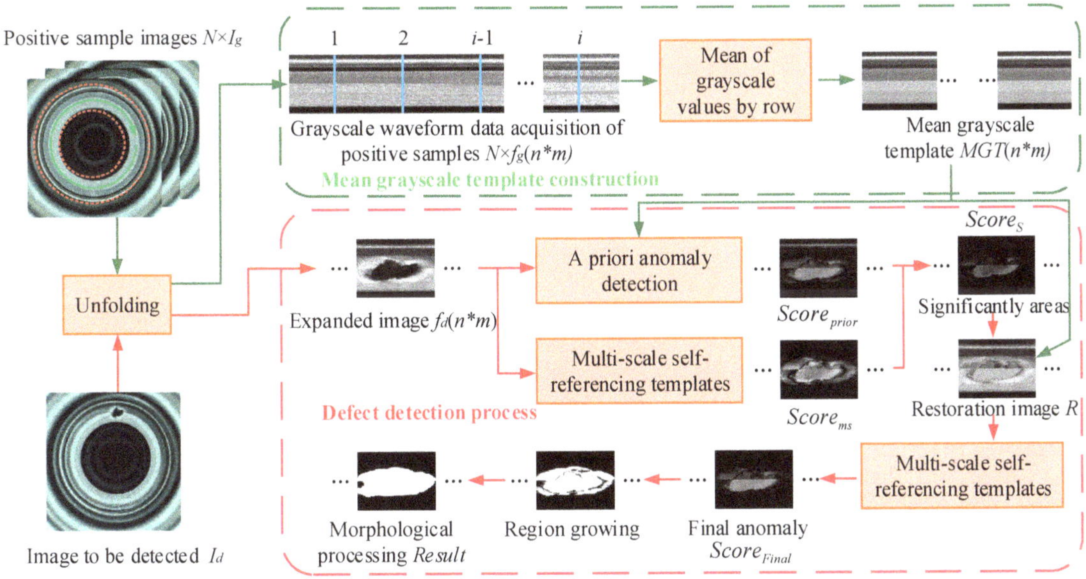

Figure 1. Pipeline of anomaly detection algorithms based on a priori information and multi-scale self-referencing template.

2.2. Sample Acquisition

Since there are relatively few public datasets for anomaly detection in annular metal turning surfaces, in this paper, as shown in Figure 2, a dataset containing three types of annular metal turning surfaces is constructed: a nuclear fuel rod groove dataset, rotary body thin-walled port dataset, and injector nozzle valve seat surface dataset. The objects to be detected in the datasets are all annular workpieces (yellow area), and their unfolded images are all long-bar images; the nuclear fuel rod groove has an unfolded size of 100 × 7000 pixels, the rotary thin-walled port has an unfolded size of 120 × 8700 pixels and the injector valve seat surface has an unfolded size of 440 × 10,000 pixels. The number of positive and negative samples for the three datasets is shown in Table 1.

Table 1. Composition of the annular metal turning surface dataset.

Dataset	Number of Positive Samples	Number of Negative Samples	Expanded Image Size (Pixels)
Groove	98	29	100 × 7000
Thin-walled port	89	34	120 × 8700
Valve seat	128	32	440 × 10,000

Figure 2. Annular metal turning surface dataset.

3. Algorithm Implementation

3.1. Unfolding of Annular Region of Interest

The region of interest of the annular turning surface only occupies part of the whole image. To reduce the number of data operations and make full use of the structural characteristics of the annular surface with uniform grayness in the circumferential direction, the annular region of interest needs to be extracted and unfolded into a long rectangular image.

In Figure 3, the schematic diagram of the annular region of interest unfolding is shown. The yellow area in Figure 3a is the annular region of interest, and its corresponding annular unfolding area is shown in Figure 3b. To facilitate the annular unfolding operation, two coordinate systems are constructed in the original image $I(x,y)$: the image Cartesian coordinate system $O-XY$ and the polar coordinate system $o-r\theta$ with the center of the annular region of interest as the origin. Let $P(r,\theta)$ be any point of the annular region in polar coordinates, and its corresponding point in the image Cartesian coordinate system is $p(x,y)$, and the relationship between the image Cartesian coordinate system and the polar coordinate system can be established.

$$
\begin{cases}
x = x_0 + r\cos\theta & \\
y = y_0 + r\sin\theta & \theta \in [0^\circ, 90^\circ) \\
x = x_0 - r\cos\theta & \\
y = y_0 - r\sin\theta & \theta \in [90^\circ, 180^\circ) \\
x = x_0 + r\cos\theta & \\
y = y_0 + r\sin\theta & \theta \in [180^\circ, 270^\circ) \\
x = x_0 - r\cos\theta & \\
y = y_0 - r\sin\theta & \theta \in [270^\circ, 360^\circ)
\end{cases}
\tag{1}
$$

where x_0 and y_0 are the coordinates of the polar coordinate system $o - r\theta$ in the center circle of the image coordinate system.

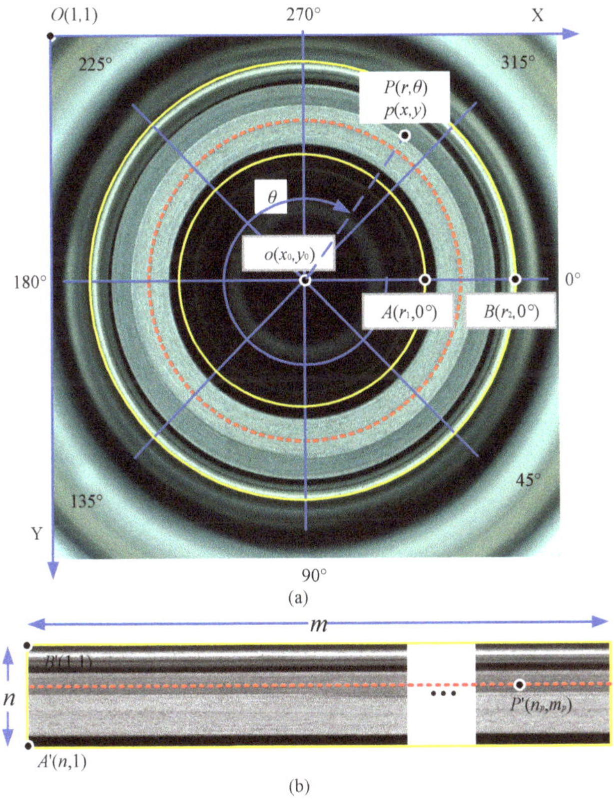

Figure 3. Annular region of interest unfolding. (**a**) The image Cartesian and polar coordinate system $I(x, y)$; (**b**) expanded image $f(x, y)$.

The inner and outer rings of the annular region of interest intersect the 0° axis of the polar coordinate system at the points $A(r_1, 0^\circ)$ and $B(r_2, 0^\circ)$, whose corresponding positions in the expanded image $f(x, y)$ are A' and B', respectively. Since the circumferences of the circles where points A and B are located are different, in this paper, the annular region of interest is unfolded along the circumference of the outer ring. Let the image $f(x, y)$ size be $n \times m$ after the annular unfolding, which is defined as follows:

$$\begin{aligned} m &= \lfloor 2\pi r_2 \rfloor \\ n &= r_2 - r_1 + 1 \end{aligned} \tag{2}$$

where n is the number of rows of the expanded image, and m is the number of columns of the expanded image; r_1 and r_2 are the radius of the inner and outer rings, respectively, and their values are integers; and $\lfloor\ \rfloor$ is a downward rounding symbol.

Let the point $P(r, \theta)$ in the polar coordinates have its corresponding position as $P'(n_p, m_p)$ in the expanded image. Then, the transformation equation of point P to P' is:

$$\begin{cases} n_p = \lfloor r - r_1 \rfloor \\ \theta = \frac{360^\circ}{m}(m_p - 1) \end{cases} \rightarrow \begin{cases} n_p = \lfloor r - r_1 \rfloor \\ m_p = \left\lfloor \frac{m\theta}{360^\circ} + 1 \right\rfloor \end{cases} \tag{3}$$

where n_p and m_p are the coordinates of point P' in the expanded image; m is the number of columns of the expanded image; r is the radius coordinate of point P in the polar coordinate system $r = r_1, r_1 + 1, \ldots, r_2$; θ is the angular coordinate of point P in the polar coordinate system $\theta = \frac{360°}{m}(t-1), t = 1, 2, \ldots, m$; and $\lfloor \rfloor$ is a downward rounding symbol.

To improve the unfolding effect of the annular turning surface and avoid the jagged effect generated by the image unfolding, the gray values in the unfolding map are calculated by bilinear interpolation when performing the unfolding. As shown in Figure 4, $p(x, y)$ is the interpolation point corresponding to point $P'(n_p, m_p)$ in the expanded image. Since the coordinates of $p(x, y)$ may not be an integer, four integer points p_1, p_2, p_3 and p_4 around p are selected. The values of point M and point N are obtained by linear interpolation using p_1 and p_2 and p_3 and p_4, respectively. Then, the grayscale value of point p is obtained by the linear interpolation of point M and point N.

$$
\begin{aligned}
p_m &= p_2 - (p_2 - p_1)(i_2 - x) \\
p_n &= p_4 - (p_4 - p_3)(i_4 - x) \\
p &= p_n - (p_n - p_m)(j_n - y)
\end{aligned}
\tag{4}
$$

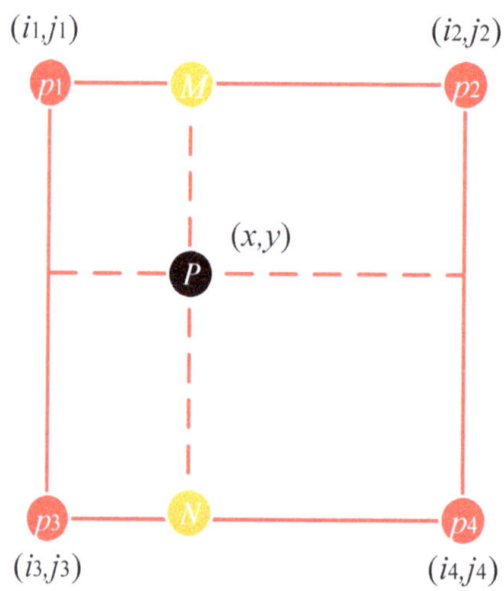

Figure 4. Bilinear interpolation method.

This completes the extraction and unfolding process of the annular region of interest. For any annular workpieces, we only need to determine the circle center $o(x_0, y_0)$, inner ring radius r_1 (in pixels) and outer ring radius r_2 (in pixels) and use the above Equations (1)–(4) in turn to achieve the counterclockwise unfolding process of the annular workpiece region of interest.

3.2. Mean Grayscale Template Construction

For computational convenience, the images are converted to single-channel grayscale maps with grayscale values ranging from 0 to 1 during the subsequent algorithm development. As shown in Figure 5, for the expanded image $f(x, y)$ of an annular workpiece without anomalies, it can be ideally considered to have equal gray values in each row (yellow region) and the same gray waveform in each column (green region).

In the actual industrial detection process, the design of the tooling fixture and imaging method can ensure that the imaging position of the annular workpiece is relatively fixed, and the imaging brightness and contrast of each position are relatively consistent, so it can be approximated that the distribution of the gray value of each row of the corresponding expanded image conforms to the normal distribution. For any annular turning surface, the mean value of each row of grayscale can be counted to construct an anomaly-free mean grayscale template MGT equal to the size of the expanded image.

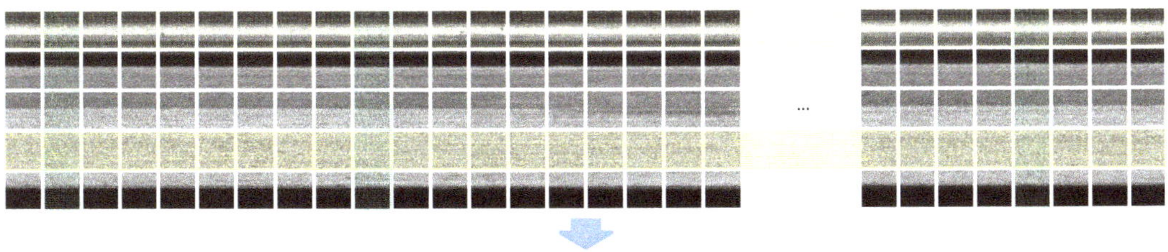

135	135	135	135	135	135	135	135	135	135	135	135	135	135	135	135	135	135	135	...	135	135	135	135	135	135	135	135
118	118	118	118	118	118	118	118	118	118	118	118	118	118	118	118	118	118	118	...	118	118	118	118	118	118	118	118
168	168	168	168	168	168	168	168	168	168	168	168	168	168	168	168	168	168	168	...	168	168	168	168	168	168	168	168
205	205	205	205	205	205	205	205	205	205	205	205	205	205	205	205	205	205	205	...	205	205	205	205	205	205	205	205
97	97	97	97	97	97	97	97	97	97	97	97	97	97	97	97	97	97	97	...	97	97	97	97	97	97	97	97

Figure 5. Image features of the annular expanded image.

N anomaly-free expanded images I_g are collected, and a positive-sample statistical matrix Ps of size $n \times (N * k)$ is constructed by randomly selecting k columns of grayscale waveforms in each expanded image. The mean grayscale value of each row of Ps is calculated, and the mean grayscale template MGT is constructed.

$$MGP(i,:) = \frac{1}{N * k} \sum_{j=1}^{N*k} Ps(i,j) \tag{5}$$

where $MGT(i,:)$ is all the grayscale values in row i of the mean grayscale template; i is the row coordinates of Ps in the range of $1, 2, \cdots, n$; and j is the column coordinate of Ps in the range of $1, 2, \cdots, N * k$.

If the MGT is directly used as a template to calculate the absolute value of the difference between MGT and the image f to be detected, the a priori anomaly score $Score_{prior}$ can be obtained.

$$Score_{prior} = abs(MGT - f) \tag{6}$$

As shown in Figure 6, the results of the a priori anomaly scores for the multiple images to be detected are shown. In Figure 6c, the a priori anomaly score has a strong response for anomalous regions, but the score map also has a high response for edges and contours. The reason for this is that the MGT template can only respond to the grayscale distribution trend of the image and cannot accurately respond to the phase shift of the image. Therefore, the a priori anomaly score cannot be directly applied to the anomaly detection task.

3.3. Anomaly Detection Algorithm Based on a Priori Information and Multi-Scale Self-Referencing Template

In practical imaging of annular turning surfaces, there are problems such as oil interference, surface texture, uneven lighting or unfolded center shift. As shown in Figure 7b, the grayscale of each row of the expanded image is the same; meanwhile, the grayscale waveforms of each column also produce phase deviation. The dashed box in Figure 7a represents the self-reference template construction area at the solid line, and the solid line is the center of the dashed box. The construction region of the self-referencing template has a total of w columns, and the row grayscale mean of the region is used as the grayscale template. Figure 7c shows the self-referencing template waveforms of an abnormal column, where the difference at the anomaly increases as the value of w is taken, but the disturbance in other regions increases. Figure 7d shows the self-referencing template waveforms of a normal column. When w is small, there is a large amplitude difference between the constructed template and the normal waveform due to the influence of the anomalous region, and false recognition occurs at this time. When w takes the value of 2000, the amplitude of the template waveform is the same as that of the anomaly-free waveform.

(a)

(b)

(c)

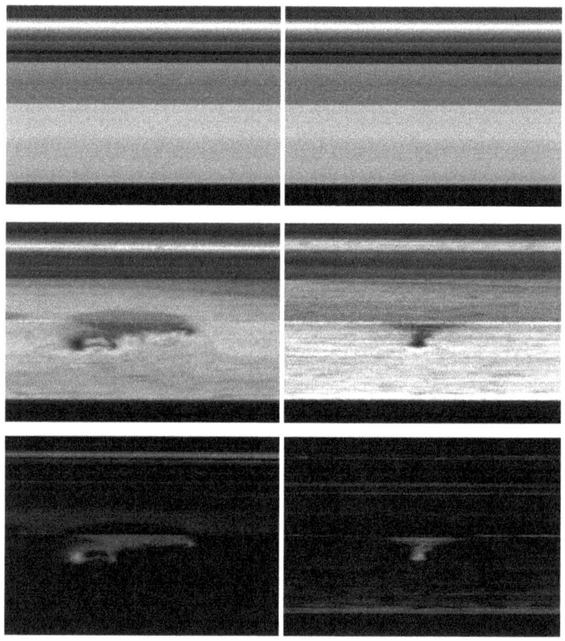

Figure 6. Results of the a priori anomaly scores. (**a**) Mean grayscale template MGT. (**b**) Image to be detected. (**c**) The a priori anomaly score $Score_{prior}$.

The above analysis shows that, in the self-referencing template algorithm, the parameter w directly determines the performance of the algorithm. Too small a w value can create problems with false recognition and missed recognition, while too large a w value can introduce too much interference. To solve the above problems, with reference to the process of manual judgment of abnormal areas, this paper proposes a highly robust anomaly detection algorithm adapted to annular metal turning surfaces by combining abnormality-free mean grayscale templates (a priori knowledge) and a multi-scale self-referencing template (own grayscale information).

First, the image to be tested f (size $n \times m$) is shrunk horizontally, and its height is kept unchanged. The shrunken image T' is obtained.

$$T'(i,j') = \frac{\sum_{j=\lfloor sk \times (j'-1) \rfloor + 1}^{\lfloor sk \times j' \rfloor} f(i,j)}{sk} \tag{7}$$

where T' is the shrunken image in the horizontal direction; i is the row coordinate of T' which takes a value in the range of $1, 2, \ldots, n$; j' is the column coordinate of T', which takes values in the range of $1, 2, \ldots, \lfloor m/sk \rfloor$; j is the column coordinate of f, which takes values in the range $1, 2, \ldots, m$; $\lfloor\ \rfloor$ is a downward rounding symbol; and sk is the scale factor, sk = (columns of the original image)/(columns of the target image). To better explain the concept of algorithm development for horizontal shrink, we introduce the sk variable in place of the w variable. The two variables have different physical meanings but can be considered numerically identical.

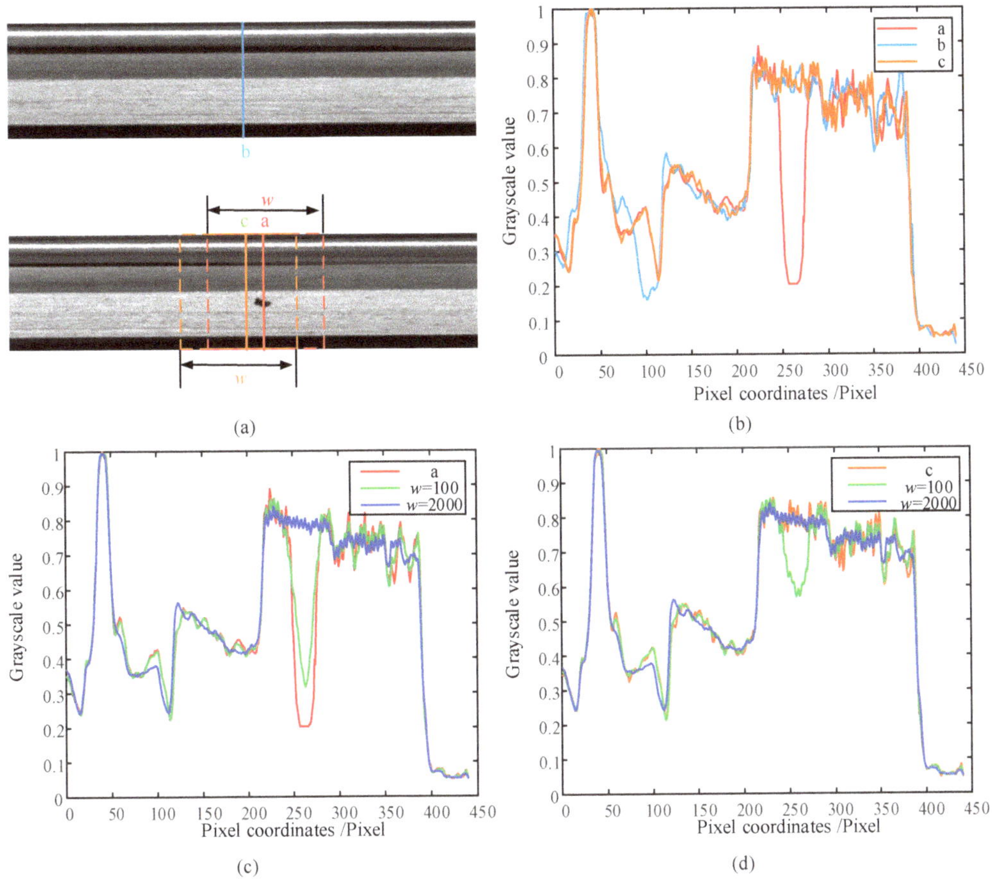

Figure 7. Grayscale waveform analysis of the expanded image. (**a**) Grayscale waveform selection; (**b**) grayscale waveforms of different columns; (**c**) self-referencing template waveforms of abnormal column; (**d**) self-referencing template waveforms of normal column.

After the operation of shrinking, the abnormal column in the original image is diluted by the data of the normal column, at which point it can be approximated that there is no abnormal data in the shrunken image T'. To ensure size consistency between the template and the expanded image f, this paper adopts the bilinear interpolation method to enlarge T' to the size of $n \times m$ to obtain the final self-referencing template T.

Calculate the absolute residual of the self-referencing template T and the expanded image f to obtain the residual map Res.

$$Res(x,y) = abs(T(x,y) - f(x,y)) \tag{8}$$

where abs is the absolute value operation.

By choosing different values of sk, a series of self-referencing algorithm residual maps of different scales can be obtained. Since different sk values have different anomaly detection effects for different sizes of anomalous regions, it can be assumed that the residual map corresponding to a series of sk values must contain all the anomalous regions. The mean value of these residual maps can be used to extract all the anomalous regions; this step is mainly to find all the anomalous regions without requiring all the anomalous regions to have a high response. The mean value of all residual maps is calculated to obtain the multi-scale self-referencing anomaly detection score $Score_{ms}$.

$$Score_{ms} = \frac{\sum_{k=1}^{Num} Res_k}{Num} \tag{9}$$

where k is the index number of the scale factor $k = 1, 2, \cdots, Num$; the scale factor takes values in the array $[sk_1, sk_2, \cdots sk_{Num}]$; and Res_k is the residual map obtained when the scale factor $sk = sk_k$.

To perform the normalization operation on $Score_{ms}$, introduce the normalization threshold t_1.

$$Score_{ms} = \begin{cases} Score_{ms}/\max(Score_{ms}) & \max(Score_{ms}) \geq t_1 \\ Score_{ms}/t_1 & \max(Score_{ms}) < t_1 \end{cases} \tag{10}$$

where $\max(Score_{ms})$ is the maximum value of $Score_{ms}$.

At this time, $Score_{ms}$ contains not only the response of the anomalous region but also the response of the non-anomalous region, but there is less phase interference in it, and the information of the mean grayscale template MGT can be introduced to suppress the non-anomalous region. Since the a priori anomaly score $Score_{prior}$ has a high response to both the significant anomaly region and the phase disturbance, the a priori anomaly scores $Score_{prior}$ are operated with the multi-scale self-referencing anomaly scores $Score_{ms}$ for the dot product, and only the significant anomaly regions can be retained to obtain the scores of the significant anomaly regions $Score_S$.

$$Score_S = Score_{prior}. \times Score_{ms} \tag{11}$$

The parameter t_2 is introduced for contrast enhancement of $Score_S$.

$$Score_S = \begin{cases} 1 & \max(Score_S) \geq t_2 \\ Score_S/t_2 & \max(Score_S) < t_2 \end{cases} \tag{12}$$

At this time, the interference caused by phase anomalies and large-size anomalies is removed from the $Score_S$, but it is less effective in detecting the anomalous regions that are close to the grayscale of MGT. Here, the strategies of image restoration and, again, multi-scale self-referencing templates are introduced to solve the above problems. The main idea is to fill the anomalous position in the original image f with the mean grayscale template MGT corresponding position data according to the significant anomaly detection score $Score_S$.

The dot product operation is performed on $Score_S$ and MGT to obtain the grayscale data MGT' for restoration.

$$MGT' = MGT. \times Score_S \tag{13}$$

The locations to be reconstructed in f are removed with $Score_S$ to obtain the anomaly region mask map f_{mask}.

$$f_{mask} = f. \times (1 - Score_S) \tag{14}$$

The final restoration map R is obtained by adding MGT' and f_{mask}.

$$R = MGT' + f_{mask} \tag{15}$$

The final anomalous region score result $Score_{final}$ is obtained by using the multi-scale self-referencing template method in Equations (7)–(9) again for the restoration map R. In $Score_{final}$, phase errors are eliminated, the effect of large size anomalies is overcome and the response to low-contrast anomalies is high.

Since the response of the anomalous region is usually not uniform, in order to achieve the complete segmentation of the anomalous region, the region growth method is introduced here, setting the region growth start point t_3 and the region growth end point t_4. The final anomalous region segmentation result $Result$ can be obtained.

4. Experiment

4.1. Bilinear Interpolation Unfolding Effect

Image unfolding is a key part of the data processing of the algorithm in this paper, where the thin-walled annular region of interest needs to be unfolded into a long strip of images to facilitate the development of the algorithm. The region of interest of the ring is converted to a rectangular region, and the area of the region of interest increases so it involves interpolation of the pixels of the increased region in the unfolded image. Since the algorithm of this paper is based on the uniformity of grayscale in the horizontal direction of the unfolded image, if the interpolation method is not well chosen, it leads to the generation of a jagged texture in the horizontal direction of the unfolded image, as shown in Figure 8a, which affects the detection effect of the algorithm of this paper.

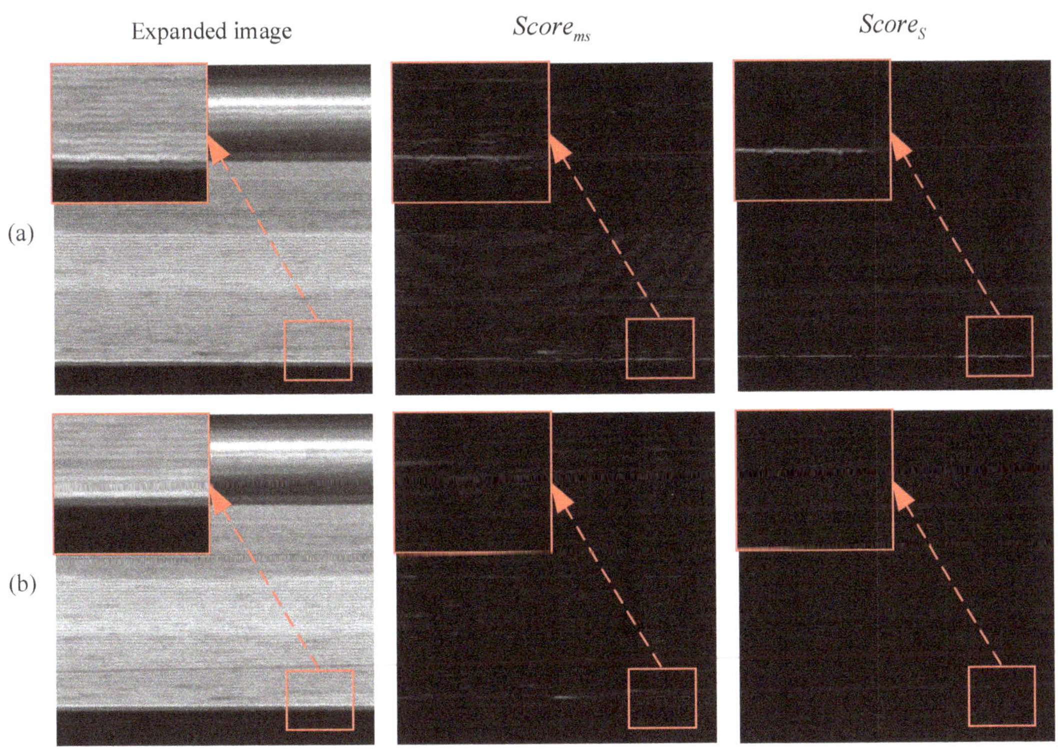

Figure 8. Bilinear interpolation unfolding effect. (**a**) Nearest-neighbor interpolation. (**b**) Bilinear interpolation.

Figure 8 demonstrates the image unfolding results of the nearest-neighbor and bilinear interpolation methods and shows $Score_{ms}$ (the multi-scale self-referencing anomaly detection score) and $Score_S$ (the significant anomaly detection score) processed by this paper's algorithm. As shown in Figure 8, the generation of jagged edges in the expanded image can be avoided by the bilinear interpolation method, and, for the algorithm in this paper, the anomalous response interference from jagged edges can also be effectively suppressed.

As shown in the comparison results, the use of the bilinear interpolation method for image unfolding can effectively increase the degree of smoothness of the unfolding map to avoid the generation of jagged edges, which ensures the normal operation of the anomaly detection algorithm in this paper.

4.2. Multi-Scale Parameter Determination

The scale factor *sk* is related to the horizontal dimension of the anomalous region to be detected in the expanded image. When the value of *sk* is too small, the algorithm becomes weak in detecting anomalies of large size, and, at the same time, it causes misrecognition, as in the green area of Figure 9b. If *sk* is too large, it introduces phase or light uneven interference. Under actual working conditions, the size of the anomaly region varies greatly, and here in this paper, the pixel range of the horizontal size of the anomalous region is defined as 5~500 pixels. As shown in Figure 9a, in this paper, the self-referencing residual map *Res* is calculated for different sizes of artificial anomalous regions with different *sk* values, and the mean response value v of the anomalous region (red region in Figure 9b) is counted to determine the choice of the scale factor *sk*.

(a)
(b)
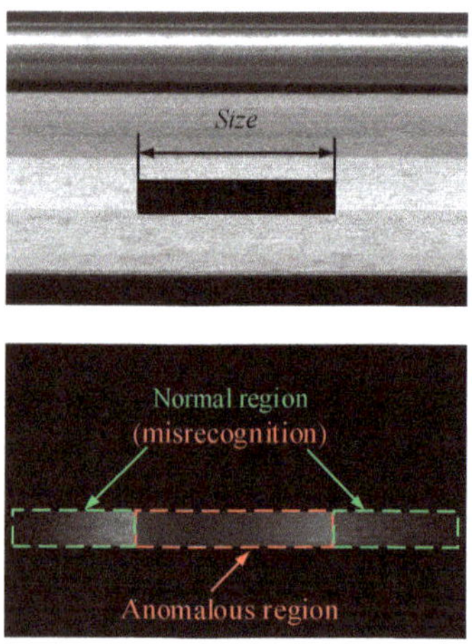

Figure 9. Artificial anomalous region and the residual maps *Res*. (**a**) Artificial anomalous regions. (**b**) Self-referencing template residual map *Res*.

The statistical results of the scale factor *sk* versus the mean response value v for the anomalous region are shown in Figure 10. The mean response value v of the anomalous region increases with the scale factor *sk*. As can be seen from the size of the anomalous region, the scale factor *sk* corresponding to the convergence position of the average response value increases as the *size* increases, while the final convergence value becomes smaller, which is consistent with the calculation principle of the self-referencing template method. From the figure, it can be seen that when *size* takes the maximum value of 500, the mean response value v takes the basic convergence at $sk = 2500$. Since the residual response is a grayscale difference, and considering that the human eye has the highest discrimination ability for grayscale display systems with a grayscale level of 8, that is, in the residual map *Res*, from 0 to 1, when the residual value $> 1/8 = 0.125$, the corresponding position can be considered as an anomalous region. To avoid the interference caused by the large-scale factor and to ensure the effect of anomaly detection, this paper takes $sk = 1000$ when the mean response value v is 0.37, which meets the reliability standard of anomaly detection.

The grayscale value of the artificial anomalous region in this paper is 0, which does not represent the full true anomaly region. However, due to the high error tolerance of the chosen average response value v, combined with a large number of supplementary experiments, this paper suggests that the algorithm in this paper has a better anomaly detection effect when the scale parameter *sk* is taken to be twice the maximum size of the anomalous region to be detected.

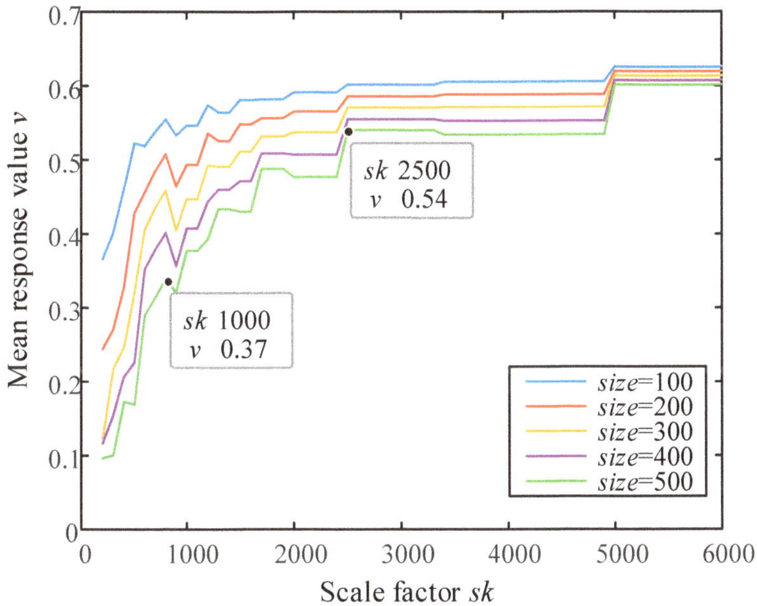

Figure 10. Experimental statistical graph of scale factor *sk*.

In summary, for the anomaly detection demand of 5~500 pixels in a horizontal direction, the scale factor *sk* selected in this paper takes the value range of 10~1000, and, to ensure that *sk* is an integer, the number of *sk* is taken as 6 by equal difference interval, and $sk = 10, 208, 406, 802, 1000$.

4.3. Determination of Algorithm Grayscale Parameters

There are four grayscale parameters in this algorithm, which are the normalization threshold t_1 for normalizing the multi-scale self-referencing anomaly detection score $Score_{ms}$, the threshold t_2 for contrast enhancement of the significant anomaly detection score $Score_S$, the anomalous region growth start point t_3 and end point t_4.

The normalization threshold t_1 serves to prevent the multi-scale self-referencing template score $Score_{ms}$ from being too low, which, in turn, amplifies the phase and noise disturbances during normalization. Studies have shown that the human eye has a high-resolution accuracy for regions with grayscale differences >0.125, i.e., the human eye resolution threshold is 0.125. Here, t_1 is set to equal the human eye resolution threshold of 0.125. The main function of contrast enhancement threshold t_2 for the significant anomalous region is to improve the contrast of anomalous regions for subsequent image restoration. If its value is too large, it affects the low-contrast abnormality recognition; if its value is too small, it introduces too much noise interference. To ensure the detection effect, its value is also taken to be the human eye resolution threshold 0.125.

The region growth parameters t_3 and t_4 are the key parameters that control the final segmentation results of the algorithm. Figure 11 shows the region growth start point t_3 comparison experiment. The starting value is taken as the human eye resolution threshold 0.125 for t_3 = 0.125, 0.1875 and 0.25, at which t_4 = 0.1. From the results of the experiment, it can be seen that if t_3 is too low, too much interference is introduced, and the noise is significantly suppressed with the increase in t_3. Therefore, this paper selects the starting point of regional growth as t_3 = 0.25.

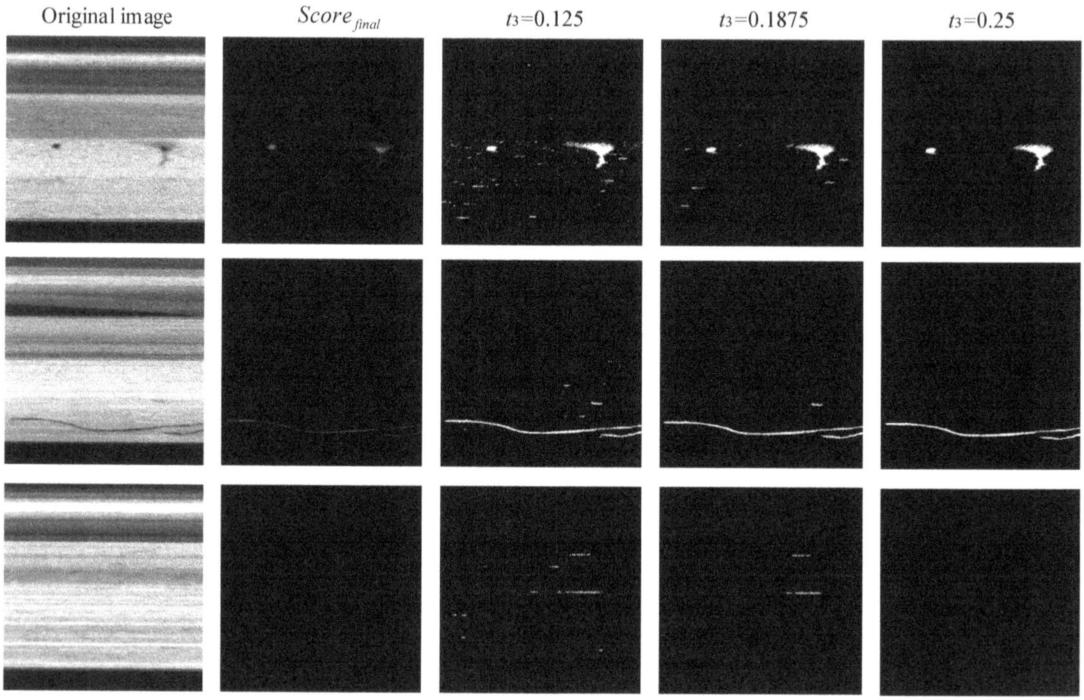

Figure 11. The region growth start point t_3 comparison experiment.

From Figure 12, it can be seen that if the region growth end point t_4 is too low, too much noise interference is introduced, and, as t_4 increases, the noise interference decreases but the integrity of the segmentation is reduced. Considering the suppression of noise interference and the integrity of the abnormal region segmentation, this paper selects the region growth end point as $t_4 = 0.1$, whose value is slightly lower than the human eye resolution threshold.

4.4. Effect of the a Priori Information on Detection

The self-referencing template method is less effective in detecting large-size anomalous regions but has a good suppression effect on phase errors, while the mean grayscale template is more effective in detecting large-size anomalies but introduces excessive phase errors. To verify the detection effect of the method in this paper for large-size anomalies and the suppression effect of phase errors, three different samples of large defects in the annular metal turning surface are selected in this section to compare the detection effect of the self-referenced template, mean grayscale template and the method in this paper.

Figure 13 shows the comparison experiment of large-size anomaly detection effect. The figure shows that the direct use of the mean grayscale template works well for detecting large-size anomalous regions, but phase interference is introduced; self-referencing methods are effective in suppressing phase interference but have a low response for large sizes, while also introducing erroneous responses in non-anomalous regions; the a-priori-information-based and multi-scale self-referencing template algorithm proposed in this paper combines the advantages of the mean grayscale template and self-referencing template to ensure the detection of large-size anomalies while effectively suppressing phase interference.

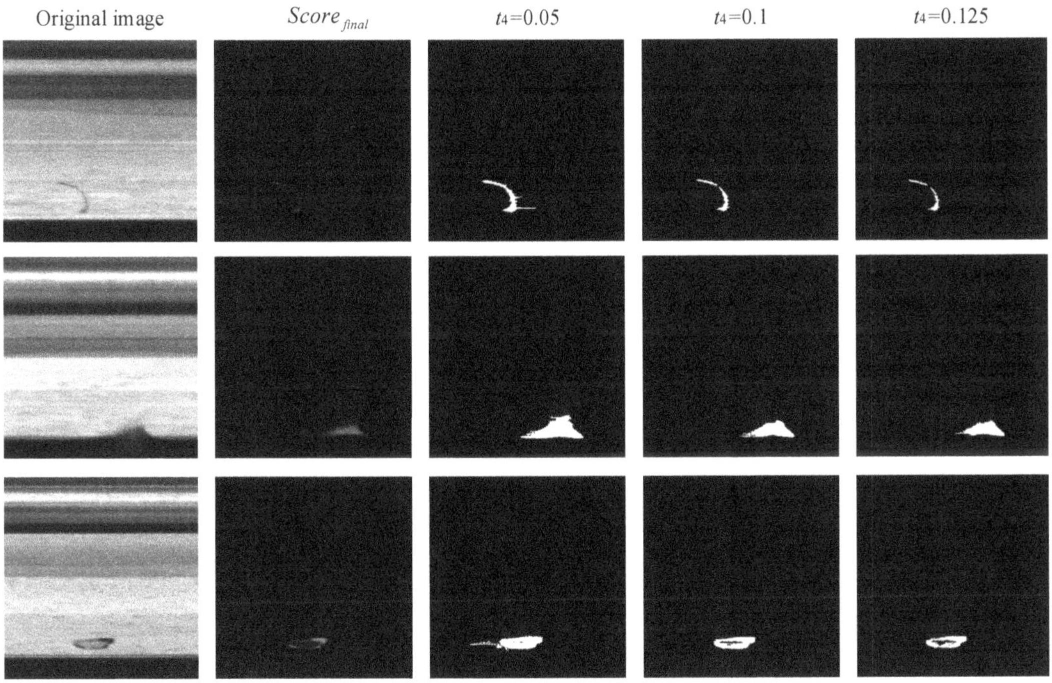

Figure 12. The region growth end point t_4 comparison experiment.

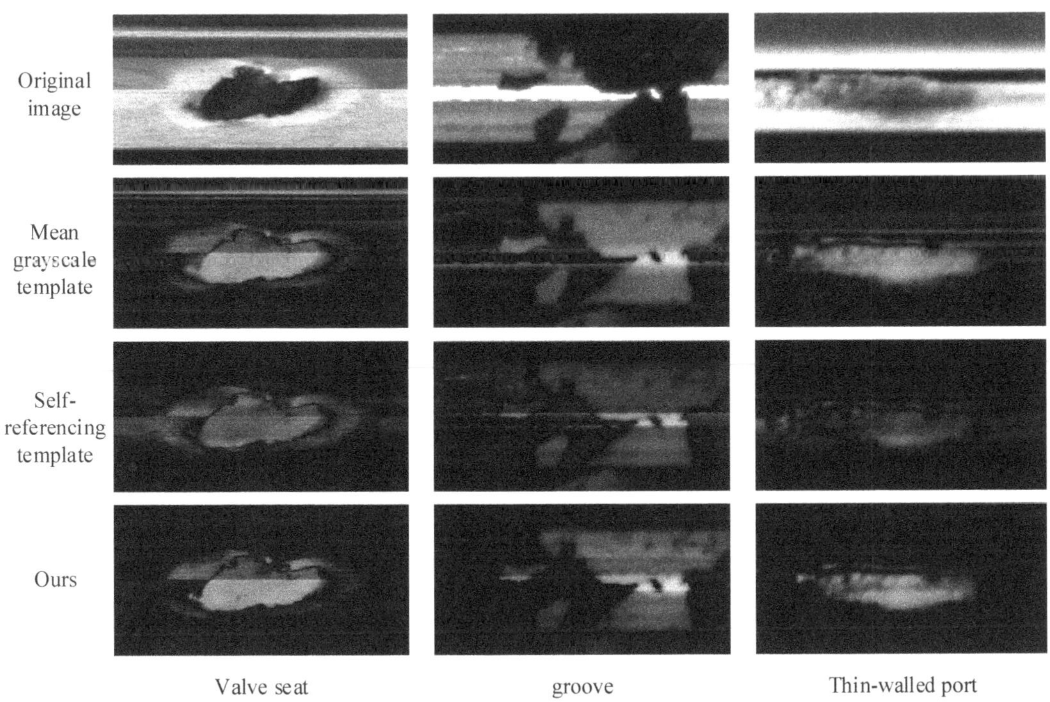

Figure 13. Comparison experiment of large-size anomaly detection effect.

4.5. Comparison of Commonly Used Unsupervised Anomaly Detection Methods

The algorithm in this paper is tested against three commonly used unsupervised anomaly detection algorithms, CFLOW-AD, PatchCore and PaDiM, which perform well in the MVTec AD dataset. Since the object processed by the algorithm in this paper is a long expanded image, and the commonly used unsupervised anomaly detection algorithms CFLOW-AD, PatchCore and PaDiM process images that are square, to compare the performance of this paper with the above commonly used methods, in this part of the experiment, this paper intercepts the expanded images as samples of equal length and width to construct the comparison experimental dataset, as shown in Table 2.

Table 2. Comparison experimental dataset.

Data	Size (Pixels)	Number of Positive-Sample Training Set	Number of Test Set	
			Positive Sample	Negative Sample
Groove	100×100	204	29	29
Thin-walled port	120×120	211	34	34
Valve seat	440×440	237	32	32

The programming environment of the algorithm in this paper is MatlabR2021b, and the programming environment of the comparison algorithm is Python, and the experiments are conducted in a Windows 10 system. The slave computer is configured with an Intel(R) Core (TM) i5-11400F @ 2.60 GHz 2.59 GHz and a memory of 16 G. The model of graphics card is an NVIDIA GeForce RTX 3060 with 12 GB memory.

Two evaluation indicators, $AUROC$ and IOU, are selected. $AUROC$ is a common indicator in the field of anomaly detection and is used to evaluate the anomaly response of each algorithm, which avoids the influence of subjective assumptions in threshold selection by summarizing the overall performance of the model under all possible thresholds. The IOU indicator is used to evaluate the anomalous region segmentation capability of each algorithm.

As can be seen from the Table 3, in terms of the $Image\ AUROC$ and $Pixel\ AUROC$ indicators, this paper's method achieves the optimal mean indicator for all three data. In terms of the $Image\ AUROC$ indicator, the advantage of this method is more obvious compared with that of other methods. The reason for this is that this method has a higher response score for anomalous regions based on the grayscale difference, but the response score is not uniform over the same anomalous region. As a result, compared with other methods, this paper is better in terms of the $Image\ AUROC$ indicator but slightly stronger in the $Pixel\ AUROC$ indicator.

Table 3. AUROC indicator performance comparison.

Item	Algorithm	Image AUROC			Mean	Pixel AUROC			Mean
		Groove	Thin-Walled Port	Valve Seat		Groove	Thin-Walled Port	Valve Seat	
1	CFLOW-AD	0.981	**0.971**	0.901	0.951	0.941	0.973	0.904	0.939
2	PaDiM	0.977	0.930	0.934	0.947	0.968	**0.975**	0.982	0.975
3	PatchCore	**1**	0.958	0.945	0.968	0.960	0.965	**0.986**	0.970
4	Ours	**1**	0.959	**1**	**0.986**	**0.976**	**0.975**	0.981	**0.977**

The goal of anomaly detection is to segment the anomalous region, and the $AUROC$ indicator more greatly reflects the divisibility of the anomalous region response score obtained by the algorithm. The ease of partitioning of the anomalous region response scores also needs to be considered when performing algorithm performance evaluation. In this section, we compare the image-level mean IOU indicator M_IOU of the anomalous region segmentation results of the method in this paper with those of other classical unsupervised methods, and the experimental results are shown in Table 4.

Table 4. **M_IOU** indicator performance comparison.

Item	Algorithm	M_IOU			Mean
		Groove	Thin-Walled Port	Valve Seat	
1	CFLOW-AD	0.573	0.595	0.52	0.563
2	PaDiM	0.582	0.567	0.575	0.575
3	PatchCore	0.66	0.617	0.584	0.620
4	Ours ($score_s$)	**0.843**	**0.691**	**0.831**	**0.788**

From the *M_IOU* indicator values in the table, it can be seen that the *M_IOU* of all algorithms exceeds 0.5, indicating that the algorithms all have excellent anomaly segmentation ability. Compared with the *Pixel AUROC* indicator, the advantage of this paper's method is more obvious in the *M_IOU* indicator. The reason is that the *Pixel AUROC* indicator is affected by the area of the anomalous region; if the large anomalous region is correctly located, the *Pixel AUROC* indicator is greatly improved, while if the small anomalous region is incorrectly located, there is only a small decrease in the *Pixel AUROC* indicator. However, when the *IOU* is counted from each image, and the mean value is calculated, the effect of anomalous region size on the indicator is effectively overcome. The experimental results further demonstrate the effectiveness of the method in this paper.

Some of the experimental results are shown in Figure 14. Comparing the anomalous response results of the algorithms, the edges of the anomalous response region of the commonly used unsupervised anomaly detection methods are fuzzy, and the anomalous response region of this paper's method is more precise, responding only to the anomalous region and not to the non-anomalous region. From the segmentation results in the figure, it can be seen that the anomaly detection method proposed in this paper achieves the optimal anomaly localization results in contrast to Ground Truth. As shown in Figure 14b,g, the method in this paper still realizes the complete segmentation of defects in the case of an inhomogeneous anomalous response due to the segmentation strategy of region growing. For large-size anomalous regions, as shown in Figure 14c,f,i, the response of this paper's method to the anomalous regions is more significant compared with that of other comparison methods, and it avoids the problem of the response of the anomalous regions not being significant. For small-size anomalous regions, as shown in Figure 14j, the method in this paper still achieves good detection results based on the grayscale difference, while the commonly used anomaly detection methods have a very low response (CFLOW-AD and PaDiM) or no response (PatchCore).

In this paper, the operational efficiency of each algorithm is also analyzed. As shown in Table 5, the image sizes processed by each algorithm differ slightly when speed tests are performed due to the different input data sizes required by the different algorithms. In terms of efficiency, the algorithm in this paper achieves the highest detection speed.

Table 5. Algorithm efficiency comparison.

Algorithm	Data	Input Size	FPS
CFLOW-AD	Groove	100 × 100	28.71
	Thin-walled port	120 × 120	22.35
	Valve seat	440 × 440	4.56
PaDiM	Groove	224 × 224	1.72
	Thin-walled port	224 × 224	1.58
	Valve seat	224 × 224	1.89
PatchCore	Groove	224 × 224	0.55
	Thin-walled port	224 × 224	0.78
	Valve seat	224 × 224	0.86
Ours	Groove	100 × 100	411.76
	Thin-walled port	120 × 120	276.51
	Valve seat	440 × 440	17.73

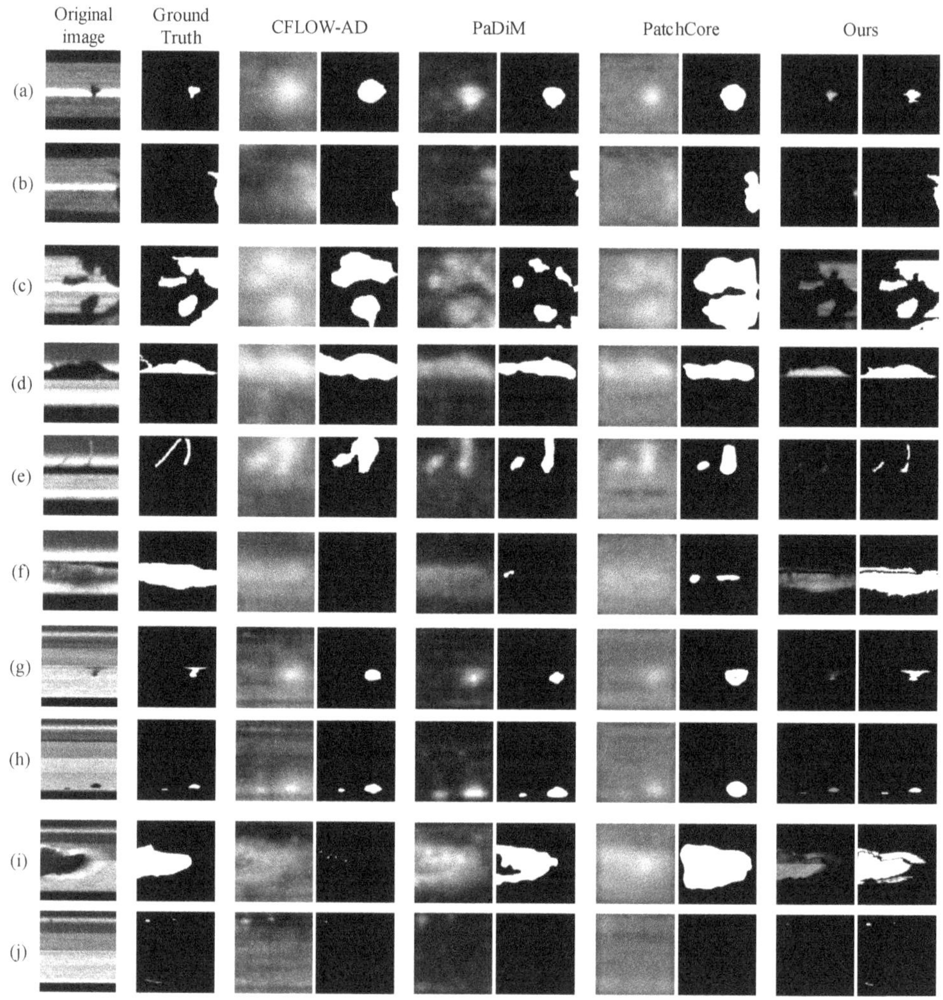

Figure 14. Comparison results of anomaly detection algorithms. (**a–c**) are grooves data. (**d–f**) are thin-walled port data. (**g–j**) are valve seat data. Each column in the figure corresponds to the results processed by the original image, Ground Truth, and 4 detection methods, with anomalous response results and final segmentation results shown under each detection method.

The experimental results in this section demonstrate that the algorithm in this paper is suitable for annular metal turning surfaces with both good anomalous region segmentation capability and high speed, while the algorithm in this paper has no excessive configuration requirements for processing units and is easy to deploy for applications.

5. Conclusions

This paper focuses on solving the problem of anomaly detection in annular metal turning surface in industrial application scenarios. In this paper, the imaging characteristics of annular metal turning surfaces are utilized to unfold the annular region of interest into a rectangular region, which reduces the amount of data processing and simplifies the development of subsequent algorithms. This paper introduces statistical grayscale a priori information, which effectively improves the detection of large-size anomalous regions.

Through the acquisition of self-referencing multi-scale templates, combining the image's own information with the a priori information, the interference caused by contours and large-size anomalous regions is effectively suppressed. The segmentation operation of the response map using the region growing method improves the completeness of anomalous region segmentation. Compared to existing unsupervised anomaly detection algorithms, the method in this paper achieves the best average detection results in the three datasets of groove, thin-walled ports and valve seat data. The experimental results show that this method achieves a mean *Pixel AUROC* of 0.977, and the mean *M_IOU* of segmentation reaches 0.788. In terms of efficiency, this method is also much more efficient than the commonly used anomaly detection algorithms. The method proposed in this paper is versatile for annular metal turning surfaces and has lower hardware requirements and superior economics for industrial applications.

It is worth noting that the anomaly detection algorithm proposed in this paper is mainly for annular turning surfaces with uniform grayscale, and the algorithm is not applicable to objects with uneven grayscale in the expanded image, such as flange surfaces with holes and gear end faces. The algorithm in this paper still relies on grayscale differences and currently performs poorly for textural anomalies with very low grayscale differences (e.g., rippling). In future research, we intend to incorporate texture information into the development of the algorithm, as well to improve the detection of texture anomalies, and we will also explore the optimization of the algorithm in this paper to accommodate objects to be detected with uneven grayscale in the horizontal direction.

Author Contributions: Conceptualization, X.S. and J.Z.; methodology, X.S.; software, J.Z.; validation, D.Y., F.Z. and J.L.; formal analysis, X.S.; investigation, J.Z.; resources, J.L.; data curation, D.Y.; writing—original draft preparation, F.Z.; writing—review and editing, X.S.; visualization, J.Z.; supervision, J.L.; project administration, J.L.; funding acquisition, J.L. All authors have read and agreed to the published version of the manuscript.

Funding: This work is supported by the China–Japan Science and Technology Joint Committee of the Ministry of Science and Technology of the People's Republic of China (grant no. 2017YFE0128400), the Key Research and Development Program of Hunan Province (2021GK2007) and the program of Leading Scientific and Technological Innovation in High-Tech Industries (2021GK4028).

Institutional Review Board Statement: Not applicable.

Informed Consent Statement: Not applicable.

Data Availability Statement: Not applicable.

Conflicts of Interest: We declare that we have no financial and personal relationships with other people or organizations that can inappropriately influence our work; there is no professional or other personal interest of any nature or kind in relation to any product, service and/or company. The data materials in this paper are all real and available.

References

1. Dongling, Y.; Xiaohui, Z.; Jianzhen, Z.; Nanxing, W. An enhancement algorithm based on adaptive updating template with Gaussian model for Si_3N_4 ceramic bearing roller surface defects detection. *Ceram. Int.* **2022**, *4*, 6672–6680. [CrossRef]
2. Cheng, S.; Lu, J.; Yang, M.; Zhang, S.; Xu, Y.; Zhang, D.; Wang, H. Wheel hub defect detection based on the DS-Cascade RCNN. *Measurement* **2023**, *206*, 112208. [CrossRef]
3. Zhao, Y.; An, X.; Sun, N. Virtual simulation experiment of the design and manufacture of a beer bottle-defect detection system. *Virtual Real. Intell. Hardw.* **2020**, *2*, 354–367. [CrossRef]
4. Suo, X.; Liu, J.; Dong, L.; Shengfeng, C.; Enhui, L.; Ning, C. A machine vision-based defect detection system for nuclear-fuel rod groove. *J. Intell. Manuf.* **2022**, *33*, 1649–1663. [CrossRef]
5. Chen, Y.; Li, Y.; Zhang, H.; Tong, L.; Cao, Y. Automatic power line extraction from high resolution remote sensing imagery based on an improved Radon transform. *Pattern Recognit.* **2016**, *49*, 174–186. [CrossRef]
6. He, Z.; Sun, L. Surface defect detection method for glass substrate using improved Otsu segmentation. *Appl. Opt.* **2015**, *54*, 9823–9830. [CrossRef]
7. Chen, J.; Liu, Z.; Wang, H.; Núñez, A.; Han, Z. Automatic Defect Detection of Fasteners on the Catenary Support Device Using Deep Convolutional Neural Network. *IEEE Trans. Instrum. Meas.* **2017**, *67*, 257–269. [CrossRef]

8. Xiang, J.; Wang, J.; Zhou, J.; Meng, S.; Pan, R.; Gao, W. Fabric defect detection based on a deep convolutional neural network using a two-stage strategy. *Text. Res. J.* **2021**, *91*, 130–142.
9. Alipour, M.; Harris, D.K.; Miller, G.R. Robust Pixel-Level Crack Detection Using Deep Fully Convolutional Neural Networks. *J. Comput. Civ. Eng.* **2019**, *33*, 04019040.1–04019040.14. [CrossRef]
10. Zhao, S.; Chen, H.; Wang, C.; Shi, S. SNCF-Net: Scale-aware neighborhood correlation feature network for hotspot defect detection of photovoltaic farms. *Measurement* **2023**, *206*, 112342. [CrossRef]
11. Cha, Y.-J.; Choi, W.; Büyüköztürk, O. Deep Learning-Based Crack Damage Detection Using Convolutional Neural Networks. *Comput. Aided Civ. Infrastruct. Eng.* **2017**, *32*, 361–378. [CrossRef]
12. Atha, D.J.; Jahanshahi, M.R. Evaluation of deep learning approaches based on convolutional neural networks for corrosion detection. *Struct. Health Monit.* **2018**, *17*, 1110–1128. [CrossRef]
13. Wang, T.; Chen, Y.; Qiao, M.; Snoussi, H. A fast and robust convolutional neural network-based defect detection model in product quality control. *Int. J. Adv. Manuf. Technol.* **2017**, *94*, 3465–3471. [CrossRef]
14. Deng, S.; Cai, W.; Xu, Q.; Bo, L. Defect detection of bearing surfaces based on machine vision technique. In Proceedings of the International Conference on Computer Application & System Modeling, Taiyuan, China, 22–24 October 2010.
15. Xiaoxun, Z.; Xinyu, H.; Xiaoxia, G.; Xing, Y.; Zixu, X.; Yu, W.; Huaxin, L. Research on crack detection method of wind turbine blade based on a deep learning method. *Appl. Energy* **2022**, *328*, 120241. [CrossRef]
16. Lu, Q.; Lin, J.; Luo, L.; Zhang, Y.; Zhu, W. A supervised approach for automated surface defect detection in ceramic tile quality control. *Adv. Eng. Inform.* **2022**, *53*, 101692. [CrossRef]
17. Zhou, Z.; Yan, L.; Zhang, J.; Zheng, Y.; Gong, C.; Yang, H.; Deng, E. Automatic segmentation of tunnel lining defects based on multiscale attention and context information enhancement. *Constr. Build. Mater.* **2023**, *387*, 131621. [CrossRef]
18. Zhang, H.; Li, H.; Chen, N.; Chen, S.; Liu, J. Novel fuzzy clustering algorithm with variable multi-pixel fitting spatial information for image segmentation. *Pattern Recognit.* **2022**, *121*, 108201. [CrossRef]
19. Tsai, D.-M.; Rivera Molina, D.E. Morphology-based defect detection in machined surfaces with circular tool-mark patterns. *Measurement* **2019**, *134*, 209–217. [CrossRef]
20. Wang, K.F. Quantitative detection of internal defects based on morphological opening, filling and binarizing operations on wrapped phase of out-of-plane deformation in digital speckle pattern interferometry. *NDT E Int.* **2019**, *103*, 173–176. [CrossRef]
21. Kong, Q.; Wu, Z.; Song, Y. Online detection of external thread surface defects based on an improved template matching algorithm. *Measurement* **2022**, *195*, 111087. [CrossRef]
22. Vaikundam, S.; Hung, T.Y.; Liang, T.C. Anomaly region detection and localization in metal surface inspection. In Proceedings of the 2016 IEEE International Conference on Image Processing (ICIP), Phoenix, AZ, USA, 25–28 September 2016.
23. Tao, X.; Zhang, D.; Ma, W.; Hou, Z.; Lu, Z.; Adak, C. Unsupervised Anomaly Detection for Surface Defects with Dual-Siamese Network. *IEEE Trans. Ind. Inform.* **2022**, *18*, 7707–7717. [CrossRef]
24. Collin, A.S.; Vleeschouwer, C.D. Improved anomaly detection by training an autoencoder with skip connections on images corrupted with Stain-shaped noise. In Proceedings of the 2020 25th International Conference on Pattern Recognition (ICPR), Milan, Italy, 10–15 January 2021; pp. 7915–7922.
25. Schlegl, T.; Seeböck, P.; Waldstein, S.M.; Schmidt-Erfurth, U.; Langs, G. *Unsupervised Anomaly Detection with Generative Adversarial Networks to Guide Marker Discovery*; Information Processing in Medical Imaging; Niethammer, M., Styner, M., Aylward, S., Zhu, H., Oguz, I., Yap, P.-T., Shen, D., Eds.; Springer International Publishing: Cham, Switzerland, 2017; pp. 146–157.
26. Gudovskiy, D.; Ishizaka, S.; Kozuka, K. CFLOW-AD: Real-Time Unsupervised Anomaly Detection with Localization via Conditional Normalizing Flows. In Proceedings of the IEEE/CVF Winter Conference on Applications of Computer Vision, Virtual Conference, 5–9 January 2021.
27. Salehi, M.; Sadjadi, N.; Baselizadeh, S.; Rohban, M.H.; Rabiee, H.R. Multiresolution Knowledge Distillation for Anomaly Detection. *arXiv* **2020**, arXiv:2011.11108.
28. Bergmann, P.; Batzner, K.; Fauser, M.; Sattlegger, D.; Steger, C. The MVTec Anomaly Detection Dataset: A Comprehensive Real-World Dataset for Unsupervised Anomaly Detection. *Int. J. Comput. Vis.* **2021**, *129*, 1038–1059. [CrossRef]
29. Defard, T.; Setkov, A.; Loesch, A.; Audigier, R. PaDiM: A Patch Distribution Modeling Framework for Anomaly Detection and Localization. In Proceedings of the Pattern Recognition, ICPR International Workshops and Challenges, Online, 10–15 January 2021.
30. Roth, K.; Pemula, L.; Zepeda, J.; Schlkopf, B.; Gehler, P. Towards Total Recall in Industrial Anomaly Detection. In Proceedings of the 2022 IEEE/CVF Conference on Computer Vision and Pattern Recognition (CVPR), New Orleans, LA, USA, 18–24 June 2022.
31. Yun, J.P.; Choi, S.H.; Seo, B.; Sang, W.K. Real-time vision-based defect inspection for high-speed steel products. *Opt. Eng.* **2008**, *47*, 685–694. [CrossRef]
32. Tsai, D.M.; Lai, S.C. Defect detection in periodically patterned surfaces using independent component analysis. *Pattern Recognit.* **2008**, *41*, 2812–2832. [CrossRef]
33. Jing, J.-F.; Chen, S.; Li, P.-F. Fabric defect detection based on golden image subtraction. *Color. Technol.* **2017**, *133*, 26–39. [CrossRef]

sensors

MDPI

Article

Design and Implementation of a Hardware and Software System for Visual Assessment of Bituminous Coating Quality

Dmitrii Kamianskii [1,*], Alexander Boldyrev [2], Nikita Vezdenetsky [1], Irina Vatamaniuk [1,*] and Marina Bolsunovskaya [1]

1 Laboratory "Industrial Systems for Streaming Data Processing", School of Advanced Engineering Studies in Digital Engineering, Peter the Great St. Petersburg Polytechnic University, Polytechnicheskaya, 29, 195251 Saint Petersburg, Russia; marina.bolsunovskaia@spbpu.com (M.B.)
2 LLC Amdor, 192148 Saint Petersburg, Russia; mail@amdor.ru
* Correspondence: kamianskii_dmitrii@mail.ru (D.K.); irina.vatamaniuk@spbpu.com (I.V.)

Abstract: Typically, the quality of the bitumen adhesion in asphalt mixtures is assessed manually by a group of experts who assign subjective ratings to the thickness of the residual bitumen coating on the gravel samples. To automate this process, we propose a hardware and software system for visual assessment of bituminous coating quality, which provides the results both in the form of a discrete estimate compatible with the expert one, and in a more general percentage for a set of samples. The developed methodology ensures static conditions of image capturing, insensitive to external circumstances. This is achieved by using a hardware construction designed to provide capturing the samples at eight different illumination angles. As a result, a generalized image is obtained, in which the effect of highlights and shadows is eliminated. After preprocessing, each gravel sample independently undergoes surface semantic segmentation procedure. Two most relevant approaches of semantic image segmentation were considered: gradient boosting and U-Net architecture. These approaches were compared by both stone surface segmentation accuracy, where they showed the same 77% result and the effectiveness in determining a discrete estimate. Gradient boosting showed an accuracy 2% higher than the U-Net for it and was thereby chosen as the main model when developing the prototype. According to the test results, the evaluation of the algorithm in 75% of cases completely coincided with the expert one, and it had a slight deviation from it in another 22% of cases. The developed solution allows for standardizing the data obtained and contributes to the creation of an interlaboratory digital research database.

Keywords: asphalt mixtures; adhesion; boiling water test; image analysis; surface segmentation; gradient boosting; U-Net

1. Introduction

Throughout the world, paved roads are mainly constructed with asphalt or asphalt concrete consisting of a mixture of stones and bitumen, which is an oil substance that serves as a gluing component for stones. In order to ensure the surface robustness and prevent the stones from detachment, it is crucial to establish a stable bond between the stones and bitumen. In the case of pavement delamination, repair or repaving may entail substantial costs. Hence, prior to approval of the bituminous mixture use, preliminary laboratory tests are executed to determine its resistance to physical impact on a particular type of gravel base.

For this purpose, two test methods, namely, the rolling bottle and boiling water tests, are widely implemented [1]. The resistance of the bituminous coating to contact interaction is tested by the rolling bottle method, whereas its resistance to temperature exposure and water erosion is studied by means of boiling. After the experiments, the test samples are subjected to an assessment of the residual bitumen adhesion.

Currently, the main assessment method involves visual analysis conducted by a group of experts who independently determine discrete quality estimates for each stone in the test sample based on visual analysis of the residual coating level. The resulting quality assessment of the coating and the bituminous mixture is the product of averaging the estimates, initially by sample and then by a group of experts.

The rolling bottle method provides a simpler visual analysis of residual coating level in comparison with the boiling water method [2]. This is due to the fact that it only requires consideration of the bare gravel area resulting from chipping of the bituminous coating. When examining the stone surface after boiling, it should be noted that the results may not be entirely conclusive, as the bitumen tends to detach unevenly. Its final thickness in different areas can vary greatly, thereby affecting the assessment. In this regard, the following four surface types are identified, which categorize certain ranges of bitumen layer thickness based on their visual attributes and proportionate areas on the stone surface (see Figure 1):

- Bare gravel, which is the surface that lacks or has very little bituminous coating;
- Insufficient layer, which is the surface with a thin bitumen layer that allows the gravel underneath to be visible or has very few patches of the coating;
- Thin layer, which is the surface with the gravel base obscured by the bitumen layer but with clearly visible relief;
- Thick layer, which is the surface with the gravel base relief either completely invisible or extremely smoothed.

Figure 1. Gravel surface coating types. Average stone size: 3 × 2 cm. Average stone image size: 378 × 402 pt.

However, the expert assessment method has a number of significant disadvantages:

Subjective assessments: There are no strict criteria for assigning a stone coating to one or another assessment class; the assessment from one expert for the same stone may vary at different times;

Slow assessments: To compensate for the subjective factor, an expert group is involved for independent assessment, which greatly slows down the rate of obtaining results;

Insufficient assessments: During the expert assessment of a stone, only four core quality estimates are distinguished: 2, 3, 4, and 5 with the possibility of assigning an intermediate estimate, thus the stones are distributed by residual coating in seven classes. Due to the subjectivity of the assessment, the distribution into a larger number of classes appears to be extremely difficult. This makes it impossible to more accurately select the emulsion mixture ingredients in the absence of Evident visible differences among various samples.

The process automation removes the subjective factor from the assessment procedure, provides a percentage integral estimate, and significantly accelerates obtaining a result. Moreover, it allows standardization of studies across different laboratories. Our study strives to devise a hardware and software system for automated quality assessment of residual bituminous coating for testing by the boiling water method, capable of both providing a discrete estimate comparable to the expert one and a more general integral estimate of the coating.

2. Related Works

It is noteworthy that quite a few researchers in the emulsion mixtures field provide automated visual assessment algorithms for residual adhesion after physical impact resistance testing. As a rule, such an algorithm is divided into two stages: recognizing stones in the analyzed image and semantically segmenting the stone surface.

The pixel-by-pixel filtering algorithm for selecting stones in the image is employed in all the papers reviewed, wherein there are variations in the filtering method used: some of them imply filtering regarding a stipulated background color value, whereas others ascertain the background color boundary by scrutinizing the color component histogram. As the rolling bottle test features in the majority of these studies, the surface is divided into two classes: bare gravel and bitumen layer.

Källén et al. [3] proposed a pixel filtering method for identifying stones in a picture. This method defines pixels with a small Euclidean distance value to white as part of the background and those with a large distance as stones. Subsequently, areas with identified stones are filtered to remove potential classification errors. Stones that are too small or possess significant differences between their area, as well as the ones with convex hulls, are excluded from the analysis. In general, in order to solve the problem of surface semantic segmentation, various approaches are utilized, from threshold filtering methods to deep convolutional networks. Let us consider some of them.

In order to segment the stone surfaces into zones of bituminous coatings and bare gravel, Källen's team uses a graph-based segmentation algorithm, which includes the image being represented as a graph with pixels as nodes. Each pixel has a connection to each neighboring pixel, with a weight inversely proportional to the distance between these pixels in color space. Additionally, each pixel has an edge with nodes of typical bituminous coating and bare gravel colors. The edge weights are determined based on the proximity of the pixel color to the typical color. Typical colors, in turn, were determined through the clustering procedure by the k-means method on both completely bare stones and the ones completely covered with bitumen. The segmentation of the stone surface is performed through the procedure of removing faces from the graph until each pixel node of the image has a connection with only one node of a typical color while minimizing the sum of the weights of the removed faces.

In their further research [4] conducted in the same field, Källen's group proposed a much simpler and faster segmentation algorithm. By means of capturing photos with varying illumination angles, the minimum and maximum illumination values for each pixel can be determined, and the difference between them can then be calculated. Two sets of stones, one completely covered with bitumen and the other completely bare, were used as samples to calculate histograms of distribution for intensity differences. These histograms were then used to derive the probabilistic intensity distribution for the two surface types. When analyzing the image of a stone based on these distributions, the probability of it having a bituminous coating is calculated for every pixel. The resulting quality assessment of the coating for the stone is calculated as the mean of these probabilities for all stone pixels.

Mulsow [5] employed akin methodology scrutinizing laser reflections from the stone surfaces in his publication. The stones were laser-illuminated both horizontally and vertically, and a composite image was then compiled. This image facilitated calculation of an illumination histogram used for establishing threshold values that discriminate between bare gravel and bituminous coating. Similarly, Källen et al. [6] investigated the distinction between natural illumination analysis and highlight analysis on stone surfaces. This approach is rooted in the fact that bitumen-coated surfaces tend to produce particularly strong highlights. Previous research provided the basis for their conceptual development.

The identification of bare gravel areas through pixel classifications according to their color characteristics is a commonly employed technique due to its straightforwardness. The ImageJ program for image processing is frequently applied in such scenarios. For instance, Blom and De Boek [7] used ImageJ for image binarization by Otsu's method, in which the separation of two surface classes is made by the color intensity distribution histogram since a black-and-white image of a stone is considered. In a similar manner, Lantieri et al. [8] applied ImageJ to classify pixels according to their color characteristics in the YUV color scheme. The surface division into bitumen and gravel is made via pixel filtering through threshold values for each of the YUV's components. Riekstins et al. [9] applied nearly the same approach involving the use of ImageJ. The key distinction of their research lies in the application of initial image processing methods, which imply enhancing hue, saturation, and brightness. This, in turn, amplifies the contrast between the gravel foundation and bitumen and increases the gap between them in the feature space, thereby simplifying the separation process.

Trejbal et al. [10] proposed an entropy segmentation algorithm to segregate bitumen-coated and uncoated stone zones. The algorithm evaluates surface smoothness using the image local entropy estimate. The algorithm consists of three steps: filtering by intensity, filtering by entropy, and smoothing. At the first step, the image is converted to grayscale, after which a primary threshold filtering of background is performed. To distinguish the pixels in areas with bitumen coating, it is necessary to take into account the texture of the area in which the pixel is located. To perform this step, the authors propose to calculate the local entropy for each pixel of the stone in a small radius neighborhood. Then, the obtained values are normalized based on the maximum entropy for the stone. After this step, threshold filtering is performed. At the final stage, the segmentation result is smoothed using a Gaussian filter to eliminate local errors. High-entropy areas indicate the bare gravel zone, whereas low-entropy areas indicate the bituminous coating zone.

Xiao et al. [11] described an algorithm to assess the residual adhesion of bitumen, which was specifically tailored for boiling water tests. The authors compared images of the gravel before and after the test procedure. In order to obtain a better assessment, an additional image sampling was applied: the color space of each RGB component in the image was represented as a union of consecutive non-overlapping intervals of a given width (the authors used an interval of width 20). For each pixel, the values of its color components were reduced to the mean value of the interval in which they fell. Thus, a significant reduction in the color heterogeneity of stone surface areas belonging to the same class was achieved, which simplified the segmentation procedure. Afterwards, the main color options characterizing the gravel base were determined from the histogram of the image of the original gravel. Accordingly, using these options, the number of pixels depicting bare surface in the image of boiled stones was determined, and the ratio of bare gravel pixels to the number of all stone pixels was calculated. Although this method was described specifically for the case of testing bitumen adhesion by the boiling method, the proposed approach deviated slightly from the surface segmentation methods using color characteristics that were applied in studies involving the rolling bottle technique. However, it did not take into account the factor of thinning the bitumen film, rather than its complete separation; as a result, the color characteristics of bare gravel could be greatly distorted.

In a study by A. Chomicz-Kowalska [12], residual adhesion was likewise explored after the boiling water method, utilizing conventional surface segmentation. Although the test stones were also segmented into two categories, the HSV color scheme was applied instead of the RGB scheme used in Xiao's study. An additional distinctive characteristic of their methodology involved performing surface segmentation on both sides of the stones being studied, with the final values of surface area classes being calculated using images from both perspectives.

In the majority of the reviewed works that rely on threshold filtering, the thresholds themselves are set manually by the researchers based on their perception, which significantly reduces the objectivity of the method and the level of its automation. Moghaddam et al. [13] applied the most complex approach, avoiding this disadvantage. After preprocessing, the pixels in the image are grouped into clusters based on color similarity via the k-means algorithm. A class and the mean value of the members of this class are assigned to each pixel. Afterward, to determine whether each class belongs to bitumen or bare gravel, a classification algorithm (SVM or decision tree) is applied to their averages. For the training, average scores corresponding to the bitumen surface were prepared in advance, and the average values for bare gravel were taken from the images before the test. Thus, the algorithm needs to be pretrained for each run. This algorithm automates the coating evaluation process for the rolling bottle method. However, as it is entirely based on using the color characteristics of the gravel before testing and does not consider the surface relief, this method can be ineffective for the boiling method application.

In his work [14], ZhenZhou Wang proposes a universal algorithm for segmenting surface areas in an image based on the threshold filtering approach. In order to determine the threshold values by which the image is segmented, the derivative of the slope difference function of the intensity distribution is used. The original RGB image is converted to grayscale, and a normalized intensity histogram is calculated for it. The resulting histogram is then converted into frequency space via the Fourier transform method. The frequencies are filtered then to retain only the very low and very high frequencies. After frequency filtering, the histogram is inversely transformed back to the original intensity space. Then, the histogram slope to the right and left of each intensity value is calculated. This is performed by finding linear models through regression over N adjacent points on each side. The difference in the slope coefficients of the constructed models gives the value of the difference in slopes at a point. The extremum points are then found by calculating the derivative function. The points where the slope difference function reaches a local minimum are considered class separation points. They are selected as thresholds for segmentation. The same algorithm is applied to calculate threshold values for the image gradient map. Then, depending on the task, either segmentation by initial intensities, gradients, or a combination of both are used as the final segmentation map.

Such methods can be further improved, for example, by a finer selection of threshold values. In particular, ZhenZhou Wang et al. propose a method for flexible and robust threshold selection based on the slope difference distribution [15]. The threshold is selected based on the analysis of the peaks and valleys of the slope difference distribution. The pixel classes are separated according to the parameters calibrated for each specific image type. However, the method requires manual parameter adjustment and comparison with a benchmark obtained for each type of image being studied.

One of the key drawbacks of the algorithms reviewed is that they segment the surface into just two classes. While it is acceptable for the rolling bottle method, where the bitumen layer chips away, the process is unsuitable for boiling water tests, as their results do not correctly represent the observed coating quality. Furthermore, the pixel-by-pixel filtering relative to threshold values, along with the analysis of a single pixel on the surface, disregards the surface relief. As a result, this method lacks the ability to accurately identify areas with a thin coating layer. If it is necessary to take into account the relief of the analyzed surface and some more complex interconnections between neighboring pixels, it is advisable to employ more compound and advanced techniques of semantic image segmentation. In particular, such problems are well solved by deep learning methods, as shown by recent studies in the field of object detection and image segmentation [16–21]. Let us consider some state-of-the-art approaches.

A neural network with an encoder–decoder architecture is the dominant technique in the field of semantic segmentation [22]. This deep learning architecture comprises two major parts: an encoder and a decoder. The encoder converts the input data into a compressed representation, also known as a latent space. The decoder then generates the output data based on this compressed representation. The encoder and decoder are jointly trained using error backpropagation to minimize the gap between the predicted and actual outputs.

One of the key strengths of the CNN encoder–decoder model is its capability to process complex and multidimensional data, such as images and time series data. This is achieved by using convolutional layers in both the encoder and decoder to extract the relevant functions from the input data. Furthermore, employing latent space enables the model to apprehend the main data patterns and interconnections, which can prove useful for tasks like anomaly detection and data clustering.

The U-Net architecture is one of the most efficient and extensively utilized structures employing this technique. It was firstly introduced in the paper [23] by Ronneberger et al., in which its effectiveness for semantic segmentation in biomedical data is demonstrated.

The principal drawback of the basic encoder–decoder architecture is the problem of positional information loss when overcompressing, which subsequently results in a decline in segmentation quality during restoration to the original dimension. Therefore, the U-Net architecture is a significant innovation that incorporates direct connections between the encoder outputs and the decoder inputs at the same level. Firstly, this feature allows the decoder to use positional information when restoring dimensionality, and, secondly, it enables the usage of the convolution results and therefore the identified dependencies at every encoder level.

In [24], Enshaei et al. presented a study showcasing the effectiveness of the U-Net-like architecture in semantically segmenting material surface defects with varying material textures. They compared the efficiency of defect detection on a DAGM dataset of textured surfaces by two implementations of the U-Net architecture. The first model was the basic U-Net version with the only change in the order of the normalization and activation layers. In the second model, the max-pooling layer was replaced with a 3×3 convolution layer with strides 2×2. According to the test results, the second model showed slightly better results. It should be noted that this method is quite general and requires further development for application to bitumen analysis problems.

The U-Net architecture is also widely used in road surface image segmentation problems. For example, Majidifard et al. [25] utilize the U-Net network to determine the condition of pavements. To assess the deterioration of the asphalt, they perform localization of the defected road area by the YOLO model and then apply directly pixel-by-pixel segmentation of the defect by the U-Net model.

Similarly, Robet et al. [26] used a U-Net-based neural network to tackle the task of semantically segmenting the roadway for pavement type determination and defect detection. The specialty of the approach is that images for analysis are captured by a road surveillance camera (the Road Traversing Knowledge (RTK) Dataset is used for training). The neural network trained on this dataset made it possible not only to detect the road surface in the image but also to determine its type (asphalt, paved, and dirt) and to identify elements on the road such as markings, tire tracks, patches, potholes, and puddles.

However, in addition to the neural network approach, other machine learning techniques can be used in the task of semantic segmentation, which, due to greater simplicity and a smaller number of parameters, allow one to achieve relatively good results in less computing time [27,28].

One of these techniques is gradient boosting. It is an ensemble learning method that combines several weak models to create a general model capable of describing complex dependencies. The main intention of gradient boosting is to iteratively add new models to the ensemble, whereas each new model corrects the errors of the previous ones. The use of gradients, which are mathematical functions that describe how the model's predictions should be adjusted to minimize the total error, is crucial to this approach. At each iteration, the gradient boosting algorithm first calculates the loss function gradient in relation to the predictions of the current model. It then establishes the appropriate direction in which to adjust the model's predictions to minimize the error. The new model is adjusted to the anti-gradient using a decision tree, correcting the previous model's errors. The new model is then added to the ensemble, and the process is repeated until the stop criterion is met.

The effectiveness of the gradient boosting algorithm is demonstrated by Dai et al. in their paper [29], wherein they successfully solve the task of detecting an object in an image. To achieve this goal, they perform several convolutional transformations with different kernels over the original image, thus forming a matrix of additional features. The gradient boosting algorithm (XGBoost) is then utilized to classify each image pixel as belonging to either the background or target object. The advantage of using XGBoost is that gradient boosting is more resistant to overfitting on small samples.

The algorithms considered provided the basis for the software component of the device for automated quality assessment of residual bituminous coating.

Given the applied nature of the present research, we did not strive to compare all known approaches to the image segmentation problem. Based on the analysis of modern surveys [19–22], we have chosen the Fully Convolutional Networks (FCN) approach as the most appropriate for the described problem. One of the most popular, relatively simple, and effective options for the FCN architecture is the U-Net architecture, which was chosen to implement semantic segmentation. In order to extend the study, XGBoost was chosen as an alternative approach with sufficient generalization capability. For comparison with "traditional" techniques of image segmentation, we implemented an algorithm based on threshold histogram analysis as the most common approach in the field of residual adhesion analysis.

Our contribution is as follows. We have developed an automated bituminous coating quality assessment device that provides the required reproducible conditions for image shooting and analysis. This eliminates the influence of light situation changes when analyzing samples. As well, we have compared the two most common approaches to image segmentation and identified their features with regard to the considered application task. The most suitable segmentation method is refined and incorporated into the industrial laboratory prototype.

3. Materials and Methods

3.1. Hardware and Software System Structure

As part of the research, a hardware and software system (HSS) was developed to perform the basic determination of the adhesion level of bitumen to gravel, as shown in Figure 2.

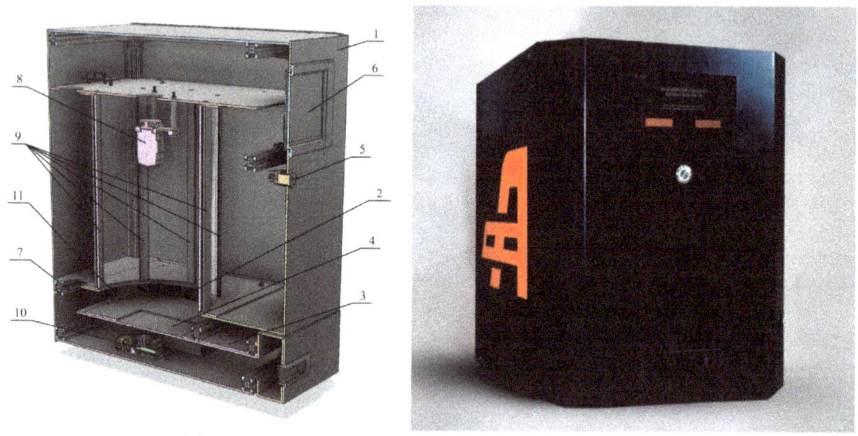

(a) (b)

Figure 2. (a) Device diagram; (b) Implemented prototype.

The developed HSS has the following characteristics. The device has an aluminum h519 × 394 × 445 mm frame (1). It is equipped with an internal cavity (2) for gravel samples to be analyzed. On the front side of the device, there is a sliding tray for the analyzed sample placement (3). The sliding tray has a hole for fixing the substrate (4), where the loaded samples are located. In order to operate this device, the power button (5) and a color LCD display (6) are located on the HSS's front side. An Ethernet port on the back of the device for connecting to a PC acts as a setup and debugging interface (7). A USB camera (8) and a set of eight LED sources (9) are installed in the internal cavity of the device to analyze the adhesion level of bitumen to a stone. An Nvidia Jetson Nano microcomputer (10) is provided in the device to control the equipment and analyze samples. A three-pin voltage supply connector is located on the back panel to power the HSS (11). In the sample tray, there is a substrate marked, as shown in Figure 3.

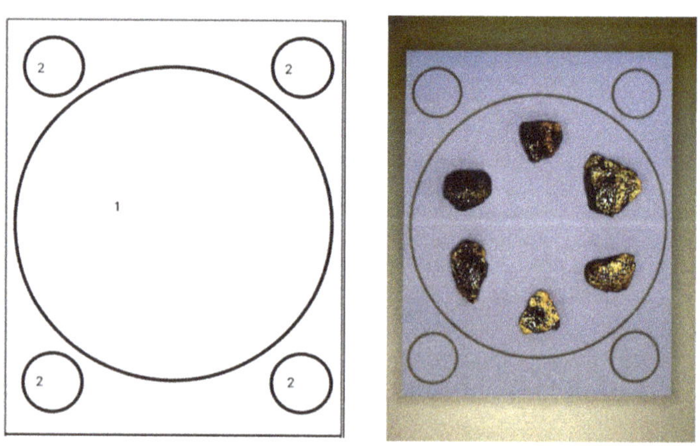

(a) (b)

Figure 3. (a) Substrate template, where 1 is the area for the samples, and 2 is the area for color adjustment; (b) Samples placed on the substrate.

The samples are put in the zone of the largest circle, which is the sample location zone. To be correctly processed, the samples must be placed as follows: the distance between the samples should not be less than 5mm, and they should not cross the line of the sample location zone or be located outside this zone. To put the samples into the device, the substrate containing the samples must be placed within a designated slot in a specially designed tray.

The device has several operational stages:

1. Running a program: At this stage, the program displays a welcome message and waits for the analysis start signal from the operator;
2. Hardware verification: At this stage, the program checks for camera connected to the device and, when it is detected, proceeds to shooting images. If, after fifteen seconds from the beginning of the transition to this stage, the camera is not detected, the program proceeds to the results stage, displaying the camera absence error;
3. Shooting images: At this stage, the LED lighting is switched on alternately, and the pictures of gravel samples put into the HSS are taken. Upon successful completion of shooting at all illumination angles, the images are transferred to the processing stage. If an error occurs during shooting, the transition to the results stage occurs with the display of the shooting error;
4. Processing: This stage involves processing and analysis of the obtained images of gravel samples in accordance with the specified requirements. If an error occurs, the corresponding error message is transmitted to the results stage instead of the analysis values;
5. Results: The stage includes displaying on the HSS's screen the obtained analysis values and images of stone samples in the outline in case of successful completion of the analysis. If an error occurs at one of the described software operation stages, the corresponding message is displayed on the HSS's screen. In any case, this stage displays a button to start a new analysis;
6. Device shutdown: The device shutdown occurs when the power button is held for several seconds while the device is operational. Its primary purpose is to confirm the operator's intention to shut down the device. In case of confirmation, the main program is stopped, followed by the device logout and shutdown. Otherwise, there is a return to the executed stage of the main program.

The following requirements are imposed on the software component of the HSS:

- Input data should comprise images of gravel samples within the inner cavity, captured at various illumination angles;
- The software must determine the total number of stone samples present in the input data;
- For each found sample, the contours of the borders and a delineative rectangle should be defined;
- The adhesion level of each found sample should be analyzed independently;
- The analysis results should be presented in both integral and discrete forms, using a value that is applicable to all samples;
- In addition to the general values, the software tool should provide an image of the analyzed stone samples in a special color outline.

3.2. Software Component Overall Structure

Consider the implementation of the HSS's software component. It is represented by a shooting module and an algorithm for assessing adhesion level. When the hardware system is run, and the program is loaded, the user puts the tested sample of stones into the device and runs the program through the user interface. Once the program is run, the shooting module captures a series of eight photos of samples at different illumination angles. The resulting images are processed by an adhesion assessment algorithm.

The adhesion assessment algorithm includes a stone detection module, a semantic segmentation module, and a final classification module. These modules are run sequentially and perform all the required image analysis steps (see Figure 4).

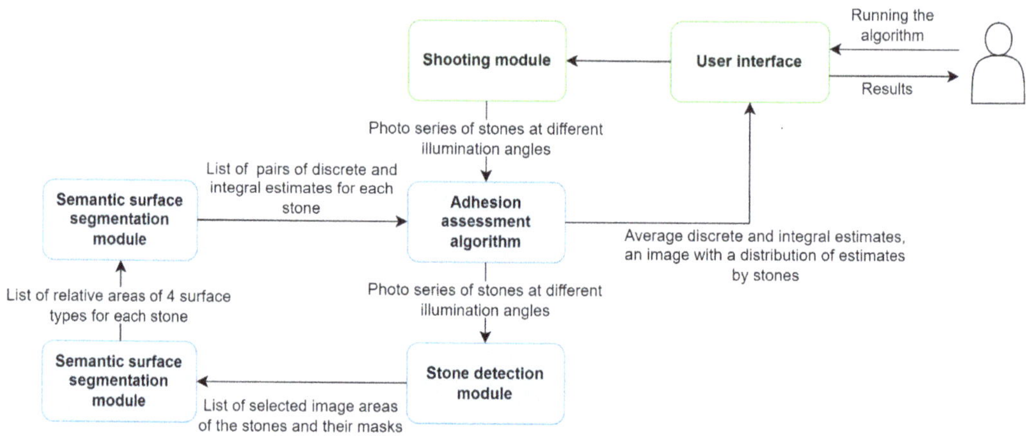

Figure 4. Hardware system overall operation scheme.

Upon completion of the final classification module, each stone from the test sample is assigned a discrete and integral (percentage) estimate. Then, the average integral and median discrete estimates for the sample are calculated. Afterwards, the resulting values are delivered to the user as final estimates of the adhesion quality in this sample.

The flow diagram for software operation is presented in Figure 5. The device is turned on by briefly pressing the button on the front side. After the system starts, the operator can select the analysis method (XGBoost or U-Net). Before starting the analysis, the operator must open the loading tray, fix the substrate, place the gravel samples on it, and then close the tray. To start the analysis, the operator should press the button on the LCD display, and then the camera launches. If the camera is connected successfully, the analysis is carried out automatically. The progress of the analysis is shown on the display. Once the analysis is complete, the results obtained will be displayed. In case of an error, the corresponding message will be shown. To shut down the device, the operator should hold down the power button for a couple of seconds.

3.3. Stone Detection

The stone detection module is responsible for detecting all the stones in a photo series to form a mask and a delineative square for each stone, followed by cutting it out of the overall image.

Classical algorithms of the Canny operator [30] and Suzuki contour search [31] are used for this task.

The Canny Edge Detection Algorithm is a widely applied boundary detection method in computer vision and image processing. The algorithm comprises four main stages: smoothing the image with a Gaussian filter, calculating the magnitude and gradient orientation, suppressing non-maximal boundaries, and applying hysteresis threshold screening to obtain the final boundary map.

A known limitation of the Canny algorithm is its sensitivity to color intensity gradients, often leading to blurring in color areas. Shadowed regions, where the stone casts a shadow, show black as the dominant color, thereby triggering blurring. To remove shadows from the image required for stone detection, the maximum value for each pixel across all images is calculated (see Figure 6).

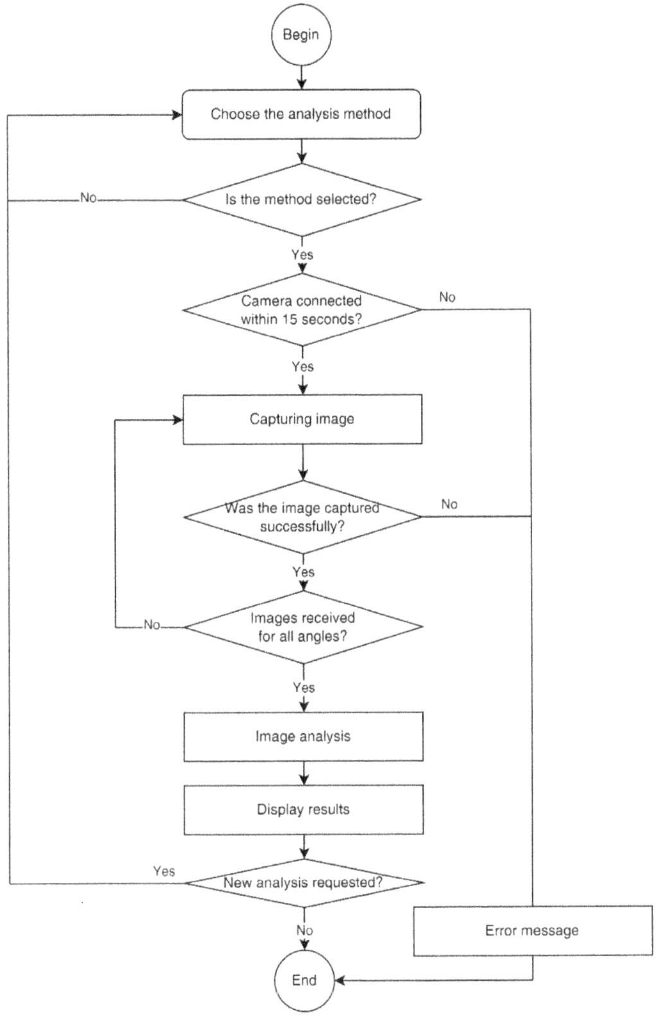

Figure 5. The flow diagram for software operation.

To improve the Canny filter operation, contrast enhancement is applied to the maximum image in order to increase the visibility of the blue component characterizing the background, followed by the removal of all pixels with a dominant, relatively large blue component (see Figure 7).

After these operations, Canny easily finds all the necessary stone boundaries (see Figure 8).

Suzuki's algorithm is a contour detection algorithm based on the Freeman chain code used in computer vision and image processing, which represents the boundary of an object as a sequence of directions. The algorithm first identifies the starting point of the contour and then uses the Freeman chain code to follow the object's boundary. The algorithm uses a stack to track the current position and direction of the contour and an array of visits to mark pixels already visited. If the algorithm encounters a pixel that has already been visited, it returns until it finds an unvisited pixel and continues to follow the boundary from there. When the contour is completely delineated, it is added to the list of contours.

Figure 6. Example of the maximum for all frames.

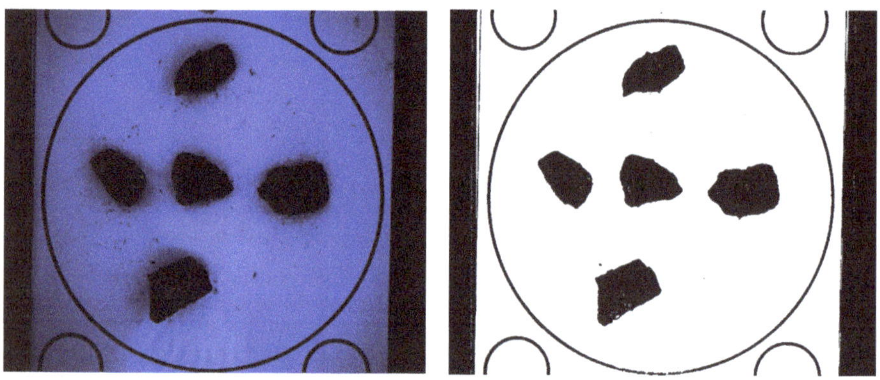

Figure 7. Increased contrast and removed blue background.

Figure 8. The contour search algorithm is used for the boundaries obtained via Canny operator.

As the algorithm does not provide any requirements for the number of stones in the analyzed set, all possible contours are initially selected. Then, a contour with a maximum area is selected, relative to which all other contours are filtered. A contour is considered a contour of a stone if its area is at least 20% of the area of the largest stone. All other contours are regarded as dirt on the substrate.

After filling in all the remaining contours, we get the resulting masks of stones (see Figure 9).

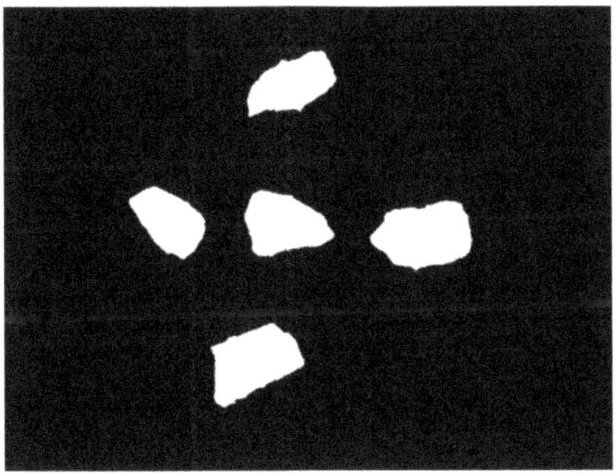

Figure 9. Resulting masks of stones.

Then, a delineative square is determined for each stone by its mask, and this part of the image is cut out of each photo. This set of images of the stone with the mask is passed on to the semantic segmentation module.

3.4. Preparing a Dataset for Model Training

Training a neural network to achieve a high-quality result requires a sufficiently large set of qualitatively marked up images in the training sample. Let us further consider preparing a dataset for training and a neural network.

In this research, we use a dataset consisting of 300 stones. Each stone's discrete estimate was determined by an expert, and half of them were additionally categorized by the surface types indicated in Figure 10. The surface types include areas of bare gravel indicated in red, insufficient layers indicated in green, and thin layers indicated in blue. Moreover, areas of the stone lacking a color mask were defined as the ones having a thick layer.

Of the set of marked up stones, 70% was allocated as a training sample for the segmentation algorithm, and the remaining 30% was allocated for a test sample. In such a way, we get a fairly limited sample for training a deep neural network, which can easily lead to the model retraining.

Accordingly, as an alternative approach, it makes sense to try a simpler machine learning algorithm that would rely on the experience of existing research's algorithms, in which image segmentation primarily relied on the color characteristics of individual pixels. To study the difference between the surface types under consideration (thick layer, thin layer, insufficient layer, and bare gravel) in the color space, we construct distribution histograms of color characteristics of stones across all images with a marked-up surface.

In order to view the image of the stone without strong distortions that arise due to the illumination angle, 8 photos with different illumination angles are taken for each stone. To eliminate the influence of excessive illumination, highlights, and shadowing, an aggregated image is applied in the analysis. As the highlight on the bitumen surface introduces too much bias into the average value, instead of simple averaging, 30% of the percentile is used to obtain an aggregated image. The image obtained by this aggregation method will hereinafter be referred to as the average image.

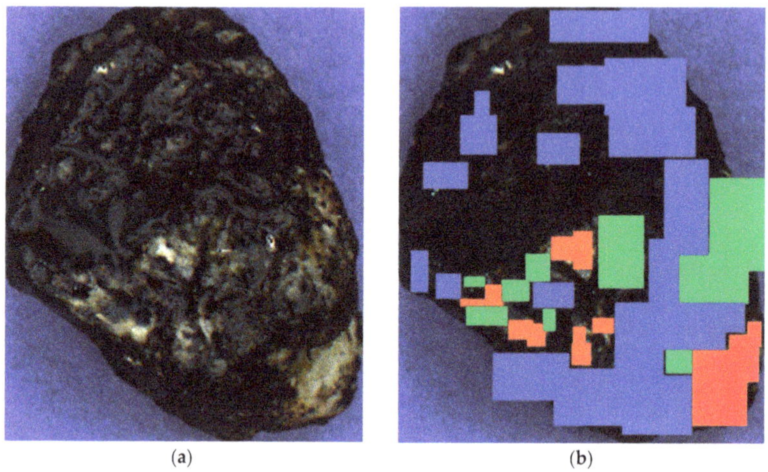

Figure 10. Stone markup by surface types: (a) The original photo; (b) The marked-up image. Each color represents regions belonging to the same class.

4. Experiment Results

4.1. Analyzing the Color Components in the Image

We consider the distribution of color components in the three most common color spaces, specifically RGB (Figure 11), HSV (Figure 12), and YUV (Figure 13). If there are significant differences in the distributions of color components for various surfaces, we can conclude that the conditional classification by color components is sufficiently effective and also determines the best color spaces to use.

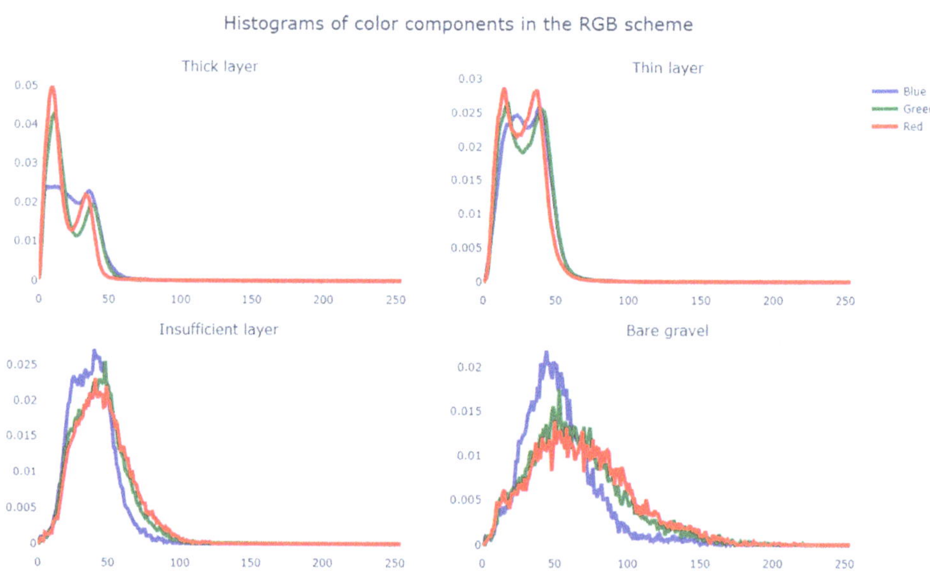

Figure 11. Histograms of color components in the RGB scheme.

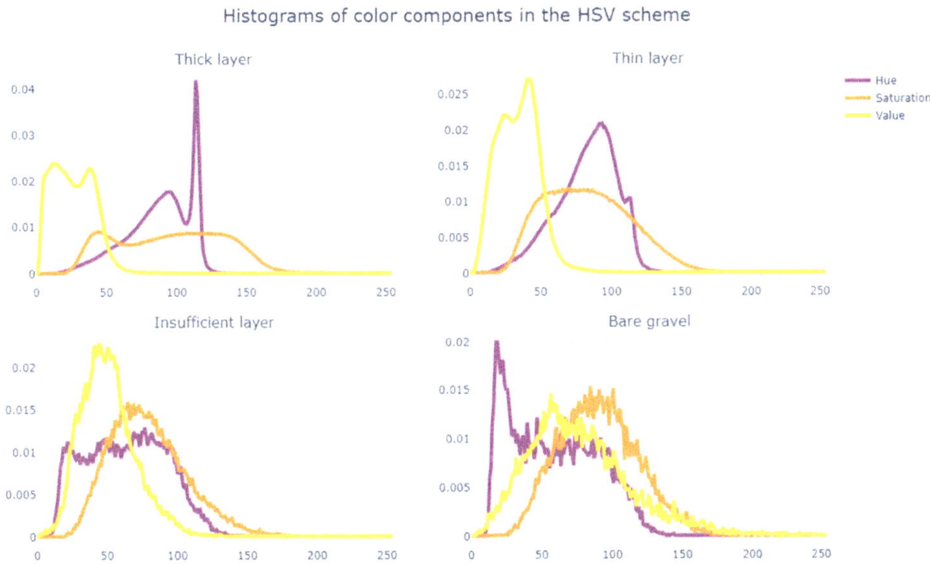

Figure 12. Histograms of color components in the HSV scheme.

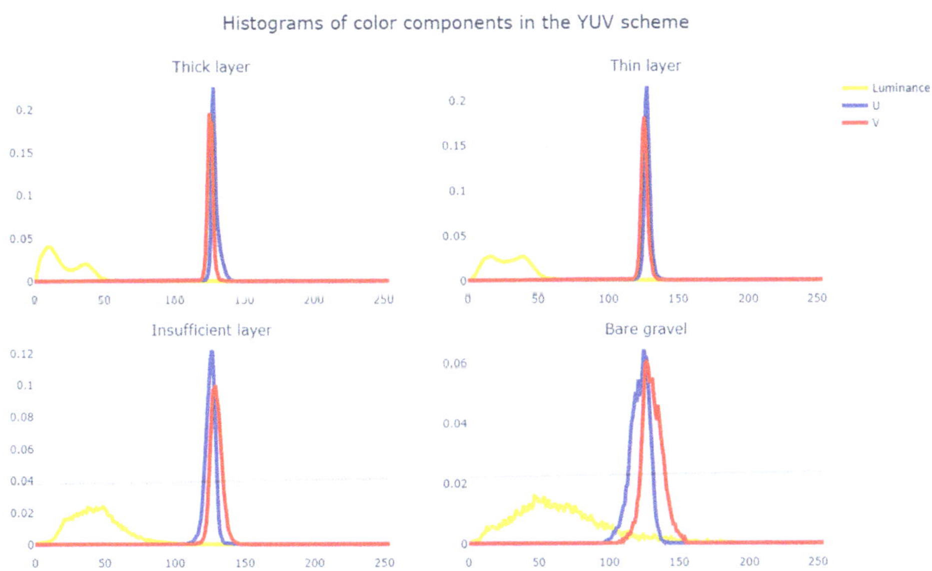

Figure 13. Histograms of color components in the YUV scheme.

Based on histogram analysis, it can be concluded that each of the four surface types has a distinctive distribution pattern for color characteristics. This is apparent mostly in the HSV color scheme and least so in YUV. Intersections in the distributions are normal since, with a sufficiently rough markup, zone parts of one type can affect zone parts of another type. One of the most challenging aspects of intersections is the insufficient coating layer since it very often delimits areas with another coating type.

It is worth examining the characteristics of the color component distributions in the HSV space for various surface types, namely, the values of the median, lower, and upper quartiles (see Table 1). These values were obtained by averaging the values of a sample of test images for samples of clearly distinguishable class of bitumen coating (see Figure 12).

Table 1. Statistical characteristics of color component distributions in HSV space for different surface types.

Surface Type	Lower Quartile	Median	Upper Quartile
Bare gravel	29, 70, 51	56, 90, 72	84, 109, 96
Insufficient layer	41, 59, 37	64, 76, 49	86, 96, 62
Thin layer	70, 59, 23	86, 81, 34	98, 104, 44
Thick layer	74, 62, 14	92, 97, 26	108, 127, 38

Table 1 shows that the value component is one of the most effective criteria for separation. On the whole, we can conclude that it is possible identify suitable threshold values for components, resulting in the satisfactory separation of sections with differing surface types.

However, to attain higher quality segmentation, the algorithm should consider more intricate pixel color dependencies with neighboring pixels, as well as the relief and highlights of the surface. In this instance, employing a gradient boosting algorithm represents the optimal choice.

In order to select a suitable method for assessing the quality of bitumen coating, we implemented three semantic segmentation methods, namely, threshold segmentation, gradient boosting, and a U-Net-based neural network.

4.2. Gradient Boosting

According to the stone images received at the module input, a dataset of feature vectors is formed, in which each vector corresponds to a specific pixel of the stone surface. The features used for classification are presented in Table 2.

Table 2. The features used for classification in gradient boosting.

Feature	RGB	HSV
Value in the average image	+	+
Average value of the average image around the (15, 15) and (21, 21) pixels	+	+
Median value of the average image around the (15, 15) and (21, 21) pixels	+	+
Value in the minimum image	+	+
Average value of the minimum image around the (15, 15) pixel	+	+
Median value of the minimum image around the (15, 15) pixel	+	+
Standard deviation of the pixel component for all frames	+	+
Average value of the standard deviation for all frames around the (15, 15) and (21, 21) pixels	+	+
Standard deviation value of the component of the average image around the (15, 15) and (21, 21) pixels	+	
Maximum value of the average image around the (11, 11) pixel	+	

The gradient boosting algorithm implementation employed XGBoost. The model was trained by combining datasets of feature vectors from all stones in the training set into a single large dataset. A 10% validation sample was used to determine the best model by selecting optimal hyperparameters through random search.

Random search grid parameters:

- Tree depth in the [3, 14] interval;
- Number of trees in the ensemble in the [50, 500] interval;
- Training parameter: [0.1, 0.05, 0.01, 0.005, 0.001];
- Tree subsample proportion: [1, 0.75, 0.5, 0.25];
- Tree features proportion: [1, 0.75, 0.5, 0.25].

Best model parameters:

- Tree depth: 12;
- Number of trees in the ensemble: 480;
- Training parameter: 0.1;
- Subsample proportion: 0.75;
- Features proportion: 1;
- Validation accuracy: 93.43%.

After classifying each pixel of the stone, a primary map of the surface type distribution is obtained (see Figure 14a). To neutralize possible errors and smooth out the zone limits, a median filter with a kernel (7, 7) is applied to this map (see Figure 14b).

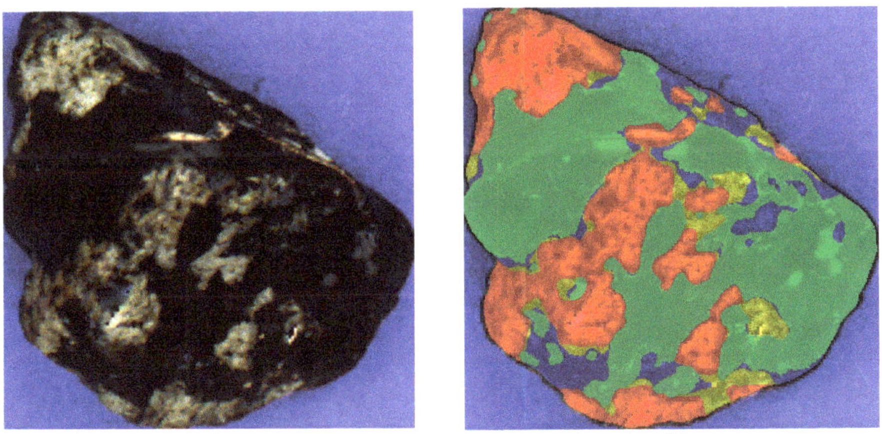

(**a**) (**b**)

Figure 14. The average image (**a**) and the final map (**b**) of the surface type distribution, where red is bare gravel, yellow is an insufficient layer, blue is a thin layer, and green is a thick layer.

4.3. U-Net Architecture

Unlike gradient boosting, where the model analyzes the image pixel by pixel so that it considers only the pixels belonging to the stone, a rectangular matrix must be input to the convolutional neural network, i.e., in addition to distinguishing four surface types, the neural network must also determine the background in the image. The average image is used for the analysis.

The standard approach for semantic segmentation of different dimension images using neural networks is to reduce images to the same size. However, in this case, a different approach involves splitting the original image into tiles of the same small size, segmenting all the tiles and then merging them into an image of the original size. This approach was chosen in order to increase the training sample size to achieve better model generalization.

For the training dataset of stones, the average image sizes are (376, 403), so (128, 128) was chosen as the one tile size. Firstly, this tile size enables the use of a neural network with a relatively small number of parameters, and, secondly, it allows to significantly increase the training sample size through synthetic instances obtained through rotation operations, whereas the tile size is large enough to capture patterns of textures and relief of various surface types.

As part of the search for the neural network optimal configuration, the following parameters varied:

- Compression depth: [2, 5];
- Base number of filters: [2, 16];
- Number of convolution blocks at the level: [1, 3];
- Kernel size in the convolution block at the level: [2, 6];

- Activation function used in the convolution block: [ReLU, Leaky ReLU, ELU, Sigmoid];
- Max pooling kernel size: [2, 4];
- Training parameter: [0.0001, 0.1] with logarithmic progression.

In order to determine the optimal network configuration, a Bayesian optimization algorithm was applied with a maximum number of 800 tested models and with 200 starting points (see Figure 15).

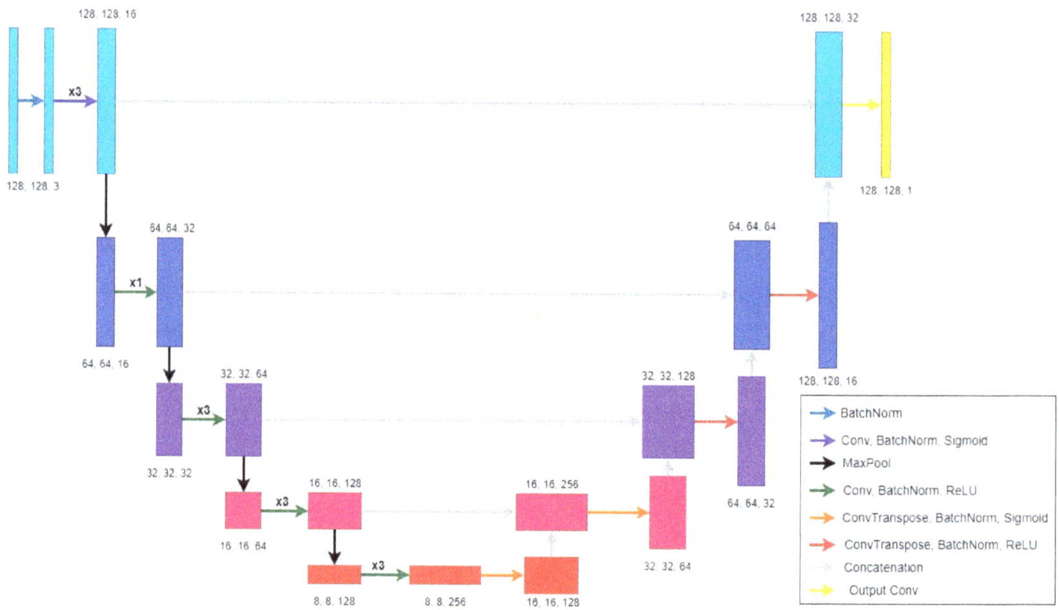

Figure 15. The final neural network architecture with the optimal configuration found.

MSE was used as a loss function to ascertain that not only does the surface type match its true value but also that the falsely defined class is as close to the authentic one as possible when misclassified.

To optimize the parameters, we applied the Adam algorithm with 50 epochs, a validation sample size of 10%, the possibility to initiate early stopping in the absence of MAE metric improvement during validation, and a patience parameter of 5 epochs (see Table 3).

Table 3. The resultant error of the optimal found model in training.

Sample	MSE	MAE
Training	0.0307	0.0534
Validation	0.1112	0.1074

For the ten models with the highest quality obtained through Bayesian optimization, further training has been carried out for an extra 1000 epochs, with a decrease in the training parameter and the possibility of an early stopping with a patience parameter of 10 epochs.

At the output of the neural network, we get an assessment matrix of surface classes for pixels in the tile. As the estimates obtained are real, in order to obtain a map of the surface type distribution, after the tile merging procedure, the procedure of rounding values to integers is applied to the complete matrix, as well as the truncation of estimates in the [0, 4] interval.

The final map is then smoothed using the median filter (7, 7), similar to the case of gradient boosting (see Figure 16).

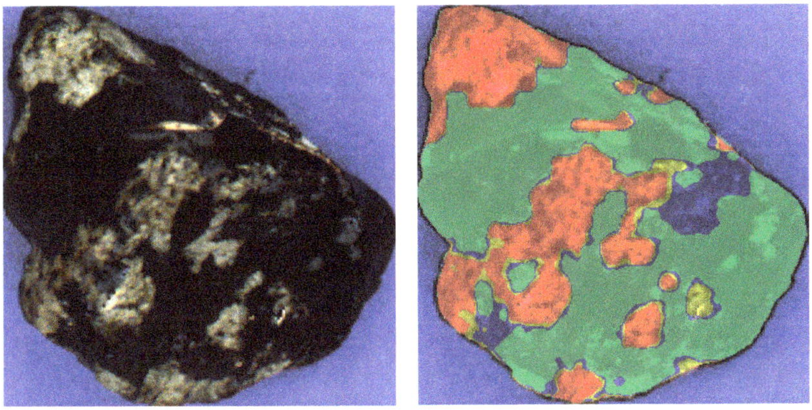

Figure 16. The average image and the final map of surface type distribution. Green color indicates thick coating; red—bare gravel; yellow—insufficient layer; blue—thin layer.

4.4. Comparison of Segmentation Accuracy

In order to perform an unbiased image segmentation quality assessment, the results of the developed models have been compared with the accuracy in naive segmentation. During this process, the entire stone surface is determined by one surface type for all the stones in the test sample (see Tables 4 and 5). Metrics utilized include the average segmentation accuracy of the test sample and the average absolute classification error.

Table 4. Naive segmentation results.

Surface Type	Average Accuracy, %	Average MAE
Bare gravel	1.86	2.41
Insufficient layer	14.36	1.45
Thin layer	24.86	0.77
Thick layer	58.92	0.59

Table 5. Model segmentation results.

Model	Average Accuracy, %	Average MAE
TS	49.79	0.68
XGboost	77.03	0.26
U-Net	77.04	0.27

Also, a simple and quick semantic segmentation method was implemented to compare the results obtained with a widely used threshold segmentation approach (TS). It is based on filtering pixels by a threshold value and is used in the majority of the reviewed works. Therefore, we adapted the algorithm for image segmentation for all surface types under consideration. In addition, we applied a number of known effective practices: image analysis in the HSV [12], in which surface types are best separated, and the use of additional image sampling [11], in order to simplify the task of component filtering of image pixels. The implemented approach includes the following steps:

- Converting all frames from RGB to HSV;
- Determining the mean of the image for H and V components;
- Sampling the image into areas of width 20;
- Partitioning the stone surface into regions where gravel surface is not detected (class 1) and where gravel surface is detected (class 2) based on comparison with threshold values;

- Partitioning the regions of class 1 into areas with insufficient coverage and bare gravel;
- Partitioning the regions of class 2 into areas with a thick and thin layer of coating.

The threshold values of H and V components were determined on a training set of images with a marked surface. For each stone, the optimal values for dividing the zone were determined according to the rule described above from the point of view of maximizing the Kullback–Leibler divergence. Then, the obtained threshold values were averaged over all stones.

It is apparent that both XGboost and U-Net demonstrate comparable results, significantly better than naive segmentation and TS. The U-Net shows marginally superior precision in identifying surface types, but XGBoost deals with surface type classification errors better than U-Net.

4.5. Final Classification Module

Having passed the semantic segmentation module, the final distribution map of the stone surface types is transferred to the final classification module. Here, the areas of each surface type are calculated in relation to the total area of the stone.

The integrated (percentage) quality assessment of the stone coating is calculated as a weighted sum of data:

$$R = (0.2Weak + 0.5Thin + Thick) \times 100, \quad (1)$$

where R is an integral estimate, *Weak* is an insufficient layer relative area, *Thin* is a thin layer relative area, and *Thick* is a thick layer relative area.

For identifying the discrete estimate of the stone coating, in order to simulate the expert's logic while determining the estimate and prevent retraining, a classifier based on a decision tree is used, with a maximum depth of 5 and with a minimum number of 7 elements in the list. The features in the classification of a stone are its values relative to the surface type areas.

All stones with marked up surfaces are utilized to train the classifier for discrete estimates.

The final test studied the correspondence of discrete estimates of stones determined by the algorithm to expert's discrete estimates for all stones in the sample. The results obtained using the XGBoost and U-Net segmentation models, as well as threshold segmentation, are compared (see Table 6).

Table 6. Comparison of test results.

Metric	TS, %	XGboost, %	U-Net, %
Accuracy	62.96	75.56	73.7
Estimates less than the expert one by 0.5	12.59	12.22	15.19
Estimates greater than the expert one by 0.5	15.19	9.63	7.78
Estimates less than the expert one by 1	2.96	1.11	0.74
Estimates greater than the expert one by 1	5.56	0.74	1.11
Estimates that differ from the expert one by more than 1	0.74	0.74	1.48
MAE	0.239	0.139	0.156
Maximum deviation of the estimate	2.5	1.5	1.5

As can be seen from the results, the accuracy of a simple threshold segmentation is significantly lower. Approximately in 8.52% of cases, the TS method assigns a stone to a different class relative to expert assessment. The main disadvantage of this algorithm is that, due to the classification of an isolated pixel, it disregards the surface topography. Therefore, it is not able to qualitatively distinguish the class 5 and class 4 (among the errors with overestimation of stone, the share of class 4 is 53%). In both cases, the surface of the stone is completely covered with bitumen, but there is a difference in layer thickness. This can be significant when industrially testing samples of asphalt mixtures.

Assessing the quality of the developed solution, it can be concluded that when using the XGBoost-based segmentation model, the algorithm shows better results, compared with U-Net, although the difference in quality is negligible. The XGBoost-based model is more likely to overestimate the stone value, whereas U-Net tends to underestimate it more frequently.

Generally, the algorithm's estimate coincided with the expert's one in over 70% of cases. In an additional 22% of cases, the algorithm's estimate differed from the expert's by only 0.5 points, which falls within the threshold value. The algorithm's estimate differed from the expert's one by more than 1 point in only 2% of cases, with the maximum error of 1.5 points. Therefore, a stone with an expert estimate of "5" has never received an algorithm estimate of "3" or lower, and vice versa.

4.6. Performance Analysis

During image processing, the algorithm combines the received images into one, segments the bitumen samples in the combined image, and processes each region of the segmented samples individually. As the developed HSS has restrictions on the hardware installed, the running time of the algorithm varies depending on the properties of the set of samples under analysis. To reveal dependencies between the algorithm's performance and the parameters of the samples, some tests on gravel of various numbers and sizes have been carried out. For the experiment, sets of 1, 3, and 5 gravel samples were used. For each sample size, five sets of available samples were applied. Before running tests for a new sample size, the application was restarted. Also, for each sample, the experiment was averaged over three runs. For the test, random samples of various sizes were used (see Figure 17). The average operating times of the algorithm are given in Table 7.

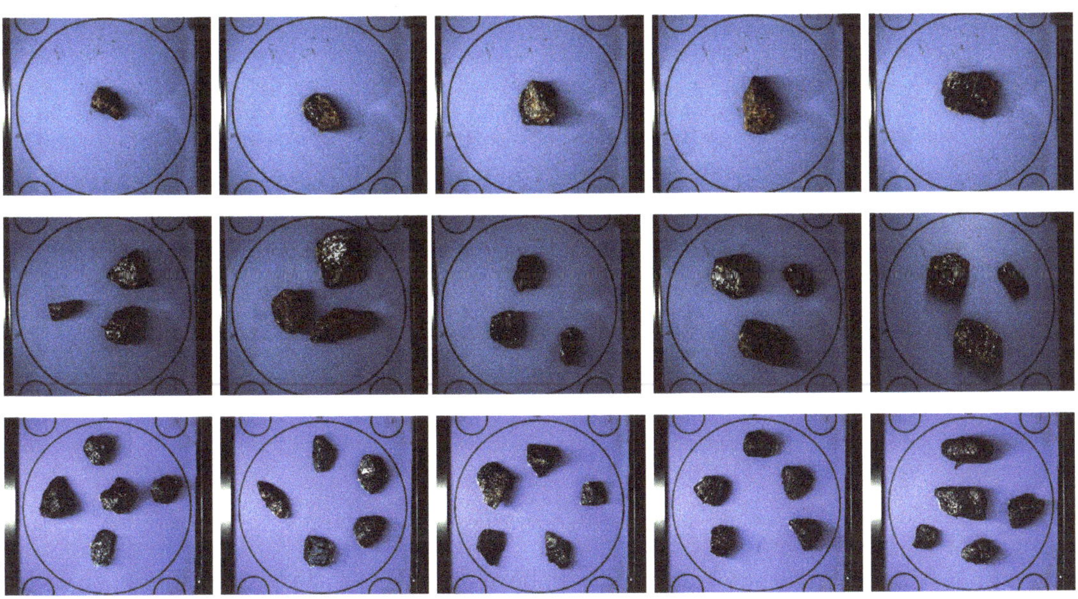

Figure 17. Sets of samples.

Table 7. Average processing time.

Set Size	Combining Images, s	Search for Stone Masks, Average by Sample, s	Surface Analysis, Average by Sample, s	Total Time, s
1	0.8	3.62	59.28	63.88
3	0.8	3.22	56.09	172.74
5	0.8	3.18	38.94	198.92

The results show that the operation of combining images and searching for stone masks takes approximately the same time. However, due to the performance degradation after initializing the XGBoost model on the prototype, there was an increase in processing time. In addition, the impact of the initialization of the XGBoost model on the execution time of the algorithm is noticeable. The experiment showed a great variability in the time of individual processing of stones in the sample: from 22 s to 75.9 s. It should be noted that as the number of samples in a set increase, the processing time for an individual sample decreases slightly due to the peculiarities of initialization and testing of the model. Analysis execution time is affected by throttling of the device's CPU after the XGBoost model is initialized.

5. Conclusions and Further Development

The adhesion properties of bitumen determine the quality and durability of road surfaces. According to existing practices, they are determined subjectively by experts, which entails known disadvantages. The aim of this work was to develop a methodology able to replace the human in assessment of bitumen quality regardless of the conditions under which the analysis is performed. For this purpose, hardware that "works out of the box" was developed, facilitating sample processing without requiring preliminary preparation and configuration.

To ensure the assessment accuracy, two versions of semantic surface segmentation algorithms have been developed by the authors. After comparing the results, it is evident that the gradient boosting approach outperforms the U-Net architecture for the purposes of the research. In general, both algorithms show a fairly high coherence of their estimates with expert discrete estimates. The significant discrepancy (error in determining the class of the sample) with the expert's assessment was less than 2%.

A possible explanation for the reason why the U-Net architecture has shown worse quality compared to the gradient boosting method is the insufficiency of the training sample, from which the model fails to extract general patterns. Therefore, the most straightforward and evident direction for further development is to increase the number of marked up samples.

Another important direction of research is the reduction in image processing time. As this work is devoted to a laboratory prototype, the main requirement for it is the accuracy of assessing the quality of bitumen coating, expressed in approximation to expert assessment. At the same time, sample processing time was not a key requirement. With further development of the proposed methodology, a conveyer analyzer can be implemented with greater computing resources, parallelization of calculations, and optimization of calculation time.

It should be taken into account that the thickness of the covering bitumen layer is a significant factor in the expert evaluation, against which the accuracy of the developed algorithms was assessed. A significant contribution of our work compared to binary classification of the known methods is that the developed solution is capable of classifying the gravel samples considering the thickness of the covering bitumen layer. It is also worth noting the general differences, both in terms of hardware implementation (number of images, shooting angle, and distance to the lens) and in the methodological part (calculating the accuracy of clustering for all stones in the image or for each one separately and the number of stones in the image, as precise surface classes have been formed). Unification of these conditions allows us to implement a more sophisticated methodology that makes our developments applicable at the industrial level.

The authors have developed and registered a patent for a hardware and software system and a methodology for automated assessment of the adhesion of bitumen to gravel. These developments may further lead to the establishment of pavement durability and quality standards. Our methodology minimizes testing errors and can be used to create a database of bituminous mixtures tests. In addition, the developed methodology allows standardizing the data obtained and contributes to the creation of digital research database compiled of various sources. If the HSS application is successful, we will consider its launch into mass production. Incorporating the developed solutions into existing adhesion level test standards may result in improved primary tests for asphalt quality.

Author Contributions: Conceptualization, A.B.; methodology, A.B. and D.K.; software, D.K. and N.V.; validation, D.K.; investigation, D.K. and N.V.; data curation, D.K.; writing—original draft preparation, D.K. and N.V.; writing—review and editing, I.V.; supervision, I.V.; project administration, M.B. All authors have read and agreed to the published version of the manuscript.

Funding: This research is partially funded by the Ministry of Science and Higher Education of the Russian Federation as part of the World-class Research Center program: Advanced Digital Technologies (contract No. 075-15-2022-311 dated 20 April 2022).

Institutional Review Board Statement: Not applicable.

Informed Consent Statement: Not applicable.

Data Availability Statement: No new data were created or analyzed in this study. Data sharing is not applicable to this article.

Conflicts of Interest: Author Alexander Boldyrev was employed by the company LLC Amdor. The remaining authors declare that the research was conducted in the absence of any commercial or financial relationships that could be construed as a potential conflict of interest.

References

1. Caputo, P.; Miriello, D.; Bloise, A.; Baldino, N.; Mileti, O.; Ranieri, G.A. A comparison and correlation between bitumen adhesion evaluation test methods, boiling and contact angle tests. *Int. J. Adhes. Adhes.* **2020**, *102*, 102680. [CrossRef]
2. Paliukaitė, M.; Vorobjovas, V.; Bulevičius, M.; Andrejevas, V. Evaluation of different test methods for bitumen adhesion properties. *Transp. Res. Procedia* **2016**, *14*, 724–731. [CrossRef]
3. Kallen, H.; Heyden, A.; Astrom, K.; Lindh, P. Measurement of Bitumen Coverage of Stones for Road Building, Based on Digital Image Analysis. In Proceedings of the 2012 IEEE Workshop on the Applications of Computer Vision (WACV), Breckenridge, CO, USA, 9–11 January 2012; pp. 337–344. [CrossRef]
4. Kallen, H.; Heyden, A.; Lindh, P. Measuring Bitumen Coverage of Stones Using a Turntable and Specular Reflections. In Proceedings of the International Conference on Computer Vision Theory and Application, Barcelona, Spain, 21–24 February 2013; pp. 333–337. [CrossRef]
5. Mulsow, C. Determination of the degree of gravel aggregate-bitumencoverage by multi-directional reflectance measurements. *Int. Arch. Photogramm. Remote Sens. Spat. Inf. Sci.* **2012**, *39*, 39–43. [CrossRef]
6. Källén, H.; Heyden, A.; Åström, K.; Lindh, P. Measuring and Evaluating Bitumen Coverage of Stones Using Two Different Digital Image Analysis Methods. *Measurement* **2016**, *84*, 56–67. [CrossRef]
7. Blom, J.; De Boeck, B.; Soenen, H. Determination of bitumen stone coverage by digital image processing. In Proceedings of the 7th International Conference "Bituminous Mixtures and Pavements", Thessaloniki, Greece, 12–14 June 2019; pp. 172–178.
8. Lantieri, C.; Lamperti, R.; Simone, A.; Vignali, V.; Sangiorgi, C.; Dondi, G.; Magnani, M. Use of Image Analysis for the Evaluation of Rolling Bottle Tests Results. *Int. J. Pavement Res. Technol.* **2017**, *10*, 45–53. [CrossRef]
9. Riekstins, A.; Haritonovs, V.; Balodis, A. Evaluation of Adhesion Between Bitumen and Aggregate with the Digital Image Processing Method. In Proceedings of the 4th International Conference on Innovative Materials, Structures and Technologies (IMST 2019), Riga, Latvia, 25–27 September 2019; p. 012047.
10. Trejbal, J.; Valentová, T.; Horová, T. Assessment Of Adhesion Between Bituminous Binder And Mineral Aggregate Using Digital Image Analysis. *Acta Polytech. CTU Proc.* **2019**, *21*, 38–41. [CrossRef]
11. Xiao, R.; Polaczyk, P.; Huang, B. Measuring Moisture Damage of Asphalt Mixtures: The Development of a New Modified Boiling Test Based on Color Image Processing. *Measurement* **2022**, *190*, 110699. [CrossRef]
12. Chomicz-Kowalska, A. A Study of Adhesion in Foamed WMA Binder-Aggregate Systems Using Boiling Water Stripping Tests. *Materials* **2021**, *14*, 6248. [CrossRef]

13. Sahari Moghaddam, A.; Rezazadeh Azar, E.; Mejias, Y.; Bell, H. Estimating Stripping of Asphalt Coating Using K-Means Clustering and Machine Learning–Based Classification. *J. Comput. Civ. Eng.* **2020**, *34*, 04019044. [CrossRef]
14. Wang, Z. A New Approach for Segmentation and Quantification of Cells or Nanoparticles. *IEEE Trans. Ind. Inform.* **2016**, *12*, 962–971. [CrossRef]
15. Wang, Z.; Xiong, J.; Yang, Y.; Li, H. A Flexible and Robust Threshold Selection Method. *IEEE Trans. Circuits Syst. Video Technol.* **2017**, *28*, 2220–2232. [CrossRef]
16. Wang, Z.; Wang, E.; Zhu, Y. Image Segmentation Evaluation: A Survey of Methods. *Artif. Intell. Rev.* **2020**, *53*, 5637–5674. [CrossRef]
17. Kütük, Z.; Algan, G. Semantic Segmentation for Thermal Images: A Comparative Survey. In Proceedings of the IEEE/CVF Conference on Computer Vision and Pattern Recognition, New Orleans, LA, USA, 18–24 June 2022; pp. 286–295.
18. Minaee, S.; Boykov, Y.; Porikli, F.; Plaza, A.; Kehtarnavaz, N.; Terzopoulos, D. Image Segmentation Using Deep Learning: A Survey. *IEEE Trans. Pattern Anal. Mach. Intell.* **2021**, *44*, 3523–3542. [CrossRef] [PubMed]
19. Kheradmandi, N.; Mehranfar, V. A Critical Review and Comparative Study on Image Segmentation-Based Techniques for Pavement Crack Detection. *Constr. Build. Mater.* **2022**, *321*, 126162. [CrossRef]
20. Mazhar, M.; Fakhar, S.; Rehman, Y. Semantic Segmentation for Various Applications: Research Contribution and Comprehensive Review. *Eng. Proc.* **2023**, *32*, 21. [CrossRef]
21. Atif, N.; Bhuyan, M.; Ahamed, S. A Review on Semantic Segmentation from a Modern Perspective. In Proceedings of the 2019 International Conference on Electrical, Electronics and Computer Engineering (UPCON), Aligarh, India, 8–10 November 2019; pp. 1–6. [CrossRef]
22. Yu, Y.; Wang, C.; Fu, Q.; Kou, R.; Huang, F.; Yang, B.; Yang, T.; Gao, M. Techniques and Challenges of Image Segmentation: A Review. *Electronics* **2023**, *12*, 1199. [CrossRef]
23. Ronneberger, O.; Fischer, P.; Brox, T. U-Net: Convolutional Networks for Biomedical Image Segmentation. In Proceedings of the Medical Image Computing and Computer-Assisted Intervention—MICCAI 2015, Munich, Germany, 5–9 October 2015; Navab, N., Hornegger, J., Wells, W.M., Frangi, A.F., Eds.; Springer International Publishing: Cham, Switzerland, 2015; pp. 234–241. [CrossRef]
24. Enshaei, N.; Ahmad, S.; Naderkhani, F. Automated Detection of Textured-Surface Defects Using U-Net-Based Semantic Segmentation Network. In Proceedings of the 2020 IEEE International Conference on Prognostics and Health Management (ICPHM), Detroit, MI, USA, 8–10 June 2020; pp. 1–5. [CrossRef]
25. Majidifard, H.; Adu-Gyamfi, Y.; Buttlar, W.G. Deep Machine Learning Approach to Develop a New Asphalt Pavement Condition Index. *Constr. Build. Mater.* **2020**, *247*, 118513. [CrossRef]
26. Robet, R.; Hasibuan, Z.A.; Arief Soeleman, M.; Purwanto, P.; Andono, P.N.; Pujiono, P. Deep Learning Model In Road Surface Condition Monitoring. In Proceedings of the 2022 International Seminar on Application for Technology of Information and Communication (iSemantic), Semarang, Indonesia, 17–18 September 2022; pp. 204–209. [CrossRef]
27. Takos, G. A Survey on Deep Learning Methods for Semantic Image Segmentation in Real-Time. *arXiv* **2020**, arXiv:2009.12942.
28. Ulku, I.; Akagündüz, E. A Survey on Deep Learning-based Architectures for Semantic Segmentation on 2D Images. *Appl. Artif. Intell.* **2020**, *36*, 2032924. [CrossRef]
29. Dai, Z.; Liu, X.; Xu, C. Image Semantic Segmentation Scheme Based on XGBoost Combination with Convolution Feature Extraction. In Proceedings of the 2022 41st Chinese Control Conference (CCC), Hefei, China, 25–27 July 2022; pp. 6334–6340. [CrossRef]
30. Canny, J. A Computational Approach to Edge Detection. *IEEE Trans. Pattern Anal. Mach. Intell.* **1986**, *8*, 679–698. [CrossRef]
31. Suzuki, S.; Be, K. Topological Structural Analysis of Digitized Binary Images by Border Following. *Comput. Vis. Graph. Image Process.* **1985**, *30*, 32–46. [CrossRef]

sensors

MDPI

Article

CLensRimVision: A Novel Computer Vision Algorithm for Detecting Rim Defects in Contact Lenses

Pawat Chunhachatrachai [1] and Chyi-Yeu Lin [1,2,*]

[1] Department of Mechanical Engineering, National Taiwan University of Science and Technology, Taipei 10632, Taiwan; pawat.chun@gmail.com

[2] Center for Intelligent Manufacturing Innovation, National Taiwan University of Science and Technology, Taipei 10632, Taiwan

* Correspondence: jerrylin@mail.ntust.edu.tw

Abstract: Automated optical inspection (AOI) plays a pivotal role in the quality control of contact lenses, safeguarding the safety and integrity of lenses intended for both medical and cosmetic applications. As the role of computer vision in defect detection expands, our study probes its effectiveness relative to traditional methods, particularly concerning subtle and irregular defects on the lens rim. In this research study, we propose a novel algorithm designed for the precise and automated detection of rim defects in contact lenses called "CLensRimVision". This algorithm integrates a series of procedures, including image preprocessing, circle detection for identifying lens rims, polar coordinate transformation, setting defect criteria and their subsequent detection, and, finally, visualization. The method based on these criteria can be adapted either to thickness-based or area-based approaches, suiting various characteristics of the contact lens. This approach achieves an exemplary performance with a 0.937 AP score. Our results offer a richer understanding of defect detection strategies, guiding manufacturers and researchers towards optimal techniques for ensuring quality in the contact lens domain.

Keywords: contact lens; computer vision; automatic optical inspection; defect detection; image processing

1. Introduction

Contact lenses, medical devices intended to be worn on the cornea, have revolutionized the world of ophthalmology and vision correction [1,2]. Their use has grown substantially over the years due to their convenience and advancements in materials and design [3–5]. Ensuring the safety and effectiveness of these lenses is crucial, not only to provide optimal vision but also to prevent potential eye-related complications [6,7]. One of the critical aspects for ensuring the quality of contact lenses is the detection and elimination of defects, particularly in areas that have a direct interface with the eye [8–10].

The rim area of a contact lens is particularly sensitive as defects in this region can lead to discomfort, tear film disruption, or even potential injury to the ocular surface. Traditional methods for defect detection, such as manual inspection, may not provide the necessary sensitivity or specificity needed for accurate and efficient quality control at an optimal cost and in an optimal time frame [11,12]. The size of a contact lens is very tiny relative to the human body. It must be in a proper liquid, comfortable, visible, transparent, soft, and curved. These are challenges for the quality control process of the product, which has strict criteria for its use in humans.

In recent times, with the advent of the computer, there has been a paradigm shift in how inspections are conducted in various industries [13–18]. Automated defect detection systems have demonstrated high levels of accuracy and consistency, surpassing human capabilities in certain cases. However, the unique challenges posed by the transparent nature of contact lenses and the subtle tiny defects that can arise in the rim area require a specialized approach.

This paper introduces a novel algorithm tailored for the robust and accurate detection of defects in the rim area of contact lenses. Leveraging advanced image processing techniques combined with mathematics and statistics, this system aims to offer superior detection capabilities, ensuring higher quality and safety standards for contact lens users worldwide. We focus on this area because it directly interacts with the rim of the human eye, making it crucial for the wearer's comfort. Defects, such as burrs, can cause irritation to wearers. Figure 1a,b illustrate examples of a clear lens and a colored and patterned contact lens edge with some defects and a lack of smoothness from our real dataset. This kind of defect is called a burr, which refers to a small, unwanted, and often sharp-edged or -ridged material that can form on the surface of a workpiece as a result of various machining or manufacturing processes. In the yellow box, display a larger view of samples showing defects in the rim area of both transparent and patterned contact lenses.

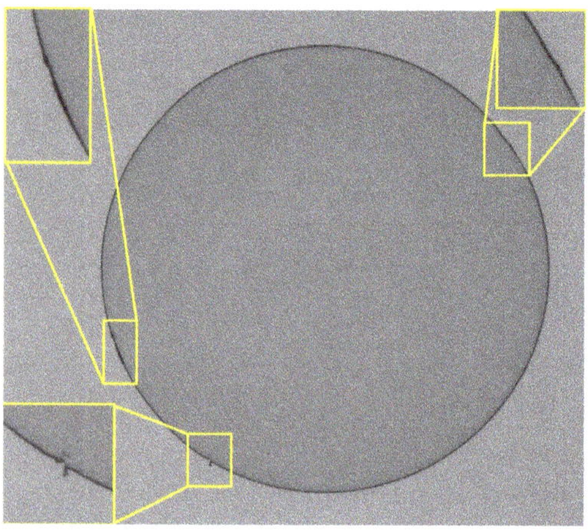

(a)

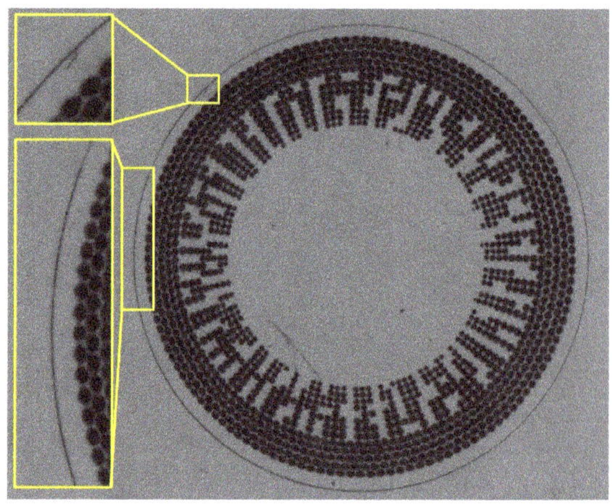

(b)

Figure 1. Sample defects on edge of contact lens. (**a**) Clear contact lens; (**b**) Colored and patterned contact lens.

Our paper's contributions are as follows.

- We proposed a novel image-processing-based algorithm to detect burr defects on the rim of contact lenses.
- We applied this comprehensively using mathematics and statistics to a computer vision algorithm.
- We presented information relevant for manufacturers and researchers in the contact lens sector, providing them with insights and methodologies that can be directly applied to improve quality control automation processes.

2. Background

Contact lenses are intricately designed medical devices that require the utmost precision during manufacturing. Defects, even if minute, not only compromise the wearer's visual clarity but also their safety. Over the years, ensuring defect-free lenses has been a priority, with multiple quality control methods put into place.

Historically, the contact lens industry has relied heavily on manual inspection methods, in which trained professionals visually inspect each lens under magnification. While this method has the advantage of human intuition, it is time-consuming and susceptible to inconsistencies and errors, especially when dealing with vast production numbers.

2.1. Quality Control in Optical Industry

In the realm of optical products, such as contact lenses and camera lens glasses, maintaining a high level of quality control is paramount to ensuring optimal transparency [7–9] and visual performance, as Rebsamen et al. stated [19]. While defects in optical products can occur, quality control systems are in place to assess whether the detected defects meet the acceptable criteria. It is important to note that different areas within the same product may have distinct criteria for evaluation [20].

For contact lenses specifically, quality control standards are outlined in ISO 18369-1:2017 and contact lens fundamental requirements in ISO 14534:2015 [19]. These standards not only address the overall quality control requirements but also delve into the criteria and responsibilities for the different areas of a contact lens. Adhering to these standards ensures that the quality of contact lenses aligns with established guidelines and contributes to the overall reliability and safety of these optical products. The most inner circle zone is called the center optical zone, Zone 2 is called the rim optical zone, and Zone 1 is called the rim connected zone. In our research, we will solely focus on Zone 1 as it is the area most sensitive to the user's safety and the area we use to develop our algorithm for the detection of edge defects.

2.2. Computer Vision for AOI of Contact Lenses

AOI is being used to shorten the process of the mass production of lenses in terms of minimizing the presence of defective products. There are several techniques and combinations inside the AOI sector. In the context of contact lens defect detection, Chang et al. use computer vision algorithms to utilize edge points to calculate the distance from the center and employ a radius abnormality function [21]. This approach involves computing the distance between the radii of adjacent parts of the circle and evaluating the absolute difference. Finally, the radius abnormality function is applied by defining an offset threshold. However, during the actual experimental setup, there may be situations in which the contact lens is not properly placed in the contact lens holder (we will explain this in Section 3.1), leading to an uneven thickness along the circular rim of the contact lens (we will discuss this in Section 4.4.1). This scenario motivates us to further develop the algorithm to accurately detect defects even under such circumstances, a consideration that Cheng et al. may not have considered.

After scoping the contact lens area, the next step is to use image processing methods to effectively detect and assess defects in contact lenses, which can contribute to overall quality control. The utilization of computer vision algorithms enhances the efficiency and accuracy of the AOI process for contact lens production. The following image preprocessing and computer vision techniques make the defect clearer and more accurately detected.

Image Processing Techniques

- Color space and morphological operations can extract information as compared to intensity or HSV color-space-based segmentation, as Dumitru Dan Burdescu et al. proved in their research [22]. The HSV (hue, saturation, and value) color space is utilized for color-based object segmentation. The hue channel represents the color category, while the saturation ranges from unsaturated to fully saturated levels. The value channel conveys the color intensity. Morphological operations, such as erosion and dilation, modify images using a structuring element [23–25]. Hasan et al. use morphological operations and prove that this technique can detect the edge of objects [26]. Moreover, the usage of erosion reduces noise and isolates elements, while dilation merges dissimilar elements and detects irregularities.
- Polar Coordinate Transformation: The conversion of Cartesian coordinates (x, y) to polar coordinates (r, θ) has various effects and facilitates specific image processing tasks. This transformation is useful for radial distortion correction, circular pattern analyses, and the transformation of images into a polar representation. The open library from CV2 [27] that supports this algorithm is Warp polar [28].

- Circle Detection: The algorithm aims to identify the best-fit circle that matches a given set of points. A circle in 3D space can be represented using a parametric equation, involving the center, radius, normal unit vector, and perpendicular vector. These parameters provide a comprehensive description of the circle in three-dimensional space. Ramlee et al. use circle detection to detect outliers in their research [29,30], which is similar to our aim to detect burr defects on the circular edge of contact lens.

3. Experimental Setup

3.1. Concept of Machine Design

This section outlines the procedural steps for configuring the traditional physical design of a contact lens. The automated optical inspection (AOI) system is designed to capture high-resolution images of contact lenses and emphasize detailed visualization of defects. The setup design is shown in Figure 2a. First, to set up the camera, position it facing downward from the top of the inspection holder. Second, set up the base, which contains the contact lens in a liquid. The container must be transparent and placed beneath the camera. Third, since contact lenses are transparent objects, a specialized photographic approach employing backlighting is crucial. A backlight is placed at the bottom of the inspection base to illuminate the object of interest and reveal all the necessary details for the camera to capture. In our study, we adopted a traditional approach inspired by the design of this contact lens AOI machine. Our camera placement was carefully calibrated to enable surface imaging, while backlighting was used to detect irregularities or foreign substances. The design of this apparatus draws from the work of Jin et al. [31,32], who used cold cathode fluorescent lamps (CCFLs) as a light source to illuminate a glass ribbon placed perpendicular to the vertical axis. The image was then captured using a CCD camera. The physical arrangement in our experiment is presented in Figure 2b. This equipment was designed and built by Intelligent Robotics Lab, Department of Mechanical Engineering, National Taiwan University of Science and Technology.

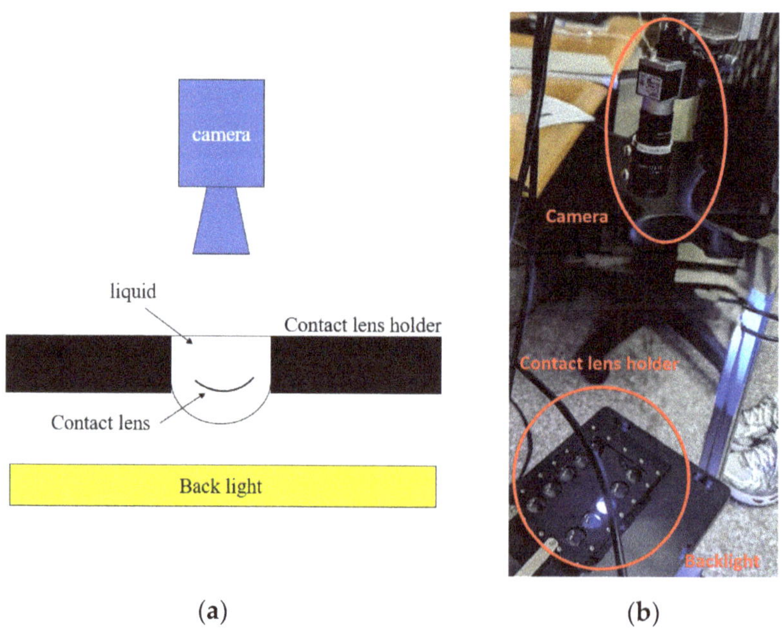

Figure 2. Design of contact lens AOI system. (**a**) Setup design; (**b**) Physical setup.

After setting up the station and successfully capturing images of an adequate quality for the experiment, the next step involves performing the experiment on a processing unit. For this experiment, we utilize a computer with the following specifications: an Intel® Core™ i9-10900X CPU running at 3.70 GHz, an RTX 1080Ti 11 GB GPU, and 16 GB of RAM.

3.2. Dataset

This dataset serves as the foundation for testing the novel algorithm using image processing approaches. Our objective is to assess the inherent learning capabilities of algorithm using raw images alongside evaluation of their corresponding labels. The raw images we use for training and evaluation are 5472 pixels in width and 3648 pixels in height. The original set of contact lens samples comprised 60 lenses; 3 images for each were captured, producing a total of 180 original images as shown in Figure 3.

(a) (b)

Figure 3. A sample of contact lens. (**a**) Original image; (**b**) Labeled image.

The dataset was composed of two types of contact lenses: a clear contact lens and colored and patterned contact lens, which are shown in Figure 1a,b, respectively. Transparent contact lens has 12 pieces and colored and patterned contact lens has 48 pieces in the dataset.

We prepared the dataset for evaluation metrics by labeling the defect area of each image. The labeled defects are shown in Figure 3. In each image, the defects are labelled in details in the green boxes. The objective is to decrease high variation in detection results' aspect ratio and size. Examples of the original field of view (FOV) and an image with the visualized labels are shown in Figure 3a,b, respectively.

3.3. Evaluation Metrics

In this section, we introduce key evaluation metrics that play a pivotal role in assessing the performance of our object detection model. These metrics provide valuable insights into the accuracy and completeness of object localization and identification, which are fundamental aspects of our research.

- Intersection over union (IoU) is a critical metric in object detection, quantifying the degree of overlap between predicted bounding boxes and ground truth bounding boxes. It is calculated as the ratio of the area of intersection between the two bounding boxes to the area of their union. IoU serves as a measure of the spatial alignment between the predicted and actual object locations.
- Precision measures the proportion of true positive predictions out of all positive predictions made by the model. In our context, it assesses how many of the objects our model detected are genuinely correct.

- Recall measures the proportion of true positive predictions out of all actual objects in the dataset. It evaluates our model's capability to identify all relevant objects in the images.
- Mean average precision (mAP) is a comprehensive metric that combines precision and recall across a range of IoU thresholds. It provides a single scalar value to gauge the overall performance of our object detection model. mAP is particularly valuable when we need to compare our model's performance against various benchmarks or evaluate its effectiveness across different scenarios. In the context of our research, the dataset exclusively comprises a single class of objects. Consequently, we have opted to utilize average precision (AP) as the primary evaluation metric. AP offers a robust means of assessing the precision/recall trade-off and model performance and is particularly well suited for single-class object detection scenarios.

In the subsequent sections, we will delve into the methodology of our experiments, the datasets used, and the results obtained, all of which are pivotal in demonstrating the effectiveness of our approach.

4. Methodology

This chapter explains the concept of CLensRimVision algorithm to detect defects on the edge of the outer circle of the contact lens. ClensRimVision algorithm comprises two main components: the offline process and the online process. The term 'offline process' refers to the setting of parameters which must be fully established prior to implementation or production. We use 'offline' to emphasize that these preparations are essential and must be completed beforehand. Conversely, the 'online process' pertains to the actual implementation of the algorithm, specifically for defect detection on a production line. The workflow of ClensRimVision algorithm is as shown in Figure 4. In this proposed algorithm, we integrate both basic image processing methods and our designed methods within the overall workflow. The description of basic image processing is provided in Section 4.1, Section 4.2, Section 4.3, Sections 4.4 and 4.5. However, the sequence for connecting each method is determined and arranged by us. This arrangement has been proven to guarantee the desired output results and performance of our experiments. The innovative part that can significantly increase our chance to fulfill our objective, as explained in Section 4.4, constitutes the core idea of this research paper. The sequencing of the system workflow is also a significant aspect contributing to the innovation of this research. Each step in the workflow plays a crucial role in enhancing the results of the preceding part, as it serves as the input for the subsequent stage.

In the offline process, various parameters are carefully configured to optimize the performance of the defect detection algorithm. During implementation, images of the contact lens may be captured under different settings, such as various methods of capture, lighting conditions, distances, and more. Therefore, proposing an algorithm that can dynamically adjust parameters according to different settings is essential. This entails setting parameters for three key aspects: image preprocessing, circle detection, and defect detection, as shown in Table 1. These parameters are predetermined and serve as the foundation for the defect detection system. In the online process, the algorithm utilizes the configured parameters to analyze and detect defects on the edges of contact lenses. This is achieved through the application of image processing procedures and the proposed algorithms. The online process takes input images or a combination of computer vision modules as input and performs defect detection based on the configured parameters from the offline process.

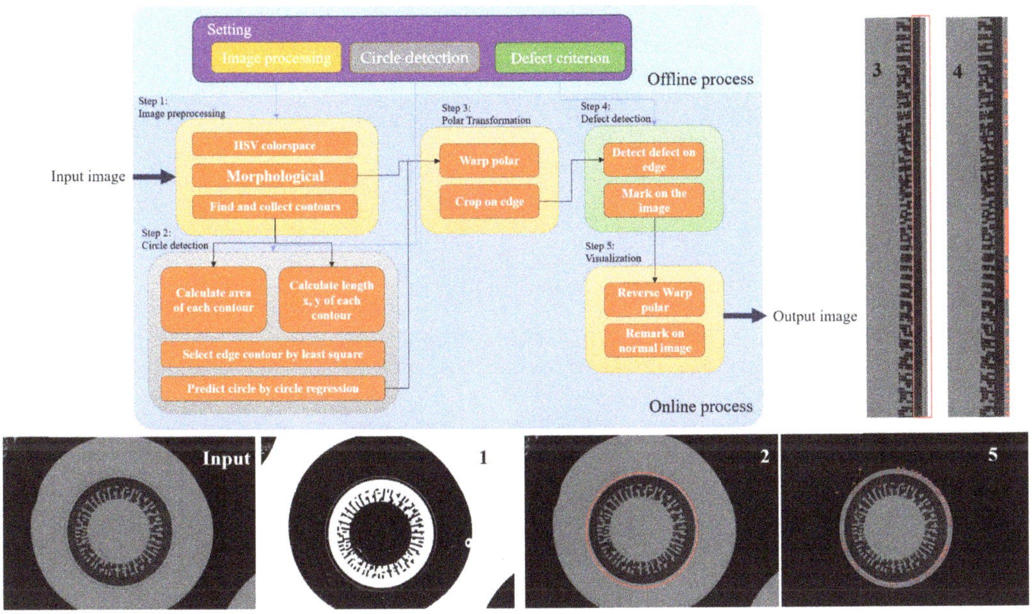

Figure 4. ClenRimVision algorithm process.

Table 1. Parameters in offline process.

Sections	Parameter Name	Domain	Default
Image processing	H color space range	0 to 180	0 to 180
	S color space range	0 to 255	0 to 18
	V color space range	0 to 255	197 to 255
	Morphological Dilate	0 to 30	5
	Morphological Erode	0 to 30	5
Circle detection	Size of radius circle in range	integer	1200 to 1350
	Center circle position	(x, y) in integer	(2010,1050)
	Max area of a contour	integer	50,000
Criterion	Size of separated image	0 to 200	10
	Number of images above	15 to 100	50
	Number of images below	15 to 100	50
	Extended thickness pixel criterion (thickness based)	0 to 30	>3
	Shrink thickness pixel criterion (thickness based)	0 to 30	>3
	Extended area ratio criterion (area based)	0 to 3.0	>0.1
	Shrink area ratio criterion (area based)	0 to 3.0	>0.1
	Factor of SD to average	0 to 3.0	0.5

Furthermore, the algorithm's workflow comprises five main execution steps as follows:

1. Image Preprocessing: This step is responsible for adjusting and preprocessing the input contact lens image to convert it into a contour image, highlighting the rich details and essential components within the image. In Figure 4, labelled "Input" shows the input image of ClensRimVision algorithm, and labelled "1" shows the output of the Image Preprocessing step.
2. Circle Detection: This step focuses on detecting the area and coordinates of the circular contact lens within the input image. The red circle in Figure 4, labelled "2" shows output of Circle Detection step.

3. Polar Coordinate Transformation: This step focuses on the transformation from circular to linear forms, stemming from the circle detection step. In Figure 4, labelled "3" shows image after Polar Coordinate Transformation is applied, and in the red box shows the region of interest for the algorithm.
4. Defect Detection: This step focuses on detecting defects at the rim area by using linear image. In Figure 4, labelled "4" shows defects in the object by printing red area in the image.
5. Visualization: This step focuses on visualizing the defects by reversing polar coordinate transformation and marking the results. In Figure 4, labelled "5" shows result of detecting defects in the original image aspect.

4.1. Image Preprocessing

We employ image processing to refine and enhance the input image, making edges sharper and more readily detectable by the circle algorithm. This enhancement utilizes two primary functions in image processing: firstly, the HSV color space, and secondly, the morphology method.

A color space is a multidimensional system by which colors can be represented, usually within three or more dimensions. Each dimension corresponds to a distinct color component. Recognizing the differences between various color spaces is pivotal in image processing, as the chosen color space can profoundly influence tasks such as segmentation, representation, and image manipulation.

Morphological image processing focuses on the shape or morphology of features within an image. Rooted in set theory, this method provides tools for extracting components of an image that are instrumental in describing and representing region shapes, like boundaries, skeletons, and the convex hull. The primary applications of morphological processing include noise reduction and sealing the contour of the circular edge of contact lenses using two core operations: dilation and erosion.

Thus, our image preprocessing involves using the HSV color space to filter and pinpoint the section within our desired color range as in Figure 5, which, in this context, is the rim of a contact lens. Subsequently, morphological image processing is employed to minimize noise and seal the circular rim of the contact lens. Following this preprocessing step, the algorithm identifies the image contours and progresses to the subsequent phase: circle detection. The final image then undergoes the image processing polar coordinate transformation.

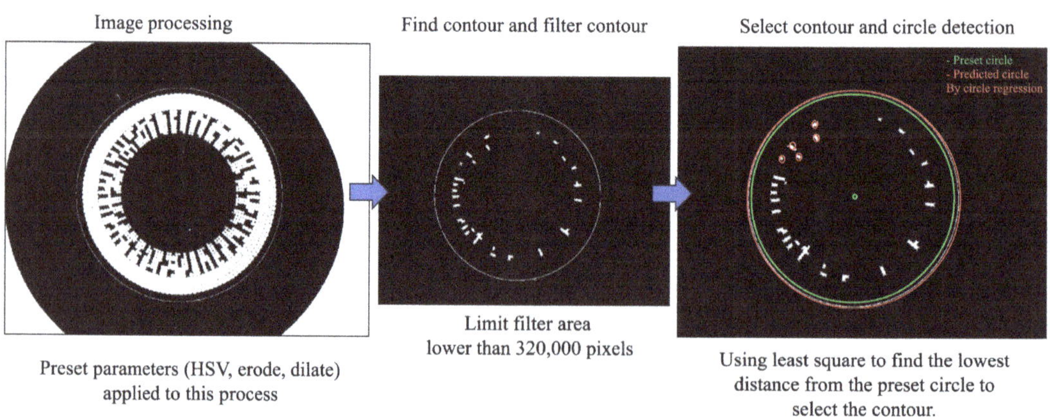

Figure 5. The workflow of image processing and circle detection steps.

4.2. Circle Detection

In this study, we opted for circle regression due to its efficiency and robustness to noise, enabling us to efficiently detect circular objects in complex, cluttered images without the need for extensive parameter tuning, thereby streamlining the image analysis process [27,33]. In this section, the input of this step is the list of contours from the first step. The algorithm needs to pick the rim contour by first filtering the size of the contour lower than parameter named "Max area of contour" which can reduce many contours from noise and container part; secondly, by selecting the final contour by using the least squares equation compared to parameters including "size of radius circle in range" and "center circle position", as shown in Figure 5; and then finally, by obtaining a very accurate circle parameter to be able to continue to the next part, which is the polar coordinate transformation. We use circle regression to fit the points from the contours. Circle fitting refers to the problem of finding a circle that best fits a set of data points with the least squares method. This method can be employed in a variety of scientific and engineering contexts, such as computer vision, robotics, and metrology.

Given a set of data points (x_i, y_i), the general equation of a circle is as follows:

$$(x_i - a)^2 + (y_i - b)^2 = r^2 \tag{1}$$

where (a, b) are the coordinates of the circle's center and r is the radius of the circle.

For the circle fitting, the goal is to determine the values of a, b, and r that minimize the sum of the squared distances or error between the data points and predicted circle. In this paper, we used linearization method to solve the circle fitting by starting with the standard circle question in Equation (2).

$$x^2 + y^2 + Dx + Ey + F = 0 \tag{2}$$

with

$$D = -2a \tag{3}$$

$$E = -2b \tag{4}$$

$$F = a^2 + b^2 - r^2 \tag{5}$$

Solve Equations (2)–(5) to find a, b, and r and perform a least squares linear regression to fit the equation.

4.3. Image Processing for Polar Coordinates' Transformation

The purpose of this section is to transform the circle image to a linear image. The polar coordinate transformation discussed in this section applies different geometrical modifications to 2D images. While they do not alter the actual content of the image, they reshape the pixel grid and project this reshaped grid onto the output image. To ensure the best quality and avoid any distortion, the projection is carried out in the opposite direction: from the output back to the input. This means that for every pixel (x, y) in the output image, the functions determine the respective original pixel in the input image and transfer its value. The linear image can apply one-dimensional window sliding, with the optimal computation cost of the algorithm for calculating information about the edges of a contact lens, such as area and thickness.

4.4. Defect Detection

The defect detection step is composed of three substeps, including contour calculation using window sliding, defining the correction criteria, and finding an acceptable range for calculation. The flowchart of defect detection is shown in Figure 6.

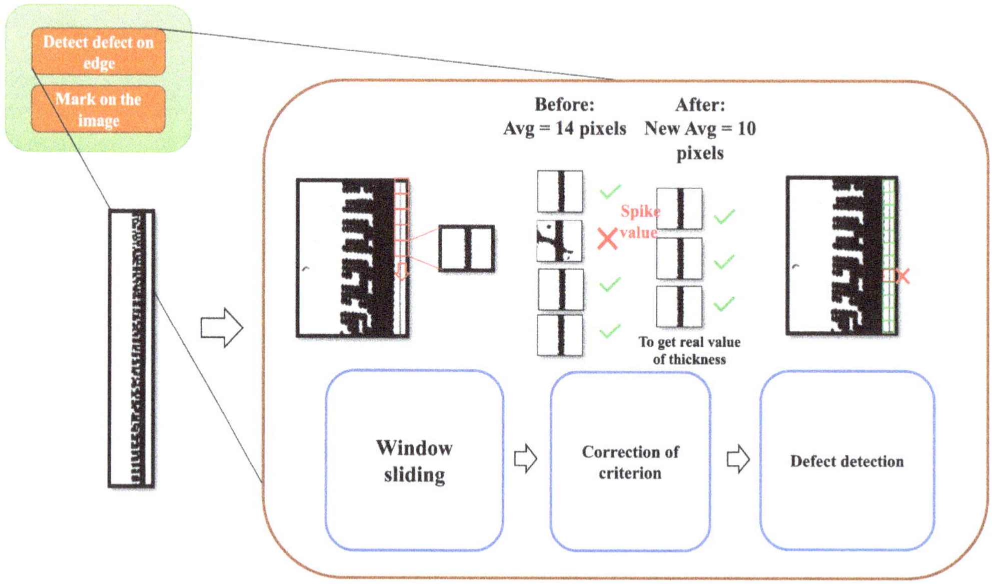

Figure 6. The workflow of defect detection.

The overall concept of defect detection in the algorithm is that it slices the long image into small pieces. Then, it selects a range of interested numbers of windows to calculate the average and standard deviation (SD). After obtaining the average and SD, the equation is used to determine if the value in the previously calculated window meets the condition. If so, then it is gathered for the second round of average value calculation. Lastly, the new average value is used for defect detection using the criteria for the acceptable range.

4.4.1. Contour Calculation Using Window Sliding

Window sliding solves the problem of the edge line of the contact lens does not have an uneven thickness along the whole circle, as shown in Figure 7. In this scenario, if we create an algorithm to obtain the data from the whole circle and calculate the statistical result, it would be biased, and unfortunately, half of the thicker circle may be detected as a defect.

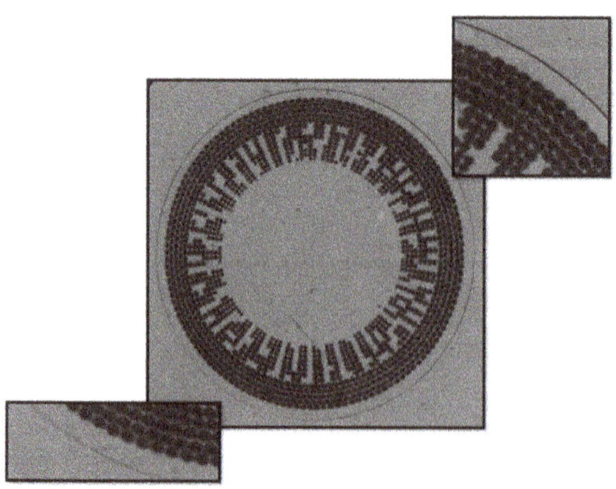

Figure 7. Edge of a contact lens with uneven thickness.

To avoid that, we perform the window-sliding technique to set a range of windows and obtain the data from the local area where it has a similar ground truth thickness. First, the algorithm starts with performing window sliding, at which point it receives the window size parameter as input along with the straightened preprocessed image. The size will be set to the window vertical height of all windows in the input image. The demonstration of window sliding in our algorithm is as shown in Figure 8.

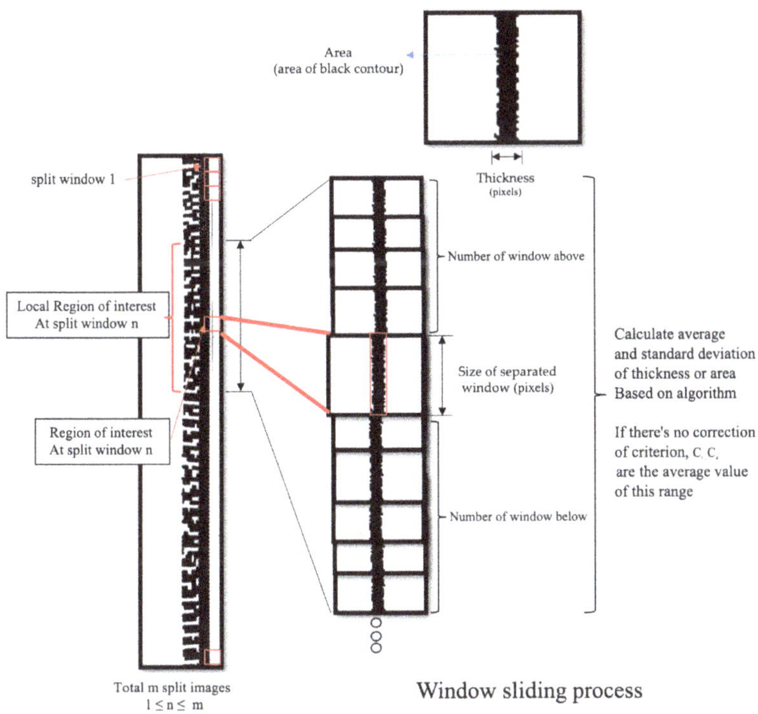

Figure 8. The sliding window process and an explanation of the parameters.

Second, we select the method to count black contours. As you can see in Figure 8, the black contours along the image are uneven with some rough edges. The red box represents a window section for the window sliding method. To determine whether these surfaces are defects, we proposed two approaches to perform the calculation, which are area-based approach and thickness-based approach.

For the area-based approach, the algorithm counts the number of black contours in pixels inside each window using OpenCV library. For a thickness-based approach, the algorithm counts the maximum width of black contour along the x-axis. This can be achieved by using OpenCV library.

The reason for proposing two different approaches is to compare their performance and check either of these approaches would be suitable for different objectives of defect detection. For example, let us assume that the average thickness in the example of 10×10 window slides is four with SD of one and the average number of black pixels (area) is 35 with SD of 10. If we consider only thickness of the black contour, there could be a case in which the thickness seems normal but somehow the defect is a slight swelling. In this case, area-based approach would detect and count the number of black pixels and return an unusually high value of black pixels inside that window. Another possibility is that, if a thin spike occurs in an image, the area-based approach may perform calculations and find that this window's black pixel value is normal; however, if we used thickness-based approach, it would immediately know that there is a spike in this window.

After calculating the value of black contours in each window, either using area-based approach or thickness-based approach, the last step is to calculate the average and SD values from a selected range of window slides. In this part, the parameter "number of windows above and below" is used to decide the range of each calculation. The reason we calculated the average is to use it as a determined value for the later step. To provide a clear explanation of each step, we present the procedure of this step in pseudocode in Algorithm 1.

Algorithm 1: Defect Criteria–Window-sliding method

```
Input
    (1) Preprocessed image
    (2) Size_window = parameter "Size of separated image"
    (3) Num_above = parameter "Number of images above"
    (4) Num_below = parameter " Number of images below"
Output
    (1) Criteria_based_list = the list contain in the local region of interest range either
        "thickness_based_list" or "area_based_list".
    (2) Criteria_avg = the average of Criteria list ("Criteria_list_value")
    (3) Criteria_sd = the standard deviation of Criteria list ("Criteria_list_value")
Begin
    1. Image_total_length = Preprocessed_image.length
    2. Number_of_window = Image_total_length/size_window
    3. Criteria_based_list = []
    4. If (area_based is True) then
    5.   For (window = 0, window=Number_of_window, window=window+size_window) then
    6.         Criteria_based_list = Area_based_list.append(find_area_of_contour(i))
    7. else then
    8.   For (window=0, window=Number_of_window, window=window+size_window) then
    9.         Criteria_based_list = Thickness_based_list.append(find_thickness_of_contour(i))
    10. //find average
    11. Window_AVG_list, Window_SD_list = [], []
    12. For (window=0, window= Criteria_based_list.length, window=window+1) then
    13.      Window_local_range = Criteria_based_list[window − Num_above : window +
         Num_below]
    14.      Window_local_AVG = Window_AVG_list.append(Average[window_local_range])
    15.      Window_local_SD = Window_SD_list.append(SD[window_local_range])
    16.    //correction of criterion (will be explained in the next section)
    17.    //defect detection (will be explained in the next section)
End
```

4.4.2. Correction of Criteria

This substep is proposed to slightly adjust the average from the previous substep to be more accurate as the thickness and area of the contact lens rims. The calculated average value from the previous step may include the value of the defects, which may lead to an unusually high value of the average. To detect this, we proposed the correction criteria method to reevaluate the average value by excluding the outliers in consideration. The first step for correction criteria is to check each window's black contour value with the set criteria that uses the average and SD values from the previous step. The equation for correction criteria is as shown in Equation (6).

$$\text{Average} - \text{F} \times \text{SD} \leq \text{qualified value} \leq \text{Average} + \text{F} \times \text{SD} \quad (6)$$

The average and SD values are specified from each set of windows for each iteration. The factor of SD to average is the required input parameter. If the window's black contour value is within the range of lower bound and upper bound, then that window's value is gathered and used for calculating a new average value. Figure 9 demonstrates how valid windows are selected for the calculation of a new average.

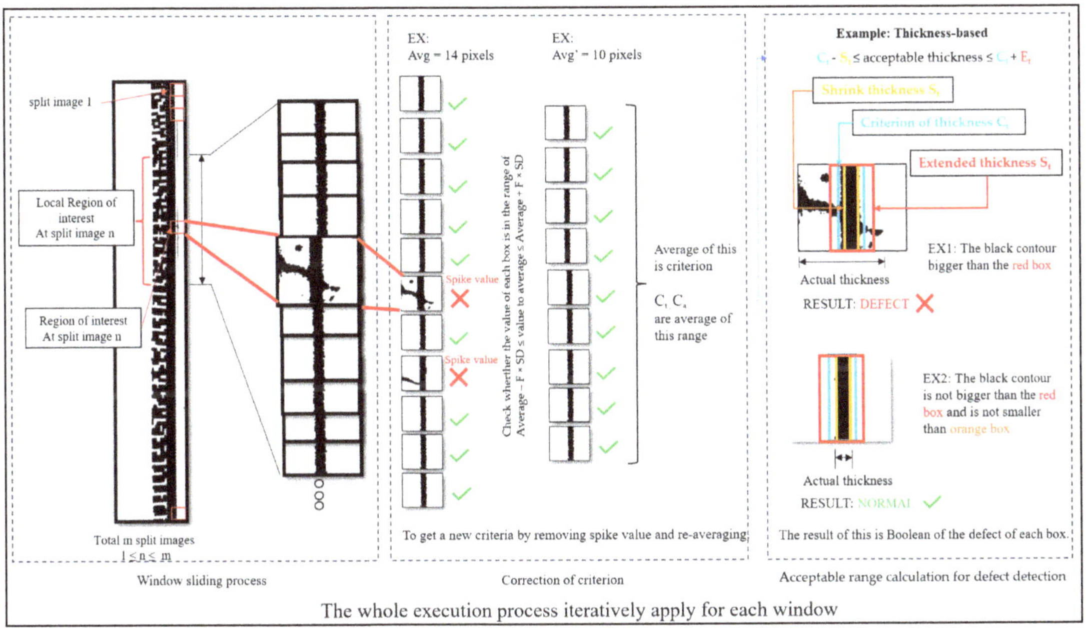

Figure 9. The execution process in a window sliding calculation.

The pseudocode used for demonstrating the concept in this step is as shown in Algorithm 2.

Algorithm 2: Defect Criteria–Correction of Criterion

Input
(1) Window_local_range = either Thickness_based_list or Area_based_list
(2) Factor_of_SD as parameter "Factor of SD to average"
(3) AVG as Window_local_AVG
(4) SD as Window_local_SD
Output
(1) Window_new_local_AVG
Begin
```
1. //find new average in correction of criterion, the for loop(line3) is the same in last loop in
   Algorithm 1
2. For (window = 0, window= Criteria_based_list.length, window=window+1) then//line 12
   in Algorithm 1
3. //window sliding return AVG, SD, Window_local_range
4.     Window_new_local_range= []
5.     For (window=0, window= Window_local_range.length, window=window+1) then
6.         If (AVG – (SD*Factor_of_SD) < Criteria_based_list[window] < AVG +
   (SD*Factor_of_SD))
7.             Window_new_local_range = Window_new_local_range.append(Criteria_based_list
   [window])
8.     Window_new_AVG = Average [Window_new_local_range]
9.     Return Window_new_AVG
```
End

4.4.3. Calculation of Acceptable Range for Defect Detection

The last step in defect detection on the circle edge of the contact lens is to perform the defect detection. In this step, the algorithm is prompted with equations that determine whether the window has a burr defect. The equations are separately designed for area-based approach and thickness-based approach, in which the lower bound is multiplied by the shrink value, and the upper bound is multiplied by extended value. These two values have different domains for area-based and thickness-based approaches. For a thickness-based approach, it is calculated with the instant value of pixel as shown in Equation (7). For area-based approach, the extended and shrink values are calculated with the percentage of the average of the black contour area as shown in Equation (8).

$$C_t - S_t \leq \text{acceptable thickness} \leq C_t + E_t \tag{7}$$

$$C_a - S_a \leq \text{acceptable area} \leq C_a + E_a \tag{8}$$

From Equations (7) and (8), C represents the new average value after being calculated from the correction criteria algorithm. S represents the shrink value and E represents the extended value. For the area-based approach, the subscription symbol a is used to represent this method, while the symbol t represents the thickness-based approach. With this equation, if any window's value has the value outside the equation range, it is considered a defect. Algorithm 3 shows the pseudocode for the calculation of the acceptable range based on 4 inputs, including the output of Algorithm 2, the shrink criterion, the extended criterion and the window value.

In this stage of the process, the algorithm undertakes a couple of significant actions. Initially, it overlays a red mask onto the image to enhance visualization and to help in assessing how well the algorithm is performing. Simultaneously, it employs a specific image processing technique, which we will delve into in the next section. In terms of performance assessment, the algorithm relies on these enhanced images to calculate evaluation metrics. This assessment involves comparing the algorithm's results to a labeled dataset, as we will discuss in detail in the following sections focusing on evaluation metrics and dataset information.

Algorithm 3: Acceptable range calculation for defect detection

Input

(1) New_AVG as Window_new_local_AVG or Window_local_AVG incase using with or without correction of criterion function respectively.
(2) Shrink_criterion is a parameter from either "Shrink thickness" or "Shrink area".
(3) Extended_criterion is a parameter from wither "Extended thickness" or "Extended area".
(4) Window_value is a thickness or area value of a certain window.

Output

(1) List Boolean of the defect as Bools_results

Begin

```
//defect detection
Bools_results = []
For (window=0, window= Criteria_based_list.length, window=window+1) then//line 12 in Algorithm 1
//window sliding: return AVG, SD, Criteria_based_list
//correction of criterion: return Window_new_local_AVG
//start defect detection
    New_AVG = Window_new_local_AVG//shorten the variable name
    If (New_AVG − Shrink_criterion < Criteria_based_list[window] < New_AVG + Extended_criterion)
        Bools_results.append(True)
    Else Bools_results.append(False)
Return Bools_results
```

End

4.5. Image Processing for Visualization

In this step, the reverse function of the polar coordinate transformation is employed. The predicted circle coordinates (x, y) and radius (r) obtained from the circle detection are used as input for this reverse function. The polar transformation equation is a reversible function used to convert the linear image back into a circular image, as illustrated in Figure 10. In the process of defect detection, the algorithm will mask the red contour on the defect following the window size in which the contour represents a rectangle size in the image before reversing polar transformation.

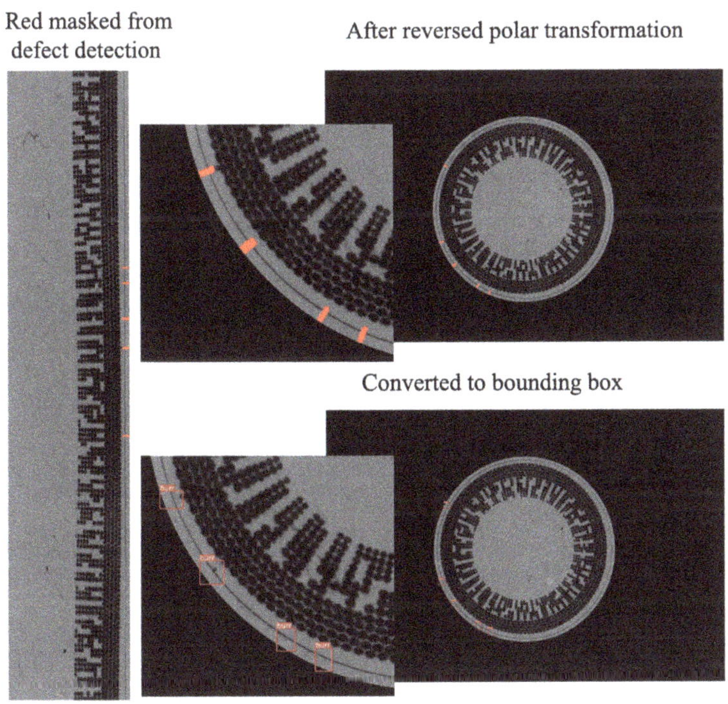

Figure 10. Image processing for visualization result.

In the output from the defect detection step, the image region containing defects is highlighted in red. The algorithm utilizes a color space method for image processing to extract the red contours, convert them into bounding areas, and present the result, as depicted in Figure 10.

5. Results

In the previous section, we proposed a novel algorithm that can perform defect detection at the rim area of contact lens bycalculating criteria by either thickness-based or area-based. In this section, we present the results of our algorithm's performance for different configurations and criteria. Table 2 summarizes the outcomes for both the "thickness-based" and "area-based" approaches. These results demonstrate the influence of varying shrink and extended thickness criteria, the number of images above and below a specified pixel size of split images, and the factor correction applied to the criteria. The values of the parameter that we set in each trial of the experiment were chosen intentionally. We carefully selected the values for each parameter through a series of experimental trials. For the shrink/extended criteria, we opted for values that were closely aligned with the original, introducing only slight variations to explore the optimal parameter settings that would be the most effective acceptable range to detect defects in the contact lens. When it came to the "Number of windows above/below" and "Factor correction of criterion", both were set as fixed parameters. The "Number of windows above" was established as 20 because this configuration involves 20 windows above and 20 windows below the window currently focused on, totaling 40 windows. This exceeds the threshold of 30, as stipulated by the central limit theorem [34]. The "Factor correction of criterion" defaults to 0.5 to maintain proximity to the average value and mitigate excessive fluctuations. Regarding the "Pixel size of split images", we employed values of 10 and 60. This choice was informed by our preliminary study of burr defects, which indicated a statistical range spanning from 8 to 52 pixels for each burr's position. Given the considerable variability in the burr sizes, we set the pixel size of the split images to 60 to encompass the largest burrs. However, we also assessed the detection performance for smaller burrs using a smaller split image size of 10. By examining the results for these two pixel sizes, we can discern meaningful contrasts and draw further conclusions from the data.

Table 2. Evaluation result with test set images for image processing algorithm.

Algorithm	Shrink/Extended Criterion	Number of Images Above/Below	Pixels Size of Spilt Image	Factor Correction of Criterion	Precision	Recall	AP@0.5
Thickness based	6	20	10	0.5	0.870	0.955	0.913
	8	20	10	0.5	0.918	0.924	
	10	20	10	0.5	0.951	0.911	
Thickness based	6	20	60	0.5	0.888	0.955	0.924
	8	20	60	0.5	0.925	0.940	
	10	20	60	0.5	0.959	0.911	
Area based	0.3	20	10	0.5	0.711	0.980	0.809
	0.4	20	10	0.5	0.821	0.971	
	0.5	20	10	0.5	0.894	0.955	
Area based	0.3	20	60	0.5	0.880	0.971	0.937
	0.4	20	60	0.5	0.951	0.955	
	0.5	20	60	0.5	0.980	0.940	

Setups of the value of shrink and extended criterion are identical, which is called the symmetric setting criterion. Similarly, symmetric setting is when the number of images above and below are identical.

5.1. Thickness-Based Approach

The "thickness-based" algorithm was evaluated using three different shrink/extended thicknesses (6, 8, and 10) and two distinct pixel sizes of the split images (10 and 60). The results indicated that for the 10-pixel split images, the algorithm achieved a high precision of 0.950 and 0.911 for the thicknesses of 6 and 10, respectively. However, as the pixel size increased to 60, the precision remained high but slightly decreased. This trend showcases the algorithm's ability to adapt to varying image characteristics while maintaining a good balance between precision and recall. Figure 11 shows an example of the result of different split image sizes; for the pixel heights of 10 and 60, the thicknesses are between 6 and 10 in 2-unit increments (6, 8, and 10 pixels). So, there are six image results for this sample. The contact lens from the examples below has inconsistent edges. The algorithm can detect but cannot cover all of the inconsistent edges. The example below represents the results of detecting burr defects on a contact lens.

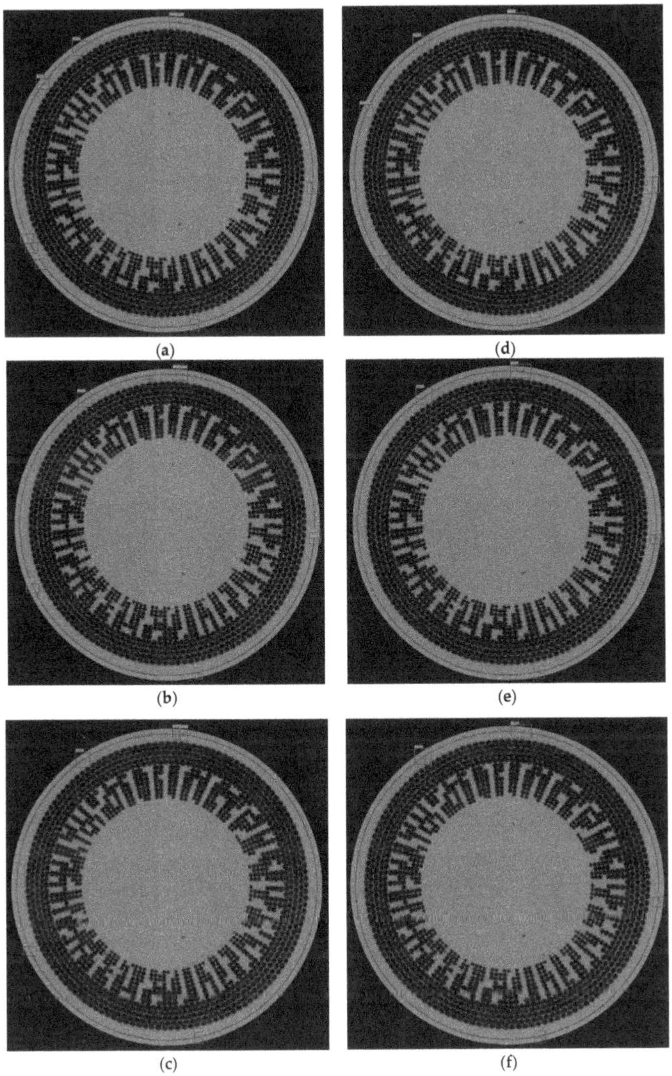

Figure 11. Results from a sample using the thickness-based method with different settings. (a) 10-pixel image splitting combined with a 6-pixel extended and shrink thickness criterion; (b) 10-pixel image splitting combined with an 8-pixel extended and shrink thickness criterion; (c) 10-pixel image splitting combined with a 10-pixel extended and shrink thickness criterion; (d) 60-pixel image splitting combined with a 6-pixel extended thickness criterion; (e) 60-pixel image splitting combined with an 8-pixel extended and shrink thickness criterion; and (f) 60-pixel image splitting combined with a 10-pixel extended and shrink thickness criterion.

In Figure 11, the 10-pixel split image performs better for the shrink defect criteria in the detection of the bottom part; the predicted results are shown for the thicknesses of 6, 8, and 10. In another setting, the 60-pixel split image cannot detect the same defect. Thus, the 10-pixels split image cannot detect the extended edge defect at the right side of the contact lens with a thickness of 10 pixels. However, the 60-pixel split image can perform well under all of the criteria settings.

5.2. Area-Based Approach

For the "area-based" algorithm, we examined different factor correction values (0.3, 0.4, and 0.5) and two different pixel sizes for split images (10 and 60). Notably, the algorithm consistently demonstrated a robust performance with precision values above 0.9 for various factor correction values. This suggests that the "area-based" method maintains its efficacy across different configurations.

Figure 12 shows a sample of the results of the area-based approach with the same samples as the thickness-based approach. This method also can adjust the parameters for the change criterion and split image size. In this method, the split image size is not set as a high value because this will affect the percentage of the area of the criteria for detection. Therefore, this parameter value is set to 10 pixels and 60 pixels, while the percentage of the area in the criteria is set as three values: 0.3, 0.4, and 0.5.

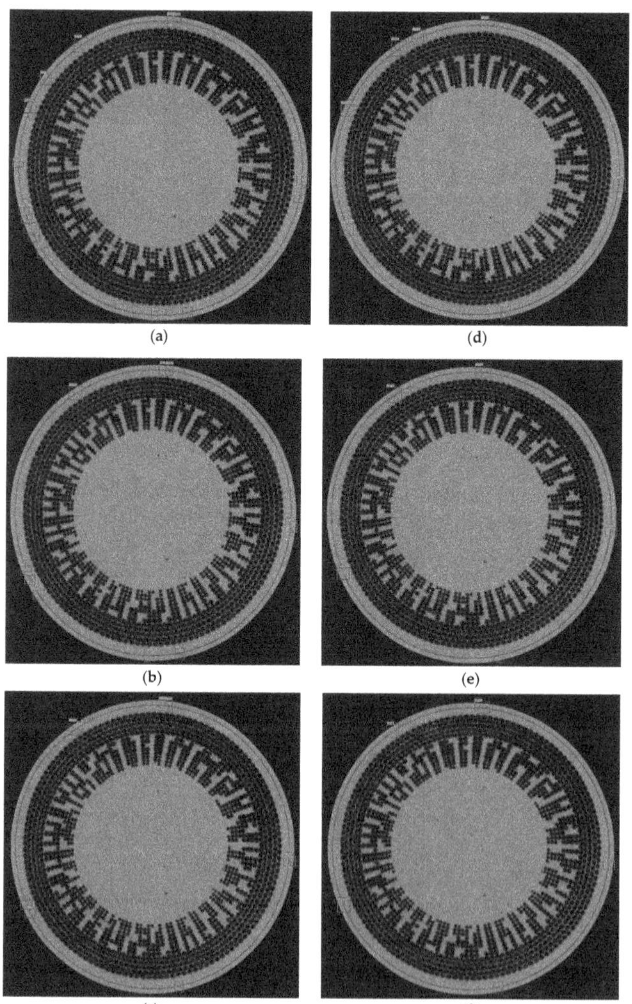

Figure 12. Results from a sample using the area-based method with different settings. (**a**) 10-pixel image splitting combined with threshold of 0.3 extended and shrink area criterion; (**b**) 10-pixel image splitting combined with threshold of 0.4 extended and shrink area criterion;

(**c**) 10-pixel image splitting combined with threshold of 0.5 extended and shrink area criterion; (**d**) 60-pixel image splitting combined with threshold of 0.3 extended and shrink area criterion; (**e**) 60-pixel image splitting combined with threshold of 0.4 extended and shrink area criterion; (**f**) 60-pixel image splitting combined with threshold of 0.5 extended and shrink area criterion.

The lower split image size of 10 pixels has more prediction detection bounding boxes. It shows that the sensitivity of the model of the setting of the smaller split image is higher, as shown in the left part of the result in Figure 12a–c. Also, the top part of the result is split into two bounding boxes, but Figure 12d–f has the one bigger box in the top part of the predicted result. In the bigger size of the split image, the number of predicted bounding boxes is obviously lower than that of the smaller split image size. This shows that the precision of the setting in which the 60-pixel size is used is greater than that of the 10-pixel split image size.

6. Discussion

6.1. Effectiveness of Window-Sliding Method

The window-sliding method is proposed with the objective of addressing the issue of uneven thickness around the entire circumference of the contact lens. The root cause of this uneven thickness on the contact lens rim stems from the imperfect placement of the contact lens in the contact lens holder. Tilt angles during its placement can lead to the conditions observed in the captured images as shown in Figure 7. Therefore, the window-sliding method is proposed to reevaluate the criteria defined by Equations (7) and (8) by calculating the average thickness within a specified local region of interest. This calculation is performed iteratively along the length of the circular rim. If the CLensRimVision algorithm is executed without the window-sliding method, the criteria defined by Equations (7) and (8) are calculated using the average value of the entire input of the contact lens. We conducted a comparative test between the results of using the window-sliding technique and tnot using it, as presented in Figure 13a,b, respectively. The testing results are based on the area-based approach, with both results having an extended and shrink area ratio set to 0.3 and a split image size of 10 pixels. As a result, the image in Figure 13a can detect fewer defects than the image in Figure 13b. Most of the detections in Figure 13b do not represent actual defects; instead, they are normal curves without defects. However, their values may fall outside the acceptable range during the defect detection stage. We can conclude that Figure 13a yields more precise results in defect detection in this case, aided by the window-sliding technique.

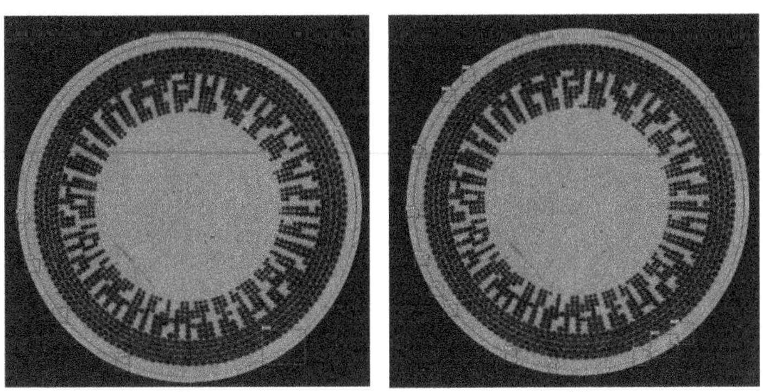

(**a**) (**b**)

Figure 13. A comparison between the results of using or not using correction criterion function in the algorithm based on the area-based algorithm. (**a**) A result from the algorithm using window-sliding method; (**b**) A result from the algorithm without window-sliding method.

6.2. Effectiveness of Correction of Criteria Function

The correction of the criterion function is used to eliminate spike values from the local regions of interest in the window-sliding method. These spikes can potentially occur due to defects in the contact lens. However, when finding the average value of the local area, we do not want to include these spikes, as they may lead to an inaccurate average value for either the thickness (in the case of the thickness-based approach) or area (in the case of the area-based approach) within the local region. The purpose of this function is to remove outliers from the average criterion value in the local region that could interfere with the overall calculated average value. The correction of the criterion function employs statistical mathematics to limit the range of the average local value, as defined in Equation (6), in which we manually set the proportion as factor F. Assuming that there are extended spikes resulting from defects, mathematically, the average will be higher than in cases without any defects. Conversely, if a local region has cuts or shrinkage defects in a linear shape within the region of interest, the average value will be lower than usual. In Figure 14a, there is a demonstration of a contact lens image that has been tested for defect detection using the CLensRimVision algorithm with the correction criteria function. In contrast, Figure 14b shows the result without using the correction criteria function. The results reveal that the example image using the correction criteria has fewer detections compared to the other image without the correction criteria. This is because it detects only the defects that appear on the rim of the contact lens. Figure 14b has some detections that are false positives for defects because they may have higher or lower average values than they should. Consequently, some values that should not be detected are also identified as defects on the rim.

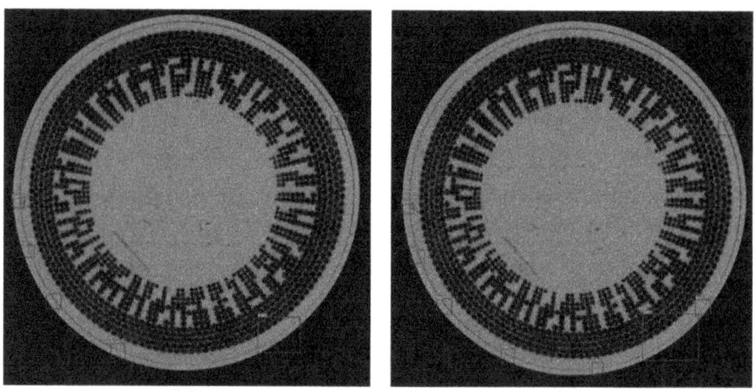

(**a**) (**b**)

Figure 14. A comparison between the results of using or not using correction criterion function in the algorithm. (**a**) A result from the algorithm using correction of criterion; (**b**) A result from the algorithm without correction of criterion.

6.3. Characteristics of Criterion-Based Methods

In the CLenRimVision algorithm, there are two criterion-based methods available: the thickness-based and the area-based approaches. Both methods serve the same purpose, which is to count the number of black pixels within the region of interest during the window-sliding process. In this section, we aim to highlight the characteristics of each approach, as they may be suitable for different applications and objectives. Each approach comes with its own set of required parameters for the shrink value and extended value used in the calculation of the acceptable range. These parameters offer the flexibility to meet the standards and specific industry needs associated with different contact lens grades, such as economy-priced beauty contact lenses, premium contact lenses, and myopia contact lenses.

When compared to physical contact lenses, the thickness-based method is more intuitive to configure within the algorithm. This is because the pixel thickness can be easily converted into the physical length, thanks to standardized quality control processes. In the premium contact lens industry, the thickness-based method has proven to be effective in quality control, as it uses micron units to precisely identify the required thickness adjustments, whether for shrinking or extending. Therefore, it can be concluded that the thickness-based method is more straightforward and suitable for most cases.

Conversely, the area-based approach may not exhibit a direct correlation with physical contact lenses. This approach relies heavily on sensitivity when configuring parameters, especially when setting the window size in the window-sliding process. This sensitivity arises from the fact that the area of black pixels in the counting method is relatively proportional to the size of the split image in the window-sliding technique. To illustrate, if the area of the split image changes from 10 to 60 pixels while assuming a constant thickness, the area difference between a 10-pixel split image and a 60-pixel split image would be six-fold. As mentioned earlier, this highlights the importance of a deep understanding of the algorithm when using the area-based method. However, it is worth noting that, in certain scenarios, the area-based method can deliver slightly superior results compared to the other approach.

6.4. Optimal Value of Pixels Size of Spilt Image

According to the results presented in Table 2, it is evident that the overall performance is significantly better with a split image pixel size of 60 pixels compared to that of 10 pixels. This leads to the conclusion that the 10-pixel size is limited in its ability to detect larger defects, resulting in a lower accuracy compared to that of the 60-pixel size. To conclude, our experiments have demonstrated that the optimal choice for the pixel size of split images is 60 pixels. This selection has been substantiated by the fact that it can effectively encompass defects of varying sizes. On the other hand, smaller sizes, such as a size of 10 pixels, do not enhance the accuracy since they are incapable of detecting larger burrs.

7. Conclusions

This research aims to provide a comprehensive understanding of the intricate technique required to detect burr defects on contact lenses. These defects may manifest as variations in thickness along the rim circle of the contact lens. Building upon these observations and our research findings, we introduce CLensRimVision, an algorithm specifically designed for detecting defects on the rims of contact lenses—an essential part that is in contact with the human eye. The proposed algorithm consists of three main components: image processing, circle detection, and defect detection. In defect detection, we developed a thoughtful window-sliding technique that overcomes the limitations posed by having an uneven thickness. Additionally, a correction criteria method was employed to reevaluate the correct average value, which is crucial for determining the acceptable range. This range calculation involves variables for shrinkage and extension, making the algorithm adaptable to various parameter inputs. Furthermore, we introduced two criteria for different detection purposes: the area-based and thickness-based approaches. As a result, we have generated prediction accuracy results based on different parameter settings and the use of these two criteria. The results of the contact lens defect detection algorithms under different parameter settings are shown, showcasing their performance in terms of precision, recall, and average precision (AP of 0.5). Notably, the "thickness-based" algorithm consistently demonstrates competitive levels of precision and recall across various settings. The choice of the shrink/extended thickness criterion significantly influences its performance, with lower criteria favoring recall. Additionally, a greater factor correction of the criterion contributes to a slightly better precision. The algorithm performs optimally with a pixel size of split images set to 20, consistently achieving a high AP of 0.5. On the other hand, the "area-based" algorithm has proven to be competitive, particularly when the shrink/extended criterion is set to 0.4 or 0.5. It offers a balanced trade-off between precision and recall, with a significantly better performance achieved with a higher size of window. Both algorithms exhibit sensitivity to parameter settings, which emphasizes the need to tailor the choice of algorithm to specific application requirements and consider the trade-off between precision and recall.

Additionally, we engage in a discussion on the proposed technique, which helps mitigate the diversity in rim thickness that can lead to errors and outliers, affecting the accuracy of the average value in each window sliding. This discussion aims to establish the meaningful and beneficial integration of each step within the algorithm. We believe that our work holds the potential to benefit the contact lens industry, offering applications in production or serving as a foundation for further implementations.

For our future work, several possibilities exist for extending this research or developing new algorithms capable of detecting other types of defects, such as hair and foreign particles. However, these defect types may not necessarily be related solely to the rim of the contact lens and may require further exploration and new techniques.

Author Contributions: Conceptualization, P.C. and C.-Y.L.; methodology, P.C.; software, P.C. and C.-Y.L.; validation, P.C.; formal analysis, P.C.; investigation, C.-Y.L.; resources, C.-Y.L.; data curation, P.C.; writing—original draft preparation, P.C.; writing—review and editing, C.-Y.L.; visualization, P.C.; supervision, C.-Y.L.; project administration, P.C.; funding acquisition, C.-Y.L. All authors have read and agreed to the published version of the manuscript.

Funding: This research received no external funding.

Institutional Review Board Statement: Not applicable.

Informed Consent Statement: Not applicable.

Data Availability Statement: Data is contained within the article.

Acknowledgments: The authors would like to express their heartfelt gratitude to Pintusorn Suttiponpisarn for his invaluable support and assistance during the preparation of the first draft. His insightful and constructive comments played a crucial role in shaping and refining this work.

Conflicts of Interest: The authors declare no conflict of interest.

References

1. Lovrec-Krstič, T.; Orthaber, K.; Maver, U.; Sarenac, T. Review of Potential Drug-Eluting Contact Lens Technologies. *Materials* **2023**, *16*, 3653. [CrossRef] [PubMed]
2. Ţălu, Ş.; Ţălu, M.; Giovanzana, S.; Shah, R. A Brief History of Contact Lenses. *Hum. Vet. Med.* **2011**, *3*, 33–37.
3. Harris, V.; Pifer, R.; Shannon, P.; Crary, M. Comparative Evaluation of *Pseudomonas aeruginosa* Adhesion to a Poly-(2-Methacryloyloxyethyl Phosphorylcholine)-Modified Silicone Hydrogel Contact Lens. *Vision* **2023**, *7*, 27. [CrossRef] [PubMed]
4. Lee, M.-J.; Park, S.-Y.; Sung, A.-Y. Ophthalmic Hydrogel Contact Lens Material Containing Magnesium Oxide Nanoparticles and 3-(Trifluoromethyl)styrene for Biomedical Application. *Micromachines* **2022**, *13*, 1897. [CrossRef] [PubMed]
5. Meretoudi, A.; Banti, C.N.; Raptis, P.K.; Papachristodoulou, C.; Kourkoumelis, N.; Ikiades, A.A.; Zoumpoulakis, P.; Mavromoustakos, T.; Hadjikakou, S.K. Silver Nanoparticles from Oregano Leaves' Extracts as Antimicrobial Components for Non-Infected Hydrogel Contact Lenses. *Int. J. Mol. Sci.* **2021**, *22*, 3539. [CrossRef] [PubMed]
6. García-Marqués, J.V.; Talens-Estarelles, C.; García-Lázaro, S.; Cerviño, A. The Effects of Soft Contact Lens Wear on The Tear Film and Meibomian Gland Drop-Out and Visibility. *Life* **2022**, *12*, 1177. [CrossRef] [PubMed]
7. Seggio, M.; Nostro, A.; Ginestra, G.; Quaglia, F.; Sortino, S. Contact Lenses Delivering Nitric Oxide under Daylight for Reduction of Bacterial Contamination. *Int. J. Mol. Sci.* **2019**, *20*, 3735. [CrossRef] [PubMed]
8. Seibel, E.J.; Trilsch, W.R.; Lee, D. Evaluating Soft Contact Lens Quality: A Manufacturer's Perspective. *Am. J. Optom. Physiol. Opt.* **1988**, *65*, 298–307. [CrossRef]
9. Sohal, A.S. Quality Control in Soft Contact Lens Manufacture—A Case Study. *Int. J. Qual. Reliab. Manag.* **1988**, *5*, 33–38. [CrossRef]
10. Torres, M.; Santhanam, K.S.V. Quality Control of Silicone Hydrogel Contact Lenses by Impedance Spectroscopy. *Mater. Res. Soc. Symp. Proc.* **2014**, *1626*, 818. [CrossRef]
11. Mana, N.; Lim, C.; Chong, Y.F.; Yazid, H.; Mohd Ali, Y. A Review on Contact Lens Inspection. *Indones. J. Electr. Eng. Comput. Sci.* **2023**, *31*, 700–712. [CrossRef]
12. Kim, G.-n.; Kim, S.-h.; Joo, I.; Kim, G.-b.; Yoo, K.-h. Center Deviation Measurement of Color Contact Lenses Based on a Deep Learning Model and Hough Circle Transform. *Sensors* **2023**, *23*, 6533. [CrossRef] [PubMed]
13. Luo, K.; Kong, X.; Zhang, J.; Hu, J.; Li, J.; Tang, H. Computer Vision-Based Bridge Inspection and Monitoring: A Review. *Sensors* **2023**, *23*, 7863. [CrossRef] [PubMed]
14. Smith, A.D.; Du, S.; Kurien, A. Vision Transformers for Anomaly Detection and Localisation in Leather Surface Defect Classification Based on Low-Resolution Images and a Small Dataset. *Appl. Sci.* **2023**, *13*, 8716. [CrossRef]

15. Chung, S.-T.; Hwang, W.-J.; Tai, T.-M. Keypoint-Based Automated Component Placement Inspection for Printed Circuit Boards. *Appl. Sci.* **2023**, *13*, 9863. [CrossRef]
16. Rudenko, M.; Plugatar, Y.; Korzin, V.; Kazak, A.; Gallini, N.; Gorbunova, N. The Use of Computer Vision to Improve the Affinity of Rootstock-Graft Combinations and Identify Diseases of Grape Seedlings. *Inventions* **2023**, *8*, 92. [CrossRef]
17. Gao, Z.; Qiu, Y.; Zhou, J. Paper Defect Detection Algorithm Based on Mathematical Morphology and Computer Vision. *J. Phys. Conf. Ser.* **2022**, *2365*, 012045. [CrossRef]
18. Zuo, D.; Hu, H.; Qian, R.; Liu, Z. An Insulator Defect Detection Algorithm Based on Computer Vision. In Proceedings of the 2017 IEEE International Conference on Information and Automation (ICIA), Macao, China, 18–20 July 2017; 2017; pp. 361–365. [CrossRef]
19. Rebsamen, M.; Boucheix, J.-M.; Fayol, M. Quality Control in the Optical Industry: From a Work Analysis of Lens Inspection to a Training Programme, an Experimental Case Study. *Appl. Ergon.* **2010**, *41*, 150–160. [CrossRef]
20. Kuo, C.-F.J.; Lo, W.-C.; Huang, Y.-R.; Tsai, H.-Y.; Lee, C.-L.; Wu, H.-C. Automated Defect Inspection System for CMOS Image Sensor with Micro Multi-Layer Non-Spherical Lens Module. *J. Manuf. Syst.* **2017**, *45*, 248–259. [CrossRef]
21. Chang, C.-L.; Wu, W.-H.; Hwang, C.-C. Automatic Optical Inspection Method for Soft Contact Lenses. In Proceedings of the International Conference on Optical and Photonic Engineering (icOPEN 2015), Singapore, 17 July 2015; SPIE: Bellingham, WA, USA, 2015; p. 9524. [CrossRef]
22. Burdescu, D.D.; Brezovan, M.; Ganea, E.; Stanescu, L. A New Method for Segmentation of Images Represented in a HSV Color Space. In *Advanced Concepts for Intelligent Vision Systems;* Lecture Notes in Computer Science; ACIVS 2009; Blanc-Talon, J., Philips, W., Popescu, D., Scheunders, P., Eds.; Springer: Berlin/Heidelberg, Germany, 2009; Volume 5807. [CrossRef]
23. Sural, S.; Qian, G.; Pramanik, S. Segmentation and Histogram Generation Using the HSV Color Space for Image Retrieval. In Proceedings of the International Conference on Image Processing, Rochester, NY, USA, 22–25 September 2002; p. II. [CrossRef]
24. Bora, D.J.; Gupta, A.K.; Khan, F.A. Comparing the Performance of LAB* and HSV Color Spaces with Respect to Color Image Segmentation. *Int. J. Emerg. Technol. Adv. Eng.* **2015**, *5*, 11.
25. Marques, O. Morphological Image Processing. In *Practical Image and Video Processing Using MATLAB*; IEEE: Piscataway, NJ, USA, 2011; pp. 299–334. [CrossRef]
26. Hasan, S.M.A.; Ko, K. Depth Edge Detection by Image-Based Smoothing and Morphological Operations. *J. Comput. Des. Eng.* **2016**, *3*, 191–197. [CrossRef]
27. OpenCV. Geometric Transformations of Images. Available online: https://docs.opencv.org/3.4/da/d6e/tutorial_py_geometric_transformations.html (accessed on 20 October 2023).
28. Puhan, N.B.; Sudha, N.; Hegde, A.S. A New Iris Liveness Detection Method against Contact Lens Spoofing. In Proceedings of the 2011 IEEE 15th International Symposium on Consumer Electronics (ISCE), Singapore, 14–17 June 2011; pp. 71–74. [CrossRef]
29. Ramlee, I.; Ibrahim, S.; Leow, W.Z.; Yusoff, M. A Review of Detecting Outliers in a Circular Regression Model. *IOP Conf. Ser. Mater. Sci. Eng.* **2020**, *767*, 012048. [CrossRef]
30. Ou, Y.; Deng, H.; Liu, Y.; Zhang, Z.; Ruan, X.; Xu, Q.; Peng, C. A Fast Circle Detection Algorithm Based on Information Compression. *Sensors* **2022**, *22*, 7267. [CrossRef] [PubMed]
31. Jin, Y.; Wang, Z.; Zhu, L.; Yang, J. Research on In-Line Glass Defect Inspection Technology Based on Dual CCFL. *Procedia Eng.* **2011**, *15*, 1797–1801. [CrossRef]
32. Huang, Y. A Novel Direct Structured-Light Inspection Technique for Contaminant and Defect Detection. *arXiv* **2020**, arXiv:2006.12186. [CrossRef]
33. Cicconet, M.; Geiger, D.; Werman, M. Complex-Valued Hough Transforms for Circles. *arXiv* **2015**, arXiv:1502.00558. [CrossRef]
34. Smith, J. Central Limit Theorem (CLT): Definition and Key Characteristics. Financial Analysis. Available online: https://www.investopedia.com/terms/c/central_limit_theorem.asp (accessed on 18 November 2023).

sensors

MDPI

Article

Anomaly Detection Based on a 3D Convolutional Neural Network Combining Convolutional Block Attention Module Using Merged Frames

In-Chang Hwang and Hyun-Soo Kang *

Department of Information and Communication Engineering, School of Electrical and Computer Engineering, Chungbuk National University, Cheongju-si 28644, Republic of Korea; inchang9507@gmail.com
* Correspondence: hskang@cbnu.ac.kr

Abstract: With the recent rise in violent crime, the real-time situation analysis capabilities of the prevalent closed-circuit television have been employed for the deterrence and resolution of criminal activities. Anomaly detection can identify abnormal instances such as violence within the patterns of a specified dataset; however, it faces challenges in that the dataset for abnormal situations is smaller than that for normal situations. Herein, using datasets such as UBI-Fights, RWF-2000, and UCSD Ped1 and Ped2, anomaly detection was approached as a binary classification problem. Frames extracted from each video with annotation were reconstructed into a limited number of images of 3×3, 4×3, 4×4, 5×3 sizes using the method proposed in this paper, forming an input data structure similar to a light field and patch of vision transformer. The model was constructed by applying a convolutional block attention module that included channel and spatial attention modules to a residual neural network with depths of 10, 18, 34, and 50 in the form of a three-dimensional convolution. The proposed model performed better than existing models in detecting abnormal behavior such as violent acts in videos. For instance, with the undersampled UBI-Fights dataset, our network achieved an accuracy of 0.9933, a loss value of 0.0010, an area under the curve of 0.9973, and an equal error rate of 0.0027. These results may contribute significantly to solve real-world issues such as the detection of violent behavior in artificial intelligence systems using computer vision and real-time video monitoring.

Keywords: video surveillance; convolutional block attention module; area under the curve; equal error rate; computer vision; UBI-Fights; 3D convolution

1. Introduction

The mandatory installation of closed-circuit television (CCTV) in daycare centers and kindergartens owing to school violence and child abuse has been extensively debated, and the revision of these laws is being promoted as a precaution. Additionally, numerous intelligent CCTVs that detect or preempt dangerous situations early by analyzing the relationship between current and past situations in real time have been released. Modern intelligent CCTVs, capable of real-time situation analysis, play an integral role not only in deterring potential criminal activities but also in aiding in their resolution. Thus, the maintenance of safety and order is improved in the community via the CCTV recognition and classification of crime scenes. Particularly, the detection of assault situations is a key task that uses video anomaly detection. Anomaly detection, which has been the focus of several studies, is an algorithm that finds abnormal instances within the patterns of a specified dataset. The biggest challenge in the study of anomaly detection is that the dataset for these abnormal situations is smaller than that for normal situations. Even if the number of datasets increases, there remains a lack of data for completely abnormal situations. Recently, anomaly detection using unsupervised learning, that is, using unannotated datasets, has been increasing because data annotation is expensive and time-consuming. Many previous studies have used unsupervised learning, such as generative adversarial network (GAN) [1] and autoencoder [2], using two-dimensional (2D) convolution. However, anomaly detection using multi-instance learning is also a popular trend. In this study, a model was constructed using a three-dimensional (3D) convolutional neural network (CNN), which is the most commonly used supervised traditional deep learning method. 3D CNNs have 3D filters and significantly more parameters; however, because fewer pre-trained models exist compared to 2D CNN, training often starts from the ground up without weights. Therefore, 3D CNNs appear unsuitable for constructing and learning sufficiently deep structures. However, space-time data can be considered simultaneously, which enables effective recognition and classification of dynamic behavior, including violent behavior. Because our model uses a 3D CNN, it can learn both spatial and temporal data, and the convolutional block attention module (CBAM) [3] is used to identify the most crucial information, thereby improving the interpretability and focus of the model. The model can extract features from various dimensions, thereby contributing to the recognition and classification of complex behavioral patterns.

While other 3D CNNs use multiple frames as input simultaneously, we merged multiple frames into one image and used multiple images as input. Therefore, even if the same number of images was input, memory usage was reduced, which was useful for environments with large datasets or memory limitations. Consequently, substantial amounts of temporal data could be used simultaneously, enabling more precise detection of anomalous behavior. Important features could be extracted and learned more effectively by aggregating multiple information frames from one location. This improved the recognition of complex behavioral patterns and could increase the accuracy of violent behavior detection. Additionally, using input data in this manner is similar to the method of light field imaging [4], which acquires 3D information by combining images captured from various angles. Light field imaging is yet to prove effective at detecting abnormalities. However, in this study, we generated data in a way similar to light-field imaging and extracted temporal patterns by merging continuous frames to generate 3D information at various points. Our method was inspired by light-field imaging, and it aims to improve precision in anomaly detection.

In contrast to traditional methods that utilize 2D images or video data, this data-merging method uses a complex input structure, in which multiple frames are contained in one image. Therefore, the form of the input data in a CNN can significantly affect the performance of the model, indicating the value of investigating various input methods in this manner.

Finally, this form of input data did not previously exist, and it may be helpful in the study of anomaly detection. Because it is similar to inputting images by dividing them into patch units, such as with a vision transformer (ViT) [5], future research could expand upon this idea. Therefore, in this study, we seek to emphasize the effect of the input data structure on the CNN performance and pave the way for anomaly detection techniques that expand on this structure. The findings of this study indicate that compared to existing state-of-the-art methods, our approach exhibits superior performance, which enables improved focus on areas of interest. Furthermore, rather than assessing a situation by observing a static scenario, such as a 2D CNN, our method can be applied to various scenarios, such as robbery, arson, and theft using the correlation between the past, present, and future. Additionally, in this paper, experiments are conducted in an environment that has relatively few obstacles and does not obscure actions, rather than a highly enclosed environment.

2. Related Works

Anomaly detection is the most popular and challenging topic in the field of computer vision, and it entails several methods, including 3D convolution, weakly supervised learning, unsupervised learning, skeleton, and optical flow. This section introduces these and other methods.

2.1. Three-Dimensional Convolution

Inspired by deep learning innovations in the rapidly developing image domain, 3D convolution (C3D) [6] can extract temporal and spatial information, in contrast to 2D convolution. The model is trained using a combination of dropout, batch normalization, and data augmentation techniques. The deep structures in the model perform better than the shallow structures, highlighting the importance of capturing long-term structures in videos. C3D has effective features for detecting anomalies, and several studies have been conducted on it [7–10].

2.1.1. Exploring Kernel Temporal Depth

In [6], C3D with a kernel size of $d \times k \times k$ was tested by changing only the kernel depth, where d is the kernel temporal depth and k is the kernel space size. In this network, eight convolution layers, five pooling layers (a pooling layer appears immediately after each convolution layer), two fully connected layers, and a softmax loss layer were used for predicting action labels. All convolution layers had the same kernel temporal depth and four networks were set with kernel temporal depths of 1, 3, 5, and 7 for testing. The kernel temporal depth with the best performance was $3 \times 3 \times 3$.

2.1.2. Spatiotemporal Feature Learning

Deconvolution is used to understand what C3D learns internally. C3D starts by focusing on the shapes in the first few frames and then tracks the prominent behavior in the following frames. Detecting shapes and movements simultaneously is the most noticeable difference between C3D and 2D CNN.

2.2. Weakly Supervised Learning

Currently, few datasets are available for anomaly detection, and the resulting annotation task requires a significant investment of time and money. Therefore, although frame-level labels have been used in supervised learning in the past, many studies have recently been conducted on anomaly detection via weakly supervised learning using labels at the video level [11–15]. Weakly supervised learning saves time and money, and the related multi-instance learning is used extensively.

2.2.1. Multi-Instance Learning

Multi-instance learning (MIL) is a variant of supervised learning in which a single class label is assigned to a set of instances such as image patches. Therefore, while each instance is not labeled for prediction, multiple instances form a bag, which is labeled for prediction. This is one of the most widely used techniques because it requires less annotation, is suitable for weakly supervised learning, and can learn the labels of each instance (snippet or frame) using annotations at the entire video level.

MIL [16–20] is useful for identifying the specific parts of various snippets in a video where an abnormality occurs. Additionally, it significantly reduces the misclassification of snippets or frames that are normal. Because it focuses on certain parts of the video, abnormalities can be detected in greater detail. In [18], we considered the video as a background, where each frame was divided into multiple instances, generating instance scores via MIL, prior to performing anomaly detection. Similarly, ref. [20] used a framework that treated a video as a bag and video clips as instances of a bag, using both normal and abnormal video clips. In [16], a self-learning framework was also used.

2.2.2. Gaussian Mixture Model

In [15], the Gaussian mixture model (GMM) modeled data distributions using probability distributions. Data distribution is identified by the combination of several Gaussian distributions. Thus, GMM expresses data distribution as the sum of the weights of several Gaussian distributions.

In video anomaly detection, the GMM models change over time at the pixel level, and changes in brightness or color values for each pixel are modeled as Gaussian distributions to distinguish between normal and abnormal patterns. An anomalous pattern is one that deviates significantly from the general Gaussian distribution. Degardin [15] proposed Bayesian filtering and GMM-based approaches for anomaly detection in video images under an MIL paradigm, similar to weakly supervised learning. This method predicts and measures the high anomaly score for a video segment in an anomalous situation. Experiments were conducted on the UBI-Fights, UCF-Crime, and UCSD Ped1 datasets, with an area under the curve (AUC) of 0.906 and equal error rate (EER) of 0.160, an AUC of 0.759 and EER of 0.302, and an AUC of 0.801 and EER of 0.236, respectively.

2.3. *Unsupervised Learning*

Unsupervised learning is a learning method that does not use any labels, in contrast to weakly supervised learning that uses a limited number of labels, owing to the constraints of available anomalous datasets on anomaly detection. Unsupervised learning methods include the k-means clustering algorithm [21], which uses Euclidean distances; one-class support vector machine (SVM) [22,23]; GAN [24–27]; and autoencoder [28–30]. Several unsupervised learning methods are described below.

2.3.1. K-Means Clustering and One-Class SVM

K-means clustering [31] is an unsupervised machine learning algorithm that divides a dataset into K nonoverlapping and distinct clusters. Doshi et al. [21] applied a fast unsupervised anomaly detection system using K-means clustering consisting of three modules responsible for pre-processing, candidate selection, and backtracking anomaly detection. This method significantly reduced the overhead during test operations and eliminated the overhead during learning operations.

A one-class SVM [23] is a model that focuses on the decision boundary of normal data when insufficient amounts of anomalous data exist, and it learns to detect anomalies based on this decision boundary. In contrast to SVM, because the hyperplane is divided based on the origin, there can be only one class regardless of the quantity of data and the severity of the normal/anomalous imbalance. The model achieved an AUC of 94.3 and EER of 10.0 at the frame level on the UCSD Ped1 dataset and an AUC of 70.3 and EER of 34.0 at the pixel level. Additionally, for the entrances of the Subway Entrances and Exits dataset, it achieved an AUC of 90.5 and EER of 16.9, whereas, for the exits, it recorded an AUC of 90.8 and ERR of 15.4.

2.3.2. Generative Adversarial Networks

GANs [1] were first introduced by Goodfellow et al. in 2014. Well-trained GANs, known for their ability to produce samples that are nearly indistinguishable from real data, have also been extensively studied for anomaly detection applications.

In 2017, Ravanbakhsh and Nabi [24] proposed a method for detecting strange events in videos using a GAN. Using this method, the generator learns the distribution of normal videos, and the discriminator detects anomalies based on this information. This study is significant for the application of GAN to video anomaly detection, and it has significantly influenced subsequent studies.

In 2018, Wang et al. [23] proposed an anomaly detection method based on future video frame prediction. The key concept of this study is a model that predicts future video frames from normal sequences and detects anomalies by comparing real and anticipated frames. This study set a new standard in the field and influenced subsequent works [32,33]. Additionally, Li et al. [33] proposed, for a given video frame, first creating continuous frames before and after using a two-branch generator network and then minimizing the prediction error between the generated frames and their real-world counterparts. They evaluated three datasets: CUHK Avenue, UCSD Ped2, and Shanghai Tech Campus. Furthermore, several other studies have been conducted, including [25,34].

2.3.3. Autoencoder

Autoencoders were popularized in 2006 by Hinton and Salakhutdinov [2]. A well-learned autoencoder can generate the same input data as that of a GAN.

Wang et al. [28] proposed a shallow generation network consisting of two neural networks to obtain models such as a Gaussian mixture that fit the distribution of real data. This model has contributed to maximizing the advantages of CNN and variational autoencoder. Experiments were conducted on UCSD, Avenue, UMN, and PETS datasets, and a high performance was achieved on all except UCSD.

Chong and Tay [29] proposed an autoencoder structure comprising 3D convolution layers. Therein, if generation was not properly performed while reproducing an image in the autoencoder, abnormalities were detected. Experiments were conducted on Avenue, UCSD Ped1 and Ped2, and Subway Entrances and Exits datasets. AUC and EER were used as evaluation indicators, and top performance was achieved on the subway exits, UCSD Ped1, and Avenue data.

2.4. *Other Methods*

In this section, we describe additional network methods such as skeleton-based, optical flow-based, ViT, and convolution long short-term memory (ConvLSTM) as well as those based on 3D CNN such as inflated 3D CNN (I3D).

2.4.1. Skeleton-Based Method

In the fields of computer vision and deep learning, the human body is mainly represented by the skeleton, that is, bones and their connections (joints). Skeletal data are used to track, analyze, recognize, and model human behavior, and human movements are represented by specific patterns or sequences. Anomalous behavior can be considered a deviation from these patterns.

Morais et al. [35] presented a new method for detecting anomalies using skeletal features in normal human behavior patterns. In [35], the authors proposed a network called Message-Passing Encoder–Decoder Recurrent Neural Network (MPED-RNN) and evaluated it using the ShanghaiTech Campus and CUHK Avenue datasets. In the ShanghaiTech dataset, 101 videos related to humans are classified as human-related (HR) ShanghaiTech. The authors achieved the highest performance on the HR ShanghaiTech and ShanghaiTech datasets with receiver operating characteristic area under the curve (ROC AUC) scores at the frame level of 0.754 and 0.734, respectively. Similarly, the Avenue dataset was divided into HR Avenue and Avenue, and ROC AUC scores were obtained at frame levels of 0.863 and 0.862, respectively.

Du et al. [36] stated that skeleton-based motion recognition is generally considered a time series problem and pointed out the deficiencies of existing methods, such as temporal pyramids (TP) and hidden Markov models (HMMs). To overcome these shortcomings, the author proposed a hierarchical recurrent neural network (RNN) that not only combined human physical structures and behavior patterns but also extracted temporal dynamic representations of skeletal sequences by maximizing deep RNN in modeling long-term context information of time sequences. Experiments on the Berkeley MHAD and Motion Capture Database HDM05 datasets showed that the best performance was achieved on both with 100% and 96.92% accuracy, respectively.

2.4.2. Optical Flow

An optical flow is a vector field that represents the movement of a pixel in an image over time by indicating the extent and direction of its movement. Assuming that the brightness of a pixel is constant over time and that adjacent pixels have similar movements, the direction and size of the pixel can be expressed as a vector by calculating its movement between two consecutive frames. Abnormalities are detected in subtle movement differences in rapid, repetitive, or complex situations, rather than in patterns of general movement.

In 2019, Nguyen and Meunier [37] proposed a model with an autoencoder structure that converts video frames into optical flows. Liu et al. [32] uses an optical flow that is optimized based on the adversarial loss between the optical flow and original output video frame, and the optical flow can be extracted directly from a single frame because FlowNet [38] is built inside the network. The model learns the spatial structure for normal data via a Conv-AE and then determines the association between the pattern of each input data in the second stream and the corresponding motion represented by the optical flow. Furthermore, optical flows extracted from FlowNet2 [39] detected discontinuous behavior well owing to clear boundaries and comparison of the optical flow prediction with the actual answer values; it achieved the highest performance on the CUHK Avenue and UCSD Ped2 datasets. However, as in [24], it has the drawback of time collisions.

Wang et al. [40] proposed a new approach to overcome these shortcomings. Owing to the significant discrepancy between the ground truth and predicted optical flow, which causes blurring in abnormal areas, optical flow with duality for both normal and abnormal data was used to predict the next frame from the current frame. Therefore, for normal behavior, the optical flow makes the predicted frame close to the real frame; however, for anomalous behavior, noise is added. They experimented with UCSD Ped2, Avenue, and ShanghaiTech datasets, achieving AUC scores of 94.2, 88.4, and 75.3, respectively, with Avenue and ShanghaiTech exhibiting the highest performance.

2.4.3. Vision Transformer

The transformer was first proposed by [41] to provide an alternative to RNN- and CNN-based natural language processing models. Additionally, in [42] they proposed a bidirectional encoder presentation from transformers (BERT), which achieved state-of-the-art performance in several natural language processing (NLP) tasks by pre-training transducers for unlabeled text in both directions.

A ViT divides the image into several equally sized patches, each of which is planarized into vectors that have position embeddings. The output of the transformer is passed to the head for classification, and during training, prediction uses the output of the last layer. Additionally, ViT can process continuous frames as time-series data and is useful for learning the relationships and patterns between frames.

Yuan et al. [43] proposed an approach for video anomaly detection using a ViT, which introduces a self-attention mechanism to overcome the inherent locality of convolutional operations and the difficulty of modeling long-range relationships when reconstructing a GAN or an autoencoder. The author used a video ViT (ViViT)- [44] and U-Net- [45] based approach called TransAnomaly, which solves vision problems based on the transformer. After decoding the features using U-Net, abnormalities were detected using the difference between the predicted and actual frames [46]. Their modified ViViT enabled the effective encoding of input images in all dimensions of space-time. The performance of the model was evaluated using the AUC. For UCSD Ped1 and Ped2 and Avenue datasets, they obtained AUCs of 0.867, 0.964, and 0.870, respectively, representing the best performance on the UCSD Ped1 and Avenue datasets.

2.4.4. ConvLSTM

ConvLSTM is a variant of long short-term memory (LSTM), an RNN effective at processing sequential data, which incorporates convolutional operations. ConvLSTM is particularly useful for processing time-series data while preserving spatial structure information by performing convolutional operations instead of matrix operations within the LSTM. It was first introduced in 2015 by Shi et al. [47] to solve weather forecasting problems. ConvLSTM is suitable for anomaly detection tasks because it can process spatial (information within frames) and time series features (information between frames) simultaneously to extract spatio-temporal features and learn dynamic patterns in video sequences.

Luo et al. [48] proposed a method for integrating ConvLSTM with an autoencoder called ConvLSTM-AE to create a common CNN capable of learning the regularity of typical shapes and movements. Similar to the other methods, anomaly detection was performed using frames regenerated in the decoder; however, abnormalities can be detected only in continuous frames. An experiment was conducted using the Moving-Mnist dataset, which was divided into application and motion parts. The results showed that the models with ConvLSTM-AE or ConvLSTM performed better than convolutional autoencoder (Conv-AE), achieving AUCs of 0.999 and 0.949, respectively, in the appearance and motion sections. Experiments were also conducted on Avenue, UCSD Ped1 and Ped2, and Subway Entrances and Exits, with AUCs of 0.70, 0.755, 0.881, 0.933, and 0.877, respectively, achieving the best performance on all datasets.

Medel and Savakis [49] proposed an approach for video anomaly detection using several stacked ConvLSTMs capable of end-to-end learning. The overall model structure was an autoencoder structure similar to the previous models, and three ConvLSTM layers were used inside the encoder. The evaluation was conducted using the USCD Ped1 and Ped2, Subway Entrances and Exits, and Avenue datasets. True positive/false positive and precision/recall were used as evaluation indicators. For each dataset, the true positive/false positive results were 40/7, 12/1, 62/14, 19/37 (anomalies: 19), 29/15 (anomalies: 30), and 40/2, and the precision/recall results were 0.851/1.00, 0.923/1.00, 0.816/0.939, 0.339/1.00, 0.659/0.967, and 0.952/0.957, respectively.

2.4.5. I3D

I3D was first proposed in 2017 by Carreira and Zisserman [50]. I3D is a 3D extension of the existing 2D CNN inflation structure where 3D represents height, width, and time, that is, the 3D convolution of the 2D CNN inflation. I3D processes information by considering the temporal continuity between video frames such that it allows existing 2D convolutional filters and pooling layers to process (inflate) video data in 3D. For example, if the size of the 2D convolution filter is $k \times k$, it is expanded to 3D and converted into $k \times k \times k$. I3D can learn the actions and patterns between consecutive frames by adding temporal dimensions. Thus, the motions and patterns of the videos are recognized more accurately. By extending the weight of the 2D filter to 3D, the initial value effectively learns the 3D network. I3D also proposes a principle that uses two streams: RGB and optical flow. The RGB stream processes the frames of the original video by focusing on the visual and color information, and the optical flow stream determines the movements and patterns of objects in a video. By combining these two streams, I3D can recognize behavior in a video more accurately. Improving the accuracy of motion recognition using both visual and motion information is a major feature of I3D. Additionally, anomaly detection studies continue to be published [51–60].

3. Proposed Method

In this section, we propose a network architecture that correctly classifies violent situations by using the spatial and temporal features of each video to detect abnormal situations, particularly violent ones, and achieves state-of-the-art performance. The proposed architecture adds a convolutional block attention module (CBAM) [3], an attention module that includes the channel attention and spatial attention modules, to a residual neural network (ResNet) [61] in the form of a 3D convolution.

3.1. CBAM (Convolutional Block Attention Module)

CBAM is a network module used to highlight the channel and spatial information within a convolutional block. This improves the performance of CNNs and is commonly used in image classification and various computer vision tasks, highlighting important information. It comprises two attention modules.

1. Channel attention: Channel attention determines the importance of each channel and plays a role in emphasizing the relationships between channels. Thus, important

features can be emphasized, and unnecessary noise can be suppressed. Channel attention is usually implemented by using the average value of each channel, calculated via global average pooling, to learn the relationships between channels. 2. Spatial attention: Spatial attention plays an important role in understanding the spatial structure of feature maps by emphasizing one area and suppressing another. To this end, the importance of the feature map calculated for each channel was determined, and a spatial weight was generated and multiplied by the feature map. If the input feature map F is multiplied by M_C, which is the channel attribute map, F' is obtained. If it is multiplied by M_S, which is the spatial attribute map, F'' is obtained, which is the final feature map. Figure 1 shows the structure of the detailed sub-module of CBAM, and the formulas for obtaining M_C and M_S, which are the channel attention map and spatial attention map, respectively, are shown as below Equations (1) and (2) in below. Figure 2 shows the structure of each attention sub-module of CBAM.

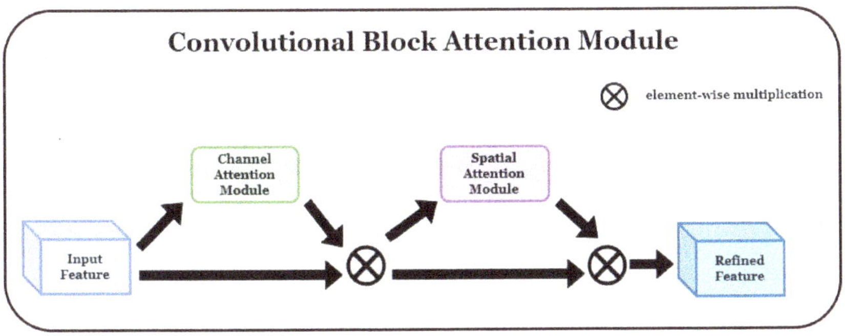

Figure 1. Overview of the convolutional block attention module (CBAM): It consists of a channel attention module and a spatial attention module [3].

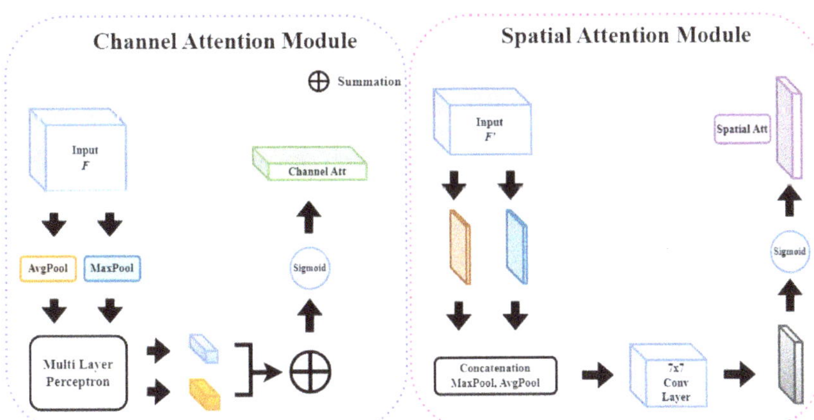

Figure 2. Structure of each attention sub-module of CBAM [3]: Divided into channel attention and spatial attention. Output is generated after passing max pooling and average pooling.

$$\begin{aligned} M_C(F) &= \sigma(MLP(AvgPool(F)) + MLP(MaxPool(F))) \\ &= \sigma(W_1(W_0(F^C_{avg})) + W_1(W_0(F^C_{max}))) \end{aligned} \quad (1)$$

$$\begin{aligned} M_S(F) &= \sigma(f^{7\times7}([AvgPool(F); MaxPool(F)])) \\ &= \sigma(f^{7\times7}([F^S_{avg}; F^S_{max}])) \end{aligned} \quad (2)$$

The channel attention map calculates the importance of each channel of an input feature map and extracts the relationships between the channels. After calculating the attention score and passing through average and max pooling, two vectors are produced, which are input into the shared multilayer perceptron to add nonlinearity. This is then encoded with a probabilistic value between zero and one using a sigmoid function to generate an M_C, one-dimensional $1 \times 1 \times C$ channel attention map. If the input feature map F is multiplied by M_C, F', which is a feature map to which the channel attachment is applied, is generated. A spatial attention map is created by encoding the pixels to be focused on per $C \times H$ pixels throughout the channel. Up until the pooling process and concatenation, it is the same as channel attention. This information is gathered for comprehensiveness because both pooling methods are carried out to extract spatial information and calculate the importance of location. Additionally, this value is placed in the 7×7 conv2d layer and a 2D spatial attention map is generated via a sigmoid function. The feature map generated in this manner is of size $H \times W \times 1$. When the resulting M_S is multiplied by F', the final attention map is obtained, that is, F'' is produced. The formulae for F' and F'' as follows Equations (3) and (4):

$$F' = M_C(F) \otimes F \tag{3}$$

$$F'' = M_S(F') \otimes F' \tag{4}$$

We introduce the proposed method by transforming the attention-based approach proposed in [62,63]. In this paper, we describe how the output of a linear classification layer can be influenced in addition to a popular CNN pipeline that has excellent feature extraction capabilities. CBAM influences the feature extraction process of the model and serves to weigh the network output for improved feature extraction.

3.2. Grid Frame

In this section, we describe how to create a grid frame for each dataset and how to assign labels accordingly.

In this study, we propose a method to improve the input data processing methods of existing 2D and 3D CNNs. In general, 2D CNNs adopt a method of processing one image at a time. Consequently, each image is processed independently by considering only the spatial features and thus does not consider the correlation between images. Therefore, to detect abnormalities using 2D CNNs, an RNN such as LSTM or a neural network structure such as ConvLSTM should be added. However, the proposed method of processing multiple images of 3D CNNs at once includes temporal and spatial features. Thus, this method has the advantage of considering correlations between images.

Additionally in this section, we present an approach that extends beyond these image-processing methods, where for each dataset, several images are processed in a single group. This group consists of continuous frames, and each image is processed considering its interrelationship with other images in the group, rather than being processed independently. This enables the model to extract more information and meaningful patterns than a single image of a 2D CNN or multiple images of a 3D CNN, thereby enabling accurate prediction and analysis. This approach aims to capture and use complex relationships between multiple images in a group (this can include, for example, multiple groups corresponding to a video) and finally detect abnormalities by identifying spatial and temporal patterns via the continuity of images in and between groups. This approach addresses complex interactions between images and groups, enabling the derivation of more accurate and meaningful results. Furthermore, this study opens up the possibility of new methods in the field of anomaly detection using an input method conceptually similar to Lightfield and ViT. In this study, we propose inputs of sizes 3×3, 4×3, 4×4, and 5×3.

For the UBI-Fights dataset, each annotation file marked fighting or nonfighting frames with a 1 or 0, respectively. Therefore, when preprocessing videos identified as fights, the annotation file was loaded and frame indices with a value of 1 were saved. The frame indices were read at intervals of 5. For example, if frames 1–20 had a value of 1, then frames

1, 6, 11, and 16 were accessed. As the video has a frame rate of 30 fps, without this interval, significant action changes may not be captured; hence, spacing between frames is necessary. Once sufficient frames were gathered for the intended composite image size (9 frames for 3×3, 15 frames for 5×3, and so on), they were combined into a single composite image. Owing to memory constraints, only ten composite images were generated per video. These ten merged images served as the input data, and depending on the size of the composite image, 90–150 frames were used from a single video. Data corresponding to the normal class were processed in the same manner to generate the input data. This method was adopted and used on all datasets, including UBI-Fights.

The UBI-Fights dataset showed an imbalance between the fight (216) and normal (784) class data. Therefore, experiments were conducted not only at the existing data ratio fight:normal = 216:784 but also by equalizing the data to fight:normal = 216:216, resulting in two experiments. Furthermore, experiments were conducted separately using ResNet depths of 10, 18, 34, and 50, with the data split in an 80:15:5 ratio for training:validation:test datasets.

The RWF-2000 dataset was balanced, and because the training to validation split was already at an 8:2 ratio, only one experiment was conducted. This dataset does not have separate annotations; therefore, the Violence and NonViolence categories were arbitrarily assigned values of 1 and 0, respectively, for experimentation. In contrast to the UBI-Fights dataset, given that the videos were 5 s long with 30 fps, after reading the video, the frames were extracted in the flow of the video to create merged images. This was performed under the assumption that sufficient actions corresponding to the classes were contained within.

UCSD Ped1 originally contained normal data in the training set and anomalous data in the test set; however, they were arbitrarily mixed for use. Similarly, Ped2 was mixed arbitrarily. For both datasets, the ratio of the training and test sets was the same as that in the original datasets, and the test set was used for validation. Furthermore, only a randomly selected 30% of the test set was used for testing. For the UCSD Ped1 and Ped2 datasets, we used all the data. A sample of each input data is described in Figure 3.

3.3. Model Architecture

Recently, various studies have been conducted on abnormal detection using CBAM, such as [62–64]. However, most research has been conducted on structures such as 2D CNN-based generative models. This study employs a different approach to propose a more effective anomaly detection model. Our work differs from other studies because it effectively combines 3D CNN and CBAM for anomaly detection. Additionally, instead of using an autoencoder structure, our approach optimizes the characteristics of 3D CNN and the effects of CBAM and proposes a more sophisticated and robust anomaly detection model. Thus, our study presents an original perspective as compared with previous studies.

3.3.1. Backbone Architecture

In this study, ResNet [61] was used as the backbone architecture. This is a suitable choice for its efficacy in learning deep neural networks. Particularly, the problem of gradient loss can be solved despite the depth of the model using the advantages of residual blocks. Residual blocks assist in the effective transfer of gradients via residual connections between inputs and outputs and minimize the loss of information, even in deep networks. Furthermore, deeper and more complex models can be developed. Subsequently, the detailed structure of the model and connections between the layers are described.

3.3.2. Proposed Architecture

The model architecture proposed in this study was constructed by adding CBAM after each bottleneck (or basic block) of 3D ResNet. This configuration may improve anomaly detection performance by effectively combining the feature extraction capabilities of the existing 3D CNN with the information emphasis capabilities of CBAM. CBAM can be placed behind each bottleneck, and the model can be extended to include additional residual blocks as required. This increases the expressiveness of the model and aids in capturing complex temporal relationships.

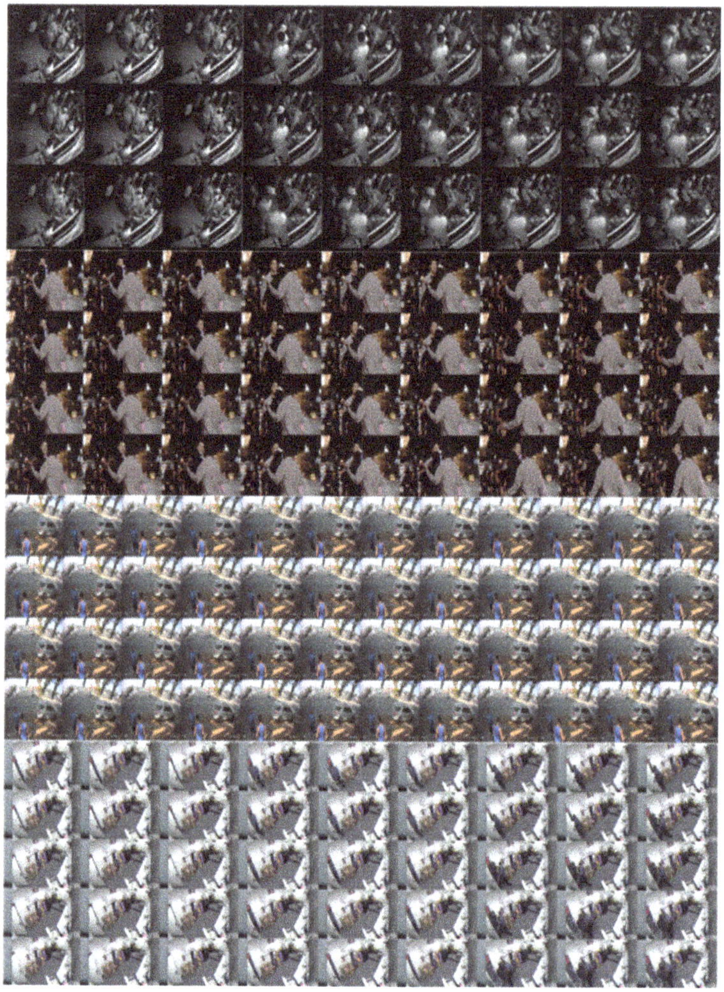

Figure 3. Example of each grid frame: Frames go from top left to top right, and from bottom left to bottom right. They have the order of 3×3, 4×3, 4×4, and 5×3 inputs in order from the top, and all are samples of three consecutive images. The total number of frames used in the example is 27, 36, 48, and 45 in order.

CBAM is an attention mechanism designed primarily for 2D data. However, as this study dealt with 3D data, the code was appropriately modified. First, 2D pooling (AvgPool2d and MaxPool2d) was replaced with AdaptiveAvgPool3d and AdaptiveMaxPool3d, which perform pooling by automatically adjusting kernel and stride sizes to fit the target output tensor sizes. They also work with inputs of various sizes and produce the same output. As channel attention aims to calculate the mean and maximum values for the entire spatial dimension, using adaptive pooling to obtain the outputs of (1, 1, 1) is appropriate. Adaptive pooling also performs pooling to the desired output size without a fixed kernel size or stride, which can help optimize the computing resources. Finally, the overall information in the input feature map was averaged to minimize information loss while maintaining a consistent output size.

(1) Channel Attention (designed to recognize and learn the importance of each channel)
   (a) Adaptive average and max pooling: This module aggregates the spatial information of 3D data. Average and max pooling each return a tensor with a size of $C \times 1 \times 1$, where C is the number of channels in the input tensor.
   (b) MLP with Conv3d: Importantly, the kernel sizes of both the first and second Conv3d layers are $1 \times 1 \times 1$. The first Conv3d layer serves to reduce the number of channels from C to $C/reductionratio$. Conversely, the second Conv3d layer increases the number of channels from the $C/reductionratio$ to C again.
(2) Spatial attention (designed to learn the importance of each 3D location)
   (a) Max and min pooling: Extracts the maximum and minimum values of the channel at each location of the input data. This results in a tensor of size $2 \times D \times H \times W$. Here, D, H, and W are depth, height, and area, respectively.
   (b) Three-dimnsional convolution: The tensor generated above passes through the 3D convolution layer, which has a kernel size of $7 \times 7 \times 7$ and a number of output filters (channels) of 1. A larger kernel size accounts for more amount of spatial information in convolutional operations, with a $7 \times 7 \times 7$ size kernel accounting for a large area of its surroundings at each 3D location, which aids in capturing a broader context. The output channel is 1 to generate a spatial attention map. This attention map has the same spatial dimension as the original input tensor but only represents the importance (or attention score) of each position. Therefore, the number of output filters is 1, which is multiplied by the original input tensor and element-wise to apply spatial attention. In summary, the setting with a kernel size of $7 \times 7 \times 7$ and a number of output channels of 1 aims to recognize a wide range of contexts, save parameters, and create a clear spatial attention map.

When the input data are provided, 3D convolution is performed, and the CBAM layer is passed through batch normalization, rectified linear unit (ReLU), and dropout. After passing through all the reserved blocks, the feature map undergoes global average pooling through avgpool. This leaves only the average value for each feature map, and the feature after pooling is a 1D vector, which passes through the fully connected layer (fc layer) and outputs the final predicted value to perform anomaly detection.

In detail, the input data go through $7 \times 7 \times 7$ convolution, batch normalization, ReLU, and MaxPooling, and they are then input to the bottleneck layer (in the case of ResNet50) or basic block (in case of ResNet 10, 18, and 34). Here, after going through $1 \times 1 \times 1$ convolution, batch normalization, ReLU, $3 \times 3 \times 3$ convolution, batch normalization, ReLU, $1 \times 1 \times 1$ convolution, and batch normalization, and lastly, reduction in computational complexity when the output and input sizes of the bottleneck are different owing to the difference in stride or number of input and output channels. Downsampling (stride = 2) is then performed for high-level feature extraction. Subsequently, CBAM is applied and channel attention and spatial attention proceed. Layers containing bottlenecks or basic blocks are repeated four times from layers 1 to 4, and CBAM is applied to each bottleneck block or basic block. In the case of ResNet 10, 18, and 34, there are (1, 1, 1, 1), (2, 2, 2, 2), and (3, 4, 6, 3), respectively, basic block layers in layer 1. In the case of ResNet50, there are 3, 4, 6, and 3 bottleneck layers in layer 1 to layer 4, respectively. After the input data have executed all the network structures, classification for anomaly detection is performed in the fully connected layer.

In the architecture of the proposed network, when an input arrives, it passes through Conv3d, batch normalization, ReLU, and AdaptiveMaxPool3d and then through the convolutional block (there are 1×1, 3×3, and 7×7 depending on the depth of the model) and the basic block (or bottleneck block). Subsequently, the process of going through CBAM is repeated four times before proceeding with the classification. In Figure 4, it is labeled as "BottleNeck" based on ResNet50, but for shallower depths (10, 18, 34), it changes to "BasicBlock".

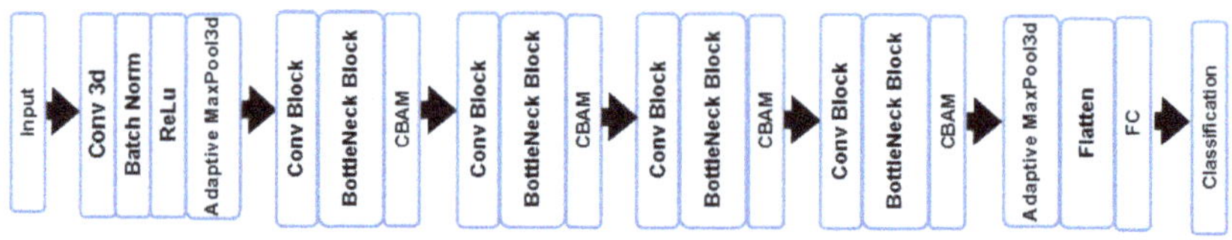

Figure 4. Proposed Architecture.

3.4. Dataset

In this section, we describe datasets that we used for experiments. When looking at existing datasets for anomalous behavior detection, few datasets contain violent actions. While there are numerous datasets like UMN [65], UCSD Ped1 and Ped2 [66,67], Subway Entrances and Exits [68], and CUHK Avenue [69], the anomalous behaviors they designate refer to actions like walking in a different direction from the majority in a crowd or riding bicycles and inline skates. There are also datasets such as [70] that occur in public transportation, especially buses. However, these datasets contain many obstacles in public transportation and there are many cases where these obstacles obscure human actions. Because we only tested frames where human actions were clearly visible, we also excluded this dataset from the experiments. Hence, a dataset that clearly distinguishes between violent and nonviolent behaviors is required. Therefore, this study used UBI-Fights [71] and three other datasets.

3.4.1. UBI-Fights

The UBI-Fights dataset was divided into fight and normal classes, consisting of 216 and 784 videos, respectively, gathered from YouTube and LiveLeak based on extensive text search results of violent incidents. The data were collected at the frame level with annotations totaling 80 h in length; however, the length of each video varied and had a frame rate of 30.00 frames per second with a fixed resolution of 640 × 360. A drawback of the videos in the fight class is that the entire duration of the video does not show fighting; hence, scenes that should be categorized as normal were included.

3.4.2. RWF-2000

Another dataset used was the RWF-2000 [72]. This dataset consists of 2000 video clips filmed in the real world by surveillance cameras, each 5 s long at 30.00 frames per second. They were also sourced from YouTube and divided into violent and nonviolent classes. Unlike in [71], the training and test sets were divided in an 8:2 ratio. The resolution of the videos varied, including 240p, 320p, 720p, and 1080p. An advantage of this dataset was that the actions were aligned with their respective classes throughout the video. Therefore, this assumption was more intuitive. The violence class included fights, robberies, explosions, shootings, bleeding, and assaults, enabling a more detailed categorization depending on the situation and purpose. However, because the resolutions varied, preprocessing through resizing could alter the original size of videos, potentially leading to a loss of information.

3.4.3. UCSD Ped1 and Ped2

We included the UCSD Ped1 and Ped2 datasets in our experiment, which considered behaviors, such as walking in the opposite direction to the majority in a crowd, riding a bicycle, or using inline skates, as anomalous. The ground truths of UCSD Ped1 and Ped2 are provided in the form of a segmentation map in which objects exhibiting anomalous behaviors are highlighted in white. The UCSD Ped1 dataset consists of images from 70 videos: 34 for training and 36 for testing. Each set contains 200 images of size 158 × 238. The training set comprises normal behaviors, whereas the testing set contains 10 anomalous behaviors. The UCSD Ped2 dataset comprises images from 28 videos, with 16 for training and 12 for testing. Similar to UCSD Ped1, each image contains 200 images of size 360 × 240. Consequently, any area in the segmentation map with a value of 1, is interpreted as representing abnormal behavior and assigned a label of 1; otherwise, it is assigned a label of 0.

3.5. Details of Training

In this section, we describe the optimizer and loss function used and the hardware and software environments and so on. As this study is a binary classification, the loss function used was BCEWithLogitsLoss. The equation for BCEWithLogitsLoss is as follows in Equation (5):

$$l(x,y) = \begin{cases} -w \cdot (1-y) \cdot \log(1-\sigma(x)) & \text{if } y = 0 \\ -w \cdot y \cdot \log(\sigma(x)) & \text{if } y = 1 \end{cases} \quad (5)$$

where σ represents the sigmoid function, ω stands for the weight for each input sample, and y indicates the actual label value (0 or 1). The loss of $y = 1$ (positive class) measures the accuracy of the model prediction using the cross-entropy term, whereas the loss of $y = 0$ (negative class) measures the inaccuracy of the model prediction. In both cases, the closer the prediction is to the actual label, the closer the loss becomes to 0. This function takes the logits (that is, the outputs before passing through the sigmoid activation function) and actual label values as inputs. This effectively combines a two-step operation (sigmoid activation function and binary cross-entropy loss). This merging method was used to achieve computational efficiency and numerical stability.

In this study, the ResNet architectures with 10, 18, 34, and 50 layers were utilized as the foundational backbones. Following the addition of the Convolutional Block Attention Module (CBAM) on four occasions within each architecture, the total number of layers increased to 14, 22, 38, and 54, respectively. For optimization, AdamW was used. AdamW, a variant of the Adam optimization algorithm, implements weight decay separately. This is particularly beneficial in preventing overfitting and enhancing the generalization performance in deep neural networks. The combination of 3D ResNet and CBAM, which forms a profound network architecture, benefits from AdamW for efficient learning and regularization. An initial learning rate of 5×10^{-3} was adopted. In complex tasks such as anomaly detection, setting the appropriate learning rate is crucial. A starting learning rate of 5×10^{-3} is high enough to facilitate rapid convergence, yet it is not so high as to risk overfitting or training instability. And the learning rate was adjusted using the ReduceLROnPlateau scheduler. This scheduler reduces the learning rate automatically when there is no improvement in validation loss. Learning rate can be optimized through training and adjusted. This prevents the model from becoming trapped in local minima during training and aids in fine-tuning performance through more delicate adjustments. Adjusting the learning rate during training is especially crucial in complex tasks like anomaly detection. Callback functions (Early Stopping) were also employed to prevent overfitting by halting the training if the validation loss did not decrease for a certain period. While the epoch count was initially configured to 100, the actual number is different due to the activation of callback functions. Owing to memory constraints, the batch size was set to one. Experiments were conducted on a Windows 10 operating system with hardware specifications of an 11th Gen Intel(R) Core(TM) i7-11700K with a 3.60 GHz CPU, 48.0 GB RAM, and an NVIDIA GeForce RTX 4090 GPU. All the proposed architectures were implemented using PyTorch.

3.6. Evaluation Metrics

In this section, we describe evaluation metrics for experiments. The evaluation metrics used were accuracy (ACC), loss, AUC, and EER. ACC is one of the most intuitive metrics used in classification problems and represents the ratio of items correctly classified by the model to the total number of items. Accuracy is follow as Equation (6):

$$Accuracy = \frac{Number\ of\ Correct\ Predictions}{Total\ Number\ of\ Predictions} = \frac{TP + TN}{TP + TN + FP + FN} \tag{6}$$

In the context of model evaluation, the loss function quantifies the discrepancy between the predicted values of the model and the actual data. A lower loss value signifies that the model's predictions are more closely aligned with the true values. Furthermore, the Receiver Operating Characteristic (ROC) curve, plotted with the False Positive Rate (FPR) on the x-axis and the True Positive Rate (TPR) on the y-axis, serves as a critical tool for assessing classification performance. The area under the ROC curve (AUC) is particularly instrumental, providing a numerical benchmark for performance evaluation. The AUC value, ranging from 0 to 1, reflects the model's effectiveness; values approaching 1 indicate superior model performance as the curve nears the upper left corner of the graph. The formulations for TPR and FPR are delineated in Equations (7) and (8), respectively.

$$TPR = \frac{TP}{TP + FN} \tag{7}$$

$$FPR = \frac{FP}{FP + TN} \tag{8}$$

EER is the error rate at the point where the false acceptance rate (FAR) and false rejection rate (FRR) intersect. In other words, in the graph, ERR denotes the error rate at the point where the FAR and FRR cross (become equal). A value closer to 0 indicates a better performance of the classification model. FRR defined as follows Equation (9):

$$FRR = \frac{FN}{TP + FN} \tag{9}$$

4. Results

In this section, we list all the experimental results of the proposed method for each dataset. Additionally, we present the results of various existing experiments and perform analyses and comparisons of these results. All figures are rounded off to the fifth decimal place. Two experiments were conducted for UBI-Fights. Before conducting the experiment, we addressed the class imbalance issue in the dataset by undersampling the fight class. Subsequently, the first experiment was conducted using the adjusted dataset, and the second experiment was performed using the original. For RWF-2000 and UCSD Ped1 and Ped2, only one experiment was conducted without altering the data ratio.

4.1. UBI-Fights

In this section, we introduce the experimental results for the UBI-Fights dataset. The UBI-Fights dataset consists of videos in two categories: fight and normal, with 216 and 784 videos, respectively. The size of the merged images used in the experiment was 800×800.

Table 1 presents the experimental results for the UBI-Fights dataset with undersampling applied. For 3×3, 4×3, 4×4, and 5×3 inputs, the networks based on ResNet 18, ResNet 34, ResNet 10, and ResNet 10, respectively, demonstrated the highest performance. This suggests that if a model, such as ResNet 50, is too deep and complex, it may face classification difficulties. We observed challenges in feature extraction when the model became excessively deep or when a certain number of frames were used on images of limited size.

Table 1. Experimental results on datasets with undersampling applied.

Input	Backbone	ACC	LOSS	AUC	EER
3 × 3	ResNet10	0.9925	0.0022	0.9970	0.0030
	ResNet18	**0.9933**	**0.0010**	**0.9973**	**0.0027**
	ResNet34	0.9858	0.0068	0.9943	0.0060
	ResNet50	0.8025	0.2368	0.8625	0.1991
4 × 3	ResNet10	0.9419	0.0174	0.9747	0.0293
	ResNet18	0.9680	**0.0009**	0.9881	0.1510
	ResNet34	**0.9762**	0.0069	**0.9905**	**0.0099**
	ResNet50	0.8625	0.1789	0.7952	0.2037
4 × 4	**ResNet10**	0.9534	**0.1804**	**0.7932**	**0.2588**
	ResNet18	**0.9595**	0.2032	0.7516	0.2707
	ResNet34	0.9535	0.3222	0.7254	0.3045
	ResNet50	0.9535	0.3778	0.6392	0.3780
5 × 3	**ResNet10**	0.9789	0.0352	**0.9363**	**0.0615**
	ResNet18	**0.9980**	**0.0286**	0.8021	0.1789
	ResNet34	0.8551	0.2153	0.8006	0.1989
	ResNet50	0.7851	0.3265	0.7978	0.2060

Table 2 lists the experimental results for the original UBI-Fights dataset. For 3 × 3, 4 × 3, 4 × 4, and 5 × 3 inputs, the networks based on ResNet10, ResNet18, ResNet10, and ResNet10, respectively, demonstrated the highest performance. Because undersampling was not performed, the amount of data increased, leading to superior performance in relatively lightweight networks. Additionally, for the 4 × 4 input, both ResNet10 and ResNet18 excelled in the two evaluation metrics. However, because the primary evaluation metrics for the UBI-Fights dataset are AUC and EER, the ResNet10-based network can be considered slightly more dominant.

Table 2. Experimental results on the original dataset.

Input	Backbone	ACC	LOSS	AUC	EER
3 × 3	**ResNet10**	**0.9916**	**0.0002**	**0.9817**	**0.0224**
	ResNet18	0.9913	0.0071	0.9345	0.0737
	ResNet34	0.9720	0.0211	0.8668	0.1418
	ResNet50	0.8344	0.2397	0.8259	0.1834
4 × 3	ResNet10	0.9419	0.0174	0.9747	0.0293
	ResNet18	**0.9680**	**0.0009**	**0.9881**	**0.0151**
	ResNet34	0.9270	0.0004	0.9647	0.0487
	ResNet50	0.8358	0.2867	0.7952	0.0950
4 × 4	**ResNet10**	**0.9799**	0.1732	**0.8411**	**0.2187**
	ResNet18	0.9799	**0.1715**	0.8406	0.2324
	ResNet34	0.9798	0.4752	0.8194	0.2481
	ResNet50	0.9799	0.4776	0.6637	0.3357
5 × 3	**ResNet10**	**0.9767**	0.1017	**0.9363**	**0.0969**
	ResNet18	0.9489	**0.0433**	0.7772	0.2231
	ResNet34	0.9500	0.0460	0.7075	0.3036
	ResNet50	0.9500	0.1480	0.6066	0.3952

As shown in Table 3, compared with other methods, our approach employed two different experimental methods based on data distribution. We chose the method with higher AUC and EER values because the remaining models were only compared based on AUC and EER. Notably, the method with 3 × 3 input achieved an AUC of 0.9973 and an EER of 0.0027. Similarly, most of our methods surpassed state-of-the-art results. Particularly, by observing the results from the 3 × 3 and 4 × 3 based on ResNet10 and ResNet18 backbone models, it is clear that fewer images were used in the merged image, and relatively lower model complexity(depth) yielded better outcomes.

Table 3. Comparison of recent studies. Table 3 compares the results of recent studies using the UBI-Fights dataset. There is only one model with an AUC exceeding 90, and in most cases, the EER is also greater than 0.2.

Method	AUC	EER
SS-Model + WS-Model + Sultani et al. [73]	0.846	0.216
SS-Model [73]	0.819	0.284
GMM [15]	0.906	0.160
Sultani et al. [19]	0.787	0.267
s2 VAE [28]	0.610	0.427
LSTM-AE [29]	0.541	0.480
Adversarial generator [24]	0.533	0.484
Grid frame (ours) (3 × 3)	**0.9973**	**0.0027**
Grid frame (ours) (4 × 3)	0.9905	0.0099
Grid frame (ours) (4 × 4)	0.8411	0.2187
Grid frame (ours) (5 × 3)	0.9363	0.0615

4.2. RWF-2000

This section presents the experimental results for the RWF-2000 dataset. Because the original RWF-2000 data ratio was split as 8:2 for the training and testing, only one experiment was conducted. Additionally, there were 2000 videos, with 800 for training and 200 for testing each for the Violence and Nonviolence categories. As shown in Table 4, similar to UBI-Fights, as the complexity of the backbone network increased, the performance tended to deteriorate. Compared with UBI-Fights, it is evident that the performance is generally higher. Table 5 presents a comparison of recent studies using the traditional RWF-2000 dataset.

Table 4. Experimental results on the RWF-2000 dataset.

Input	Backbone	ACC	LOSS	AUC	EER
3 × 3	**ResNet10**	**0.9920**	**0.0211**	**0.9962**	**0.0059**
	ResNet18	0.9884	0.0343	0.9363	0.1418
	ResNet34	0.8596	0.0386	0.8548	0.1436
	ResNet50	0.8846	0.2449	0.6006	0.3586
4 × 3	ResNet10	0.9784	0.0177	**0.9962**	**0.0487**
	ResNet18	**0.9787**	**0.0026**	0.9330	0.0677
	ResNet34	0.8764	0.3460	0.8800	0.0976
	ResNet50	0.8579	0.3524	0.8732	0.1158
4 × 4	**ResNet10**	**0.9506**	**0.0029**	**0.9071**	**0.1572**
	ResNet18	0.9015	0.0343	0.9001	0.1756
	ResNet34	0.8687	0.4059	0.8059	0.1989
	ResNet50	0.7350	0.4734	0.7000	0.2700
5 × 3	**ResNet10**	**0.9506**	**0.0029**	**0.9550**	**0.0583**
	ResNet18	0.9382	0.0071	0.9071	0.0972
	ResNet34	0.8532	0.3023	0.7734	0.1789
	ResNet50	0.8004	0.3087	0.7900	0.2139

Table 4 lists the experimental results for RWF-2000. For 3 × 3, 4 × 3, 4 × 4, and 5 × 3 inputs, the networks based on ResNet10, ResNet18, ResNet10, and ResNet10 demonstrated the highest performance. Because the primary evaluation metric for this dataset was accuracy, in the case of the 4 × 3 input, ResNet18 was considered superior.

Table 5. Comparison with other methods using the RWF-2000 dataset.

Method	ACC
Structured keypoint pooling [74]	0.934
Semi-supervised hard attention (SSHA) [75]	0.904
Human skeletons + change detection [76]	0.9025
Separable convolutional LSTM [77]	0.8975
SPIL convolution [78]	0.893
Flow-gated network [72]	0.8725
Grid frame (ours) (3 × 3)	**0.9920**
Grid frame (ours) (4 × 3)	0.9787
Grid frame (ours) (4 × 4)	0.9506
Grid frame (ours) (5 × 3)	0.9506

Table 5 compares the proposed method with other approaches using the RWF-2000 dataset. Because accuracy is the primary evaluation metric for this dataset, we prioritized it when comparing performances. Compared with other methods, the network using a 3 × 3 input configuration achieved the highest performance with an accuracy of 0.9920.

4.3. *UCSD Ped1*

In this section, we present the experimental results for the UCSD Ped1 dataset. Only the abnormal data have ground truth in the form of segmentation maps, in which the anomalies are segmented in white. If there is even one pixel with a value of 1 in the ground truth image, a label of 1 is assigned; otherwise, a label of 0 is assigned. For our experiment, we randomly added a portion of the anomalous data from the test dataset to the training set and proceeded with the training.

Table 6 lists the experimental results for the UCSD Ped1 dataset. For 3 × 3, 4 × 3, 4 × 4, and 5 × 3 inputs, the networks based on ResNet18, ResNet34, ResNet18, and ResNet10 demonstrated the highest performance. Overall, using ResNet18 as the backbone network with a 3 × 3 input achieved the highest performance, with an AUC of 0.9188 and an EER of 0.1197. However, this record is ranked second highest following an adversarial generator [24] (Table 7).

Table 6. Experimental results using the UCSD Ped1 dataset.

Input	Backbone	AUC	EER
3 × 3	ResNet10	0.8828	0.1880
	ResNet18	**0.9188**	**0.1197**
	ResNet34	0.8988	0.1606
	ResNet50	0.8787	0.1849
4 × 3	ResNet10	0.8846	0.1875
	ResNet18	0.8846	0.1890
	ResNet34	**0.9047**	**0.1597**
	ResNet50	0.8717	0.1915
4 × 4	ResNet10	0.8721	0.1913
	ResNet18	**0.8846**	**0.1836**
	ResNet34	0.8503	0.1985
	ResNet50	0.8726	0.1875
5 × 3	**ResNet10**	**0.8846**	**0.1875**
	ResNet18	0.8830	0.1884
	ResNet34	0.8824	0.1881
	ResNet50	0.8652	0.1936

Table 7. Comparison with other methods on the UCSD Ped1 dataset.

Method	AUC	EER
S2-VAE [28]	0.876	-
sRNN [79]	0.8171	-
ConvAE [29]	0.899	0.125
GMM [15]	0.801	0.236
Adversarial generator [24]	**0.974**	**0.008**
Grid frame (ours) (3 × 3)	0.9188	0.1197
Grid frame (ours) (4 × 3)	0.9047	0.1597
Grid frame (ours) (4 × 4)	0.8846	0.1875
Grid frame (ours) (5 × 3)	0.8846	0.1875

For the UCSD Ped1 dataset, the method with a 3 × 3 input achieved an AUC of 0.9188 and an EER value of 0.1197. However, this was the second-highest record, following the Adversarial generator [24].

4.4. UCSD Ped2

In this section, we present experimental results for the UCSD Ped2 dataset. The training dataset contained only normal data, whereas the test dataset contained only abnormal data. Thus, we randomly added a portion of the abnormal data from the test set to the training set for the experiment. The ground truth was provided in the same manner as UCSD Ped1; therefore, the labeling method and subsequent training were conducted in the same way.

Table 8 presents the experimental results for UCSD Ped2. For 3 × 3, 4 × 3, 4 × 4, and 5 × 3 inputs, the networks based on ResNet10 demonstrated the highest performance. The best performance was achieved with a 3 × 3 input using ResNet10 as the backbone network, resulting in an AUC of 0.9959 and an ERR of 0.0081.

Table 8. Experimental results using the UCSD Ped2 dataset.

Input	Backbone	AUC	EER
3 × 3	**ResNet10**	**0.9959**	**0.0081**
	ResNet18	0.9953	0.0111
	ResNet34	0.9000	0.1666
	ResNet50	0.7875	0.2134
4 × 3	**ResNet10**	**0.8750**	**0.2000**
	ResNet18	0.8457	0.2040
	ResNet34	0.8730	0.2006
	ResNet50	0.7969	0.2121
4 × 4	**ResNet10**	**0.8672**	**0.2025**
	ResNet18	0.8594	0.2051
	ResNet34	0.8613	0.2006
	ResNet50	0.8262	0.1895
5 × 3	**ResNet10**	**0.8750**	**0.2057**
	ResNet18	0.8574	0.2000
	ResNet34	0.8691	0.2000
	ResNet50	0.8449	0.2011

Table 9 compares the proposed approach with other methods. Because other methods primarily assessed performance based solely on AUC, we assigned the highest priority to AUC for comparison. The network using the 3 × 3 format achieved the second-best performance, with an AUC of 0.9959, slightly below that of the state-of-the-art model.

Table 9. Comparison with other methods using the UCSD Ped2 dataset.

Method	AUC	EER
Background-agnostic framework [80]	0.987	-
RTFM [14]	0.986	-
CL-VAD [81]	0.978	-
SSMTL [82]	0.975	-
FastAno [83]	0.963	-
GCN-Anomaly [84]	0.932	-
AI-VAD [85]	0.991	-
Adversarial generator [24]	0.935	0.14
DMAD [86]	**0.997**	-
Grid frame (ours) (3 × 3)	0.9959	0.0081
Grid frame (ours) (4 × 3)	0.8750	0.2000
Grid frame (ours) (4 × 4)	0.8672	0.2025
Grid frame (ours) (5 × 3)	0.8750	0.2057

5. Conclusions and Future Research

In this study, we introduce a 3D CNN-based network combined with an attention module for real-time anomaly detection, particularly for violent behavior detection and video surveillance, and a unique input data approach. This network has several unique features compared with those of previous studies and uses state-of-the-art deep learning technology and attention mechanisms to effectively extract the features of temporal and spatial information in the context of the data. A comparison of our method with state-of-the-art methods demonstrates the superior performance of our approach, which enables improved focus on areas of interest. Rather than assessing a situation by observing a static scenario, such as a 2D CNN, our method can be applied to various scenarios, such as robbery, arson, and theft, if desired, by using the correlation between the past, present, and future. Additionally, in this study, an input video with multiple consecutive frames in one image, called a grid frame, was applied to the network with cbam to simultaneously view spatiotemporal features from multiple viewpoints. This approach enabled a shift from simple 2D image analysis to dynamic video analysis.

However, there remains the challenge of memory constraints. Because of memory constraints, the number of frames applied to a single video was limited. To address this issue, we are currently exploring methods such as reducing the size of images using grid frames or diminishing the complexity of 3D convolutional models. To circumvent the memory limitations restricting the number of frames per video, it is essential to reduce either the image size during data preprocessing or the duration of each video in the dataset. For instance, segmenting a 10 s video into ten 1 s clips ensures the availability of complete data information, irrespective of memory constraints. So, addressing this issue in the data preprocessing step significantly will improve the classification performance owing to the increased input data. We believe that this network will lead to significant breakthroughs in anomaly detection. This study is expected to advance research on the application of the attention module to 2D CNN. If these models can be efficiently lightweight, it holds promise for real-time detection applications in industrial settings. Furthermore, it is anticipated that the application of this model will not be limited solely to crimes occurring between individuals. This study can serve as a basis for research on potentially hazardous areas, such as manufacturing plants, food inspection sites, and construction sites, where dangerous situations can occur instantaneously.

To improve results, it is expected that the results of all the used datasets can be improved by performing transfer learning on the network used in this paper or through data augmentation and ensemble techniques. A second approach is to use multimodal data, adding other types of data, such as audio. This is expected to improve results for this dataset by allowing the detection of anomalies that are difficult to identify using visual data alone.

Finally, future studies will apply this network to broader datasets to ensure the best classification performance, regardless of the input data method used. We will also continue to work to better detect anomalous behavior in datasets with deeper occlusion and public transportation and to achieve good results on larger and more complex datasets. We expect that this study will be used in various fields and will inspire further research directions.

Author Contributions: Conceptualization, I.-C.H.; Methodology, I.-C.H.; Software, I.-C.H.; Validation, I.-C.H. and H.-S.K.; Formal analysis, I.-C.H.; Investigation, H.-S.K.; Resources, H.-S.K.; Data curation, I.-C.H.; Writing—original draft, I.-C.H.; Writing—review and editing, I.-C.H. and H.-S.K.; Visualization, I.-C.H. and H.-S.K.; Supervision, H.-S.K.; Project administration, H.-S.K.; Funding acquisition, H.-S.K. All authors have read and agreed to the published version of the manuscript.

Funding: This work was supported by the MSIT (Ministry of Science and ICT), Korea, under the Grand Information Technology Research Center support program (IITP-2023-2020-0-01462) supervised by the IITP, and in part by the Basic Science Research Program through the National Research Foundation of Korea (NRF) funded by the Ministry of Education under Grant 2023R1A2C1006944.

Institutional Review Board Statement: Not applicable.

Informed Consent Statement: Not applicable.

Data Availability Statement: The datasets used in this paper are public datasets.

Conflicts of Interest: The authors declare no conflict of interest.

References

1. Goodfellow, I.; Pouget-Abadie, J.; Mirza, M.; Xu, B.; Warde-Farley, D.; Ozair, S.; Courville, A.; Bengio, Y. Generative adversarial nets. *Adv. Neural Inf. Process. Syst.* **2014**, *27*, 139–144.
2. Hinton, G.E.; Salakhutdinov, R.R. Reducing the dimensionality of data with neural networks. *Science* **2006**, *313*, 504–507. [CrossRef] [PubMed]
3. Woo, S.; Park, J.; Lee, J.Y.; Kweon, I.S. Cbam: Convolutional block attention module. In Proceedings of the European Conference on Computer Vision (ECCV), Munich, Germany, 8–14 September 2018; pp. 3–19.
4. Levoy, M.; Hanrahan, P. Light field rendering. In Proceedings of the 23rd Annual Conference on Computer Graphics and Interactive Techniques, New Orleans, LA, USA, 4–9 August 1996; pp. 31–42.
5. Dosovitskiy, A.; Beyer, L.; Kolesnikov, A.; Weissenborn, D.; Zhai, X.; Unterthiner, T.; Dehghani, M.; Minderer, M.; Heigold, G.; Gelly, S.; et al. An image is worth 16x16 words: Transformers for image recognition at scale. *arXiv* **2020**, arXiv:2010.11929.
6. Tran, D.; Bourdev, L.; Fergus, R.; Torresani, L.; Paluri, M. Learning spatiotemporal features with 3d convolutional networks. In Proceedings of the IEEE International Conference on Computer Vision, Santiago, Chile, 11–18 December 2015; pp. 4489–4497.
7. Dubey, S.; Boragule, A.; Gwak, J.; Jeon, M. Anomalous event recognition in videos based on joint learning of motion and appearance with multiple ranking measures. *Appl. Sci.* **2021**, *11*, 1344. [CrossRef]
8. Maqsood, R.; Bajwa, U.I.; Saleem, G.; Raza, R.H.; Anwar, M.W. Anomaly recognition from surveillance videos using 3D convolution neural network. *Multimed. Tools Appl.* **2021**, *80*, 18693–18716. [CrossRef]
9. Hu, X.; Lian, J.; Zhang, D.; Gao, X.; Jiang, L.; Chen, W. Video anomaly detection based on 3D convolutional auto-encoder. *Signal Image Video Process.* **2022**, *16*, 1885–1893. [CrossRef]
10. Koshti, D.; Kamoji, S.; Kalnad, N.; Sreekumar, S.; Bhujbal, S. Video anomaly detection using inflated 3D convolution network. In Proceedings of the 2020 International Conference on Inventive Computation Technologies (ICICT), IEEE, Coimbatore, India, 26–28 February 2020; pp. 729–733.
11. Chen, Y.; Liu, Z.; Zhang, B.; Fok, W.; Qi, X.; Wu, Y.C. Mgfn: Magnitude-contrastive glance-and-focus network for weakly supervised video anomaly detection. In Proceedings of the AAAI Conference on Artificial Intelligence, Washington, DC, USA, 7–14 February 2023; Volume 37, pp. 387–395.
12. Pu, Y.; Wu, X.; Wang, S. Learning Prompt-Enhanced Context Features for Weakly Supervised Video Anomaly Detection. *arXiv* **2023**, arXiv:2306.14451.
13. Lv, H.; Zhou, C.; Cui, Z.; Xu, C.; Li, Y.; Yang, J. Localizing anomalies from weakly labeled videos. *IEEE Trans. Image Process.* **2021**, *30*, 4505–4515. [CrossRef]
14. Tian, Y.; Pang, G.; Chen, Y.; Singh, R.; Verjans, J.W.; Carneiro, G. Weakly supervised video anomaly detection with robust temporal feature magnitude learning. In Proceedings of the IEEE/CVF International Conference on Computer Vision, Montreal, BC, Canada, 11–17 October 2021; pp. 4975–4986.
15. Degardin, B.M. Weakly and Partially Supervised Learning Frameworks for Anomaly Detection. Ph.D. Thesis, Universidade da Beira Interior (Portugal), Covilha, Portugal, 2020.
16. Feng, J.C.; Hong, F.T.; Zheng, W.S. Mist: Multiple instance self-training framework for video anomaly detection. In Proceedings of the IEEE/CVF Conference on Computer Vision and Pattern Recognition, Nashville, TN, USA, 20–25 June 2021; pp. 14009–14018.

17. Lv, H.; Yue, Z.; Sun, Q.; Luo, B.; Cui, Z.; Zhang, H. Unbiased Multiple Instance Learning for Weakly Supervised Video Anomaly Detection. In Proceedings of the IEEE/CVF Conference on Computer Vision and Pattern Recognition, Vancouver, BC, Canada, 18–22 June 2023; pp. 8022–8031.
18. Dengxiong, X.; Bao, W.; Kong, Y. Multiple Instance Relational Learning for Video Anomaly Detection. In Proceedings of the 2021 International Joint Conference on Neural Networks (IJCNN), IEEE, Shenzhen, China, 18–22 July 2021; pp. 1–8.
19. Sultani, W.; Chen, C.; Shah, M. Real-world anomaly detection in surveillance videos. In Proceedings of the IEEE Conference on Computer Vision and Pattern Recognition, Salt Lake City, UT, USA, 18–22 June 2018; pp. 6479–6488.
20. Kamoona, A.M.; Gostar, A.K.; Bab-Hadiashar, A.; Hoseinnezhad, R. Multiple instance-based video anomaly detection using deep temporal encoding–decoding. *Expert Syst. Appl.* **2023**, *214*, 119079. [CrossRef]
21. Doshi, K.; Yilmaz, Y. Fast unsupervised anomaly detection in traffic videos. In Proceedings of the IEEE/CVF Conference on Computer Vision and Pattern Recognition Workshops, Seattle, WA, USA, 14–19 June 2020; pp. 624–625.
22. Xia, X.; Gao, Y. Video Abnormal Event Detection Based on One-Class Neural Network. *Comput. Intell. Neurosci.* **2021**, *2021*, 1955116. [CrossRef]
23. Yang, M.; Rajasegarar, S.; Erfani, S.M.; Leckie, C. Deep learning and one-class SVM based anomalous crowd detection. In Proceedings of the 2019 International Joint Conference on Neural Networks (IJCNN) IEEE, Budapest, Hungary, 14–19 July 2019; pp. 1–8.
24. Ravanbakhsh, M.; Nabi, M.; Sangineto, E.; Marcenaro, L.; Regazzoni, C.; Sebe, N. Abnormal event detection in videos using generative adversarial nets. In Proceedings of the 2017 IEEE International Conference on Image Processing (ICIP), Beijing, China, 17–20 September 2017; pp. 1577–1581.
25. Jackson, S.D.; Cuzzolin, F. Svd-gan for real-time unsupervised video anomaly detection. In Proceedings of the British Machine Vision Conference (BMVC), Virtual, 22–25 November 2021; pp. 22–25.
26. Zaheer, M.Z.; Mahmood, A.; Khan, M.H.; Segu, M.; Yu, F.; Lee, S.I. Generative cooperative learning for unsupervised video anomaly detection. In Proceedings of the IEEE/CVF Conference on Computer Vision and Pattern Recognition, New Orleans, LA, USA, 18–24 June 2022; pp. 14744–14754.
27. Yang, Y.; Fu, Z.; Naqvi, S.M. Enhanced adversarial learning based video anomaly detection with object confidence and position. In Proceedings of the 2019 13th International Conference on Signal Processing and Communication Systems (ICSPCS), IEEE, Surfers Paradise, Australia, 16-18 December 2019; pp. 1–5.
28. Wang, T.; Qiao, M.; Lin, Z.; Li, C.; Snoussi, H.; Liu, Z.; Choi, C. Generative neural networks for anomaly detection in crowded scenes. *IEEE Trans. Inf. Forensics Secur.* **2018**, *14*, 1390–1399. [CrossRef]
29. Chong, Y.S.; Tay, Y.H. Abnormal event detection in videos using spatiotemporal autoencoder. In Proceedings of the Advances in Neural Networks-ISNN 2017: 14th International Symposium, ISNN 2017, Sapporo/Hakodate/Muroran, Japan, 21–26 June 2017; Part II 14; Springer: Berlin/Heidelberg, Germany, 2017; pp. 189–196.
30. Hasan, M.; Choi, J.; Neumann, J.; Roy-Chowdhury, A.K.; Davis, L.S. Learning temporal regularity in video sequences. In Proceedings of the IEEE Conference on Computer Vision and Pattern Recognition, Las Vegas, NV, USA, 27–30 June 2016; pp. 733–742.
31. Jain, A.K. Data clustering: 50 years beyond K-means. *Pattern Recognit. Lett.* **2010**, *31*, 651–666. [CrossRef]
32. Liu, W.; Luo, W.; Lian, D.; Gao, S. Future frame prediction for anomaly detection—A new baseline. In Proceedings of the IEEE Conference on Computer Vision and Pattern Recognition, Salt Lake City, UT, USA, 18–23 June 2018; pp. 6536–6545.
33. Li, D.; Nie, X.; Li, X.; Zhang, Y.; Yin, Y. Context-related video anomaly detection via generative adversarial network. *Pattern Recognit. Lett.* **2022**, *156*, 183–189. [CrossRef]
34. Zhang, Z.; Zhong, S.h.; Liu, Y. Video abnormal event detection via context cueing generative adversarial network. In Proceedings of the 2021 IEEE International Conference on Multimedia and Expo (ICME), Shenzhen, China, 5–9 July 2021; pp. 1–6.
35. Morais, R.; Le, V.; Tran, T.; Saha, B.; Mansour, M.; Venkatesh, S. Learning regularity in skeleton trajectories for anomaly detection in videos. In Proceedings of the IEEE/CVF Conference on Computer Vision and Pattern Recognition, Long Beach, CA, USA, 15–20 June 2019; pp. 11996–12004.
36. Du, Y.; Fu, Y.; Wang, L. Representation learning of temporal dynamics for skeleton-based action recognition. *IEEE Trans. Image Process.* **2016**, *25*, 3010–3022. [CrossRef] [PubMed]
37. Nguyen, T.N.; Meunier, J. Anomaly detection in video sequence with appearance-motion correspondence. In Proceedings of the IEEE/CVF International Conference on Computer Vision, Seoul, Republic of Korea, 27 October–2 November 2019; pp. 1273–1283.
38. Dosovitskiy, A.; Fischer, P.; Ilg, E.; Hausser, P.; Hazirbas, C.; Golkov, V.; Van Der Smagt, P.; Cremers, D.; Brox, T. Flownet: Learning optical flow with convolutional networks. In Proceedings of the IEEE International Conference on Computer Vision, Santiago, Chile, 7–13 December 2015; pp. 2758–2766.
39. Ilg, E.; Mayer, N.; Saikia, T.; Keuper, M.; Dosovitskiy, A.; Brox, T. Flownet 2.0: Evolution of optical flow estimation with deep networks. In Proceedings of the IEEE Conference on Computer Vision and Pattern Recognition, Honolulu, HI, USA, 21–26 July 2017; pp. 2462–2470.
40. Wang, H.; Zhang, X.; Yang, S.; Zhang, W. Video anomaly detection by the duality of normality-granted optical flow. *arXiv* **2021**, arXiv:2105.04302.

41. Vaswani, A.; Shazeer, N.; Parmar, N.; Uszkoreit, J.; Jones, L.; Gomez, A.N.; Kaiser, Ł.; Polosukhin, I. Attention is all you need. In Proceedings of the Advances in Neural Information Processing Systems 30 (NIPS 2017), Long Beach, CA, USA, 4–9 December 2017; Volume 30.
42. Devlin, J.; Chang, M.W.; Lee, K.; Toutanova, K. Bert: Pre-training of deep bidirectional transformers for language understanding. *arXiv* **2018**, arXiv:1810.04805.
43. Yuan, H.; Cai, Z.; Zhou, H.; Wang, Y.; Chen, X. Transanomaly: Video anomaly detection using video vision transformer. *IEEE Access* **2021**, *9*, 123977–123986. [CrossRef]
44. Arnab, A.; Dehghani, M.; Heigold, G.; Sun, C.; Lučić, M.; Schmid, C. Vivit: A video vision transformer. In Proceedings of the IEEE/CVF International Conference on Computer Vision, Montreal, BC, Canada, 11–17 October 2021; pp. 6836–6846.
45. Ronneberger, O.; Fischer, P.; Brox, T. U-net: Convolutional networks for biomedical image segmentation. In Proceedings of the Medical Image Computing and Computer-Assisted Intervention–MICCAI 2015: 18th International Conference, Munich, Germany, 5–9 October 2015; Part III 18; Springer: Berlin/Heidelberg, Germany, 2015; pp. 234–241.
46. Isola, P.; Zhu, J.Y.; Zhou, T.; Efros, A.A. Image-to-image translation with conditional adversarial networks. In Proceedings of the IEEE Conference on Computer Vision and Pattern Recognition, Honolulu, HI, USA, 21–26 July 2017; pp. 1125–1134.
47. Shi, X.; Chen, Z.; Wang, H.; Yeung, D.Y.; Wong, W.K.; Woo, W.C. Convolutional LSTM network: A machine learning approach for precipitation nowcasting. In Proceedings of the Advances in Neural Information Processing Systems 28 (NIPS 2015), Montreal, QC, Canada, 7–12 December 2015; Volume 28.
48. Luo, W.; Liu, W.; Gao, S. Remembering history with convolutional lstm for anomaly detection. In Proceedings of the 2017 IEEE International Conference on Multimedia and Expo (ICME), Hong Kong, China, 10–14 July 2017; pp. 439–444.
49. Medel, J.R.; Savakis, A. Anomaly detection in video using predictive convolutional long short-term memory networks. *arXiv* **2016**, arXiv:1612.00390.
50. Carreira, J.; Zisserman, A. Quo vadis, action recognition? A new model and the kinetics dataset. In Proceedings of the IEEE Conference on Computer Vision and Pattern Recognition, Honolulu, HI, USA, 21–26 July 2017; pp. 6299–6308.
51. Qi, Z.; Zhu, R.; Fu, Z.; Chai, W.; Kindratenko, V. Weakly Supervised Two-Stage Training Scheme for Deep Video Fight Detection Model. In Proceedings of the 2022 IEEE 34th International Conference on Tools with Artificial Intelligence (ICTAI), Virtual, 31 October–2 November 2022; pp. 677–685.
52. Deniz, O.; Serrano, I.; Bueno, G.; Kim, T.K. Fast violence detection in video. In Proceedings of the 2014 International Conference on Computer Vision Theory and Applications (VISAPP), IEEE, Lisbon, Portugal, 5–8 January 2014; Volume 2, pp. 478–485.
53. Bermejo Nievas, E.; Deniz Suarez, O.; Bueno García, G.; Sukthankar, R. Violence detection in video using computer vision techniques. In Proceedings of the Computer Analysis of Images and Patterns: 14th International Conference, CAIP 2011, Seville, Spain, 29–31 August 2011; Part II 14; Springer: Berlin/Heidelberg, Germany, 2011; pp. 332–339.
54. Ding, C.; Fan, S.; Zhu, M.; Feng, W.; Jia, B. Violence detection in video by using 3D convolutional neural networks. In Proceedings of the Advances in Visual Computing: 10th International Symposium, ISVC 2014, Las Vegas, NV, USA, 8–10 December 2014; Part II 10; Springer: Berlin/Heidelberg, Germany, 2014; pp. 551–558.
55. Abbass, M.A.B.; Kang, H.S. Violence Detection Enhancement by Involving Convolutional Block Attention Modules into Various Deep Learning Architectures: Comprehensive Case Study for UBI-Fights Dataset. *IEEE Access* **2023**, *11*, 37096–37107. [CrossRef]
56. Mahareek, E.A.; El-Sayed, E.K.; El-Desouky, N.M.; El-Dahshan, K.A. Detecting anomalies in security cameras with 3DCNN and ConvLSTM. *Int. J. Electr. Comput. Eng.* **2023**, 993–1004. [CrossRef]
57. Gao, Y.; Liu, H.; Sun, X.; Wang, C.; Liu, Y. Violence detection using oriented violent flows. *Image Vis. Comput.* **2016**, *48*, 37–41. [CrossRef]
58. Zhang, T.; Yang, Z.; Jia, W.; Yang, B.; Yang, J.; He, X. A new method for violence detection in surveillance scenes. *Multimed. Tools Appl.* **2016**, *75*, 7327–7349. [CrossRef]
59. De Souza, F.D.; Chavez, G.C.; do Valle, E.A., Jr.; Araújo, A.d.A. Violence detection in video using spatio-temporal features. In Proceedings of the 2010 23rd SIBGRAPI Conference on Graphics, Patterns and Images, IEEE, Gramado, Brazil, 30 August–3 September 2010; pp. 224–230.
60. Mahmoud, M.; Kang, H.S. GANMasker: A Two-Stage Generative Adversarial Network for High-Quality Face Mask Removal. *Sensors* **2023**, *23*, 7094. [CrossRef] [PubMed]
61. He, K.; Zhang, X.; Ren, S.; Sun, J. Deep residual learning for image recognition. In Proceedings of the IEEE Conference on Computer Vision and Pattern Recognition, Las Vegas, NV, USA, 27–30 June 2016; pp. 770–778.
62. Jetley, S.; Lord, N.A.; Lee, N.; Torr, P.H. Learn to pay attention. *arXiv* **2018**, arXiv:1804.02391.
63. Yong, H.L. Anomaly Detection with Attention-Based Deep Autoencoder. Ph.D. Thesis, UTAR, Kampar, Malaysia, 2022.
64. Gong, D.; Liu, L.; Le, V.; Saha, B.; Mansour, M.R.; Venkatesh, S.; Hengel, A.v.d. Memorizing normality to detect anomaly: Memory-augmented deep autoencoder for unsupervised anomaly detection. In Proceedings of the IEEE/CVF International Conference on Computer Vision, Seoul, Republic of Korea, 27 October–2 November 2019; pp. 1705–1714.
65. Raghavendra, R.; Bue, A.; Cristani, M. Unusual Crowd Activity Dataset of University of Minnesota. 2006. Available online: https://mha.cs.umn.edu/ (accessed on 26 November 2023).
66. Mehran, R.; Oyama, A.; Shah, M. Abnormal crowd behavior detection using social force model. In Proceedings of the 2009 IEEE Conference on Computer Vision and Pattern Recognition, Miami, FL, USA, 20–25 June 2009; pp. 935–942.

67. Mahadevan, V. UCSD Anomaly Detection Dataset@ONLINE. 2010. Available online: http://www.svcl.ucsd.edu/projects/anomaly/dataset.htm (accessed on 26 November 2023).
68. Adam, A.; Rivlin, E.; Shimshoni, I.; Reinitz, D. Robust real-time unusual event detection using multiple fixed-location monitors. *IEEE Trans. Pattern Anal. Mach. Intell.* **2008**, *30*, 555–560. [CrossRef]
69. Lu, C.; Shi, J.; Jia, J. Abnormal event detection at 150 fps in matlab. In Proceedings of the IEEE International Conference on Computer Vision, Sydney, Australia, 1–8 December 2013; pp. 2720–2727.
70. Ciampi, L.; Foszner, P.; Messina, N.; Staniszewski, M.; Gennaro, C.; Falchi, F.; Serao, G.; Cogiel, M.; Golba, D.; Szczęsna, A.; et al. Bus violence: An open benchmark for video violence detection on public transport. *Sensors* **2022**, *22*, 8345. [CrossRef]
71. Degardin, B.; Proença, H. Human activity analysis: Iterative weak/self-supervised learning frameworks for detecting abnormal events. In Proceedings of the 2020 IEEE International Joint Conference on Biometrics (IJCB), Abu Dhabi, United Arab Emirates, 10–13 October 2020; pp. 1–7.
72. Cheng, M.; Cai, K.; Li, M. RWF-2000: An open large scale video database for violence detection. In Proceedings of the 2020 25th International Conference on Pattern Recognition (ICPR), Milan, Italy, 10–15 January 2021; pp. 4183–4190.
73. Degardin, B.; Proença, H. Iterative weak/self-supervised classification framework for abnormal events detection. *Pattern Recognit. Lett.* **2021**, *145*, 50–57. [CrossRef]
74. Hachiuma, R.; Sato, F.; Sekii, T. Unified keypoint-based action recognition framework via structured keypoint pooling. In Proceedings of the IEEE/CVF Conference on Computer Vision and Pattern Recognition, Vancouver, BC, Canada, 18–22 June 2023; pp. 22962–22971.
75. Mohammadi, H.; Nazerfard, E. Video violence recognition and localization using a semi-supervised hard attention model. *Expert Syst. Appl.* **2023**, *212*, 118791. [CrossRef]
76. Garcia-Cobo, G.; SanMiguel, J.C. Human skeletons and change detection for efficient violence detection in surveillance videos. *Comput. Vis. Image Underst.* **2023**, *233*, 103739. [CrossRef]
77. Islam, Z.; Rukonuzzaman, M.; Ahmed, R.; Kabir, M.H.; Farazi, M. Efficient two-stream network for violence detection using separable convolutional lstm. In Proceedings of the 2021 International Joint Conference on Neural Networks (IJCNN), IEEE, Shenzhen, China, 18–22 July 2021; pp. 1–8.
78. Su, Y.; Lin, G.; Zhu, J.; Wu, Q. Human interaction learning on 3d skeleton point clouds for video violence recognition. In Proceedings of the Computer Vision–ECCV 2020: 16th European Conference, Glasgow, UK, 23–28 August 2020; Part IV 16; Springer: Berlin/Heidelberg, Germany, 2020; pp. 74–90.
79. Wang, T.; Snoussi, H. Detection of abnormal visual events via global optical flow orientation histogram. *IEEE Trans. Inf. Forensics Secur.* **2014**, *9*, 988–998. [CrossRef]
80. Georgescu, M.I.; Ionescu, R.T.; Khan, F.S.; Popescu, M.; Shah, M. A background-agnostic framework with adversarial training for abnormal event detection in video. *IEEE Trans. Pattern Anal. Mach. Intell.* **2021**, *44*, 4505–4523. [CrossRef] [PubMed]
81. Doshi, K.; Yilmaz, Y. Continual learning for anomaly detection in surveillance videos. In Proceedings of the IEEE/CVF Conference on Computer Vision and Pattern Recognition Workshops, Seattle, WA, USA, 13–19 June 2020; pp. 254–255.
82. Georgescu, M.I.; Barbalau, A.; Ionescu, R.T.; Khan, F.S.; Popescu, M.; Shah, M. Anomaly detection in video via self-supervised and multi-task learning. In Proceedings of the IEEE/CVF Conference on Computer Vision and Pattern Recognition, Nashville, TN, USA, 20–25 June 2021; pp. 12742–12752.
83. Park, C.; Cho, M.; Lee, M.; Lee, S. FastAno: Fast anomaly detection via spatio-temporal patch transformation. In Proceedings of the IEEE/CVF Winter Conference on Applications of Computer Vision, Waikoloa, HI, USA, 4–8 January 2022; pp. 2249–2259.
84. Zhong, J.X.; Li, N.; Kong, W.; Liu, S.; Li, T.H.; Li, G. Graph convolutional label noise cleaner: Train a plug-and-play action classifier for anomaly detection. In Proceedings of the IEEE/CVF Conference on Computer Vision and Pattern Recognition, Long Beach, CA, USA, 15–20 June 2019; pp. 1237–1246.
85. Reiss, T.; Hoshen, Y. Attribute-based Representations for Accurate and Interpretable Video Anomaly Detection. *arXiv* **2022**, arXiv:2212.00789.
86. Liu, W.; Chang, H.; Ma, B.; Shan, S.; Chen, X. Diversity-measurable anomaly detection. In Proceedings of the IEEE/CVF Conference on Computer Vision and Pattern Recognition, Vancouver, BC, Canada, 18–22 June 2023; pp. 12147–12156.

sensors

MDPI

Article

FR-PatchCore: An Industrial Anomaly Detection Method for Improving Generalization

Zhiqian Jiang, Yu Zhang *, Yong Wang, Jinlong Li and Xiaorong Gao

School of Physical Science and Technology, Southwest Jiaotong University, Chengdu 611756, China; 2022201064@my.swjtu.edu.cn (Z.J.); wangyonga@swjtu.edu.cn (Y.W.); jinlong_lee@126.com (J.L.); gxrr@vip.163.com (X.G.)

* Correspondence: zhangyu2@swjtu.edu.cn

Abstract: In recent years, a multitude of self-supervised anomaly detection algorithms have been proposed. Among them, PatchCore has emerged as one of the state-of-the-art methods on the widely used MVTec AD benchmark due to its efficient detection capabilities and cost-saving advantages in terms of labeled data. However, we have identified that the PatchCore similarity principal approach faces significant limitations in accurately locating anomalies when there are positional relationships between similar samples, such as rotation, flipping, or misaligned pixels. In real-world industrial scenarios, it is common for samples of the same class to be found in different positions. To address this challenge comprehensively, we introduce Feature-Level Registration PatchCore (FR-PatchCore), which serves as an extension of the PatchCore method. FR-PatchCore constructs a feature matrix that is extracted into the memory bank and continually updated using the optimal negative cosine similarity loss. Extensive evaluations conducted on the MVTec AD benchmark demonstrate that FR-PatchCore achieves an impressive image-level anomaly detection AUROC score of up to 98.81%. Additionally, we propose a novel method for computing the mask threshold that enables the model to scientifically determine the optimal threshold and accurately partition anomalous masks. Our results highlight not only the high generalizability but also substantial potential for industrial anomaly detection offered by FR-PatchCore.

Keywords: image anomaly detection; self-supervised learning; feature processing

1. Introduction

The issue of data imbalance is widely observed in industrial anomaly detection, further exacerbated by the scarcity of valuable anomalous data available for this task. Anomalies can arise from various unknown external influences [1], and they can also originate from the objects themselves. It is impractical to account for every possible exception. Unsupervised approaches address this challenge by leveraging unlabeled samples to enable the model to learn the distribution properties and features of both normal and abnormal samples. This enables effective anomaly detection while avoiding the high cost associated with labelled data. In the field of industrial anomaly detection, training autoencoders [2,3] and methods based on generative adversarial networks (GANs) [4–7] are commonly employed. Lyu et al. [8] fused GAN with a deep convolutional neural network (DCNN) to build an anomaly location model, which can find the mapping relationship between images and high-dimensional features. Son et al. [9] computed anomaly scores by training encoder–decoder-based long short-term memory (LSTM) networks to evaluate data based on the time the anomaly existed. J. Mitra et al. [10] used generative adversarial networks to identify a manifold of normal samples and, at the same time, identify abnormal patterns that fall outside the model. Xiao, K., Cao, J., Zeng, Z., and Ling, W.-K. [11] explore sample distribution by utilizing the fast-labelling characteristics of a graph structure and use an autoencoder to blur low-level features while retaining local features to achieve better anomaly detection performance. However, these methods often rely on the availability of a substantial amount of data. When the number of anomalous samples is limited, the model may struggle to distinguish between normal and anomalous images accurately.

Consequently, self-supervised approaches are highly favored in the field. Several state-of-the-art self-supervised methods, such as PatchCore [12], have achieved remarkable detection results on the widely recognized MVTec AD benchmark [13]. However, it has been observed that the PatchCore method imposes stringent requirements for image alignment. It is worth noting that in the MVTec dataset, most categories satisfy the condition of "similarity", where samples within the same category share common characteristics and can be approximately aligned at identical positions. These samples do not exhibit significant spatial differences, such as rotation or flipping. Hence, they demonstrate excellent test results and are referred to as "similar categories". In contrast, categories that fail to meet these similarity conditions are termed "spatial transformation categories". The effectiveness of their anomaly detection may be compromised as we have observed that these methods are not proficient in handling the latter type of category.

To establish that this is not a random occurrence, we also conducted experiments on the MPDD dataset [14], whose six categories also have "spatial location relationships". These experiments confirmed that the PatchCore method is more sensitive to "similar pictures" and does not pay much attention to the spatial position relationship between samples. In industrial detection, the images to be analyzed are not limited to "similar categories". They can be captured by sensors from various angles and locations. Moreover, when training data are scarce, employing data augmentation techniques to expand the dataset (e.g., rotation, mirror flipping, left, and right flipping, etc.) can introduce "spatial transformation categories", which may reduce the effect of anomaly detection, resulting in increased rates of missed detection and false positives.

To enhance the generalization performance of the method and enable it to effectively detect even in the presence of "spatial transformation categories", we introduce registration as a pretext task. We extract fused features from the feature extraction layer and store them in a database through a series of dimensionality reduction operations. By utilizing a negative cosine similarity loss, we update the features in the memory bank, eliminating overlapping features while incorporating previously neglected ones during training. As the tests adopt a pixel-by-pixel comparison approach, disregarding pixel relationships, we construct a module similar to pyramid pooling to enforce connectivity between patches. Additionally, we devise an innovative approach that enables the model to look for the optimal threshold, resulting in accurate ground truth segmentation. Our proposed method achieves an impressive AUROC performance of 98.81% on the MVTec dataset, significantly enhancing generalization and reducing false positive rates.

2. Related Works

Recently, supervised methods have exerted a significant impact on the field of anomaly detection, resulting in the assimilation of concepts from these approaches into self-supervised methods. Self-supervised anomaly detection methods can be broadly classified into reconstruction-based techniques and feature embedding-based techniques.

2.1. Reconstruction-Based Approaches

Reconstruction-based anomaly detection relies on the assumption that the model can reconstruct normal samples well but struggles with anomalous regions. Early attempts involved using generative adversarial networks (GANs) [15,16]. However, due to the strong generalization of neural networks, abnormal samples could also achieve good reconstruction, rendering the comparison unreliable. More recently, You et al. [17] introduced a new paradigm for reconstruction networks, addressing the issue where popular reconstruction networks may fall into a "same shortcut" scenario, where both normal and abnormal samples can be accurately recovered, leading to difficulties in detecting outliers. This work tackles the practical challenge of detecting anomalies across different object classes using a unified framework.

2.2. Feature Embedding-Based Approaches

The feature embedding-based approach involves feeding images into the model, extracting features, and constructing interface scoring rules in the feature space. Unlike the reconstruction method that operates in the RGB image space, feature embedding focuses on the high-dimensional feature space for anomaly detection. Notably, Bergman, L., Cohen, N. and Hoshen, Y. [18] encouraged the use of pre-trained models on large-scale datasets for anomaly detection. This recommendation gives great encouragement to such methods.

SPADE [19] initially extracts features of all normal samples (training set) in a network pre-trained on ImageNet. During testing, the K-nearest neighbors (KNN) [20] metric is applied to each pixel-wise feature of the test sample to calculate the anomaly score for each pixel point. However, the more normal images used in training, the more features are stored, resulting in higher KNN complexity during testing. Building upon SPADE, a method named PaDiM [21] improves it by eliminating the construction of a feature bank and performing KNN for anomaly detection. However, the pixels at each position are not strictly aligned, so modelling each pixel position alone may lead to inaccuracies. Y. Zheng et al. [22] performs fine-tuning based on the anomaly localization task and proposed FYD. To address the feature misalignment issue in PaDiM, they introduce the spatial transformer network (STN) [23] module for coarse alignment and use a self-supervised learning paradigm inspired by the SimSiam network (without requiring negative samples). However, certain images may still not be aligned. Furthermore, from a visualization perspective, the detection performance is not optimal, primarily due to the evaluation metric (pixel-wise AUCROC score [24]), which tolerates anomalies in small areas.

Recently, Huang et al. [25] innovated based on PaDiM and FYD and proposed RegAD, which employs registration as a proxy task to train an anomaly detection model with unknown categories. PatchCore addresses the slow testing speed of SPADE. KNN and greedy coreset subsampling are applied to select the most representative feature points for measuring anomaly scores during testing, which reduce the size of the feature bank and finally achieve excellent results.

The method FR-PatchCore (Feature-level Registration PatchCore) of ours preserves the high efficiency of PatchCore. Feature-level registration is employed as a pretext task, and the information stored in the feature database is updated using a negative cosine similarity loss. To enforce pixel relationships, a module similar to pyramid pooling is constructed. During testing, the Euclidean distance from each pixel to its corresponding feature in the database is used for scoring.

3. Methodology

In our methodology, we introduce the concept of feature-level registration and utilize it as a non-cumbersome pretext task during training. The negative cosine similarity loss is employed as the training loss, which we optimize to update the memory bank. The framework of our approach is depicted in Figure 1.

The upper section represents the registration module, while the lower section represents the memory bank module. Both sections operate concurrently. Two images from samples belonging to the same category are randomly selected, and their features are extracted using the convolutional neural network and spatial transformer network (CNN + STN) for registration purposes. The feature registration process is supervised by maximizing the absolute value of the negative cosine similarity loss. After the registration process, the extracted fusion features are stored in a dedicated memory bank and continuously updated with training losses after each registration iteration.

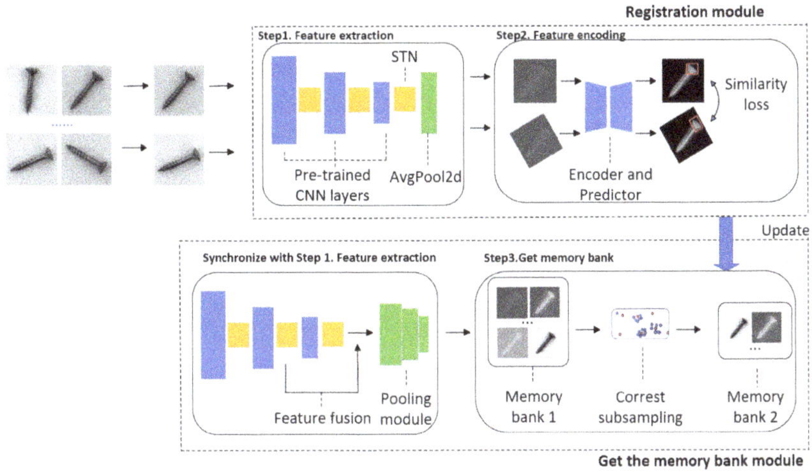

Figure 1. The architecture of the FR-PatchCore model proposed is presented. The entire framework can be divided into three main steps, and the feature extraction processes in both modules are executed simultaneously. In the registration module, randomly selected images are fed into CNN + STN to obtain pre-encoded features. Meanwhile, in the memory bank module, we combine the features passing through the second STN layer and the third STN layer to retain sufficient spatial information of features. Finally, the fusion features are stored in the memory bank.

3.1. Registration Module

Neural network training necessitates task-driven learning. Therefore, the essence of self-supervised learning lies in the thoughtful design of tasks that facilitate effective model learning. Inspired by the work of [25], which obtained a Gaussian distribution model of normal data through feature-level registration training, we leverage the registration task as a pretext task to enhance the model's understanding of features and emphasize spatial and positional differences. Accordingly, we construct the registration module, consisting of a feature extractor, feature encoder, and predictor, as illustrated in Figure 2.

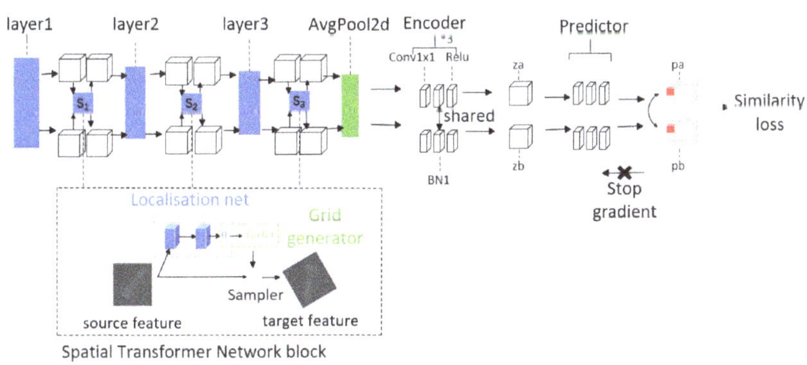

Figure 2. Registration module: two different images of the same category are inputted, and the CNN + STN combination is utilized to extract features. A Siamese structure network is employed for feature encoding and prediction. The objective is to train the registration network in such a way that the predicted features closely resemble the encoded features in the figure.

In feature registration, since spatial transformation can be represented as matrix operations, it is advantageous to allow the network to learn the generation of matrix parameters, thereby acquiring spatial transformation capabilities. Common network frameworks in deep learning include the CNN and the transformer. Additionally, the spatial transformer network, which plays a pivotal role in our research, can seamlessly integrate into any component of the CNN architecture. To ensure a fair comparison with state-of-the-art methods, we have selected the wide_resnet_50_2 [26] network as the backbone for our experiments among various CNNs commonly employed in anomaly detection tasks. This network has demonstrated exceptional performance on the ImageNet dataset, achieving an accuracy of 78.51% for Top 1 and 94.09% for Top 5. For this specific task addressed in our paper, we conducted an ablation experiment (4.5.1) to compare the feature extraction capabilities of both Resnet and VIT models, where wide_resnet_50_2 exhibited superior performance.

We incorporated the *STN* module into the wide_resnet_50_2 architecture. The overall structure of the *STN* module is illustrated in Figure 2 and consists of a localization network, a grid generator, and a sampler. In the first component, a feature map is given as Formula (1):

$$U \in R^{H\times W\times C} \tag{1}$$

After several convolutions or fully connected layers, a regression layer follows, leading to the output of regression transformation parameter $\boldsymbol{\theta}$. The dimension of $\boldsymbol{\theta}$ depends on the specific transformation type chosen by the network.

In the second component, we utilize $\boldsymbol{\theta}$ and the specified transformation mode from the localization network output, and the grid generator performs further spatial transformations of the feature map to determine the mapping of $\boldsymbol{T(\theta)}$ between the output and input features. It employs the predicted transformation parameters to create a sampling grid, which represents a set of points where the input map should be sampled to produce the transformed output. In the sampler, it utilizes the sampling grid to determine which points in the input feature map will be utilized for the transformation. It samples the input feature map against the sampling grid to obtain the final output.

In Step 1, assuming that the input RGB image has a resolution of (224,224), Formula (2) is applied.

$$U \in R^{224\times 224\times 3} \tag{2}$$

The fourth layer of the wide_resnet_50_2 network is excluded to preserve more comprehensive spatial information, and the spatial transformer network (STN) is integrated after the initial three layers of the network. The input feature $f_i(x_i^s, y_i^s)$ undergoes a transformation function τ_θ. The mapping relationship between the input and output feature mappings is defined as Formula (3).

$$\begin{pmatrix} x_i^s \\ y_i^s \end{pmatrix} = \tau_\theta(G_i) = A_\theta \begin{pmatrix} x_i^t \\ y_i^t \\ 1 \end{pmatrix} = \begin{bmatrix} \theta_{11} & \theta_{12} & \theta_{13} \\ \theta_{21} & \theta_{22} & \theta_{23} \end{bmatrix} \begin{pmatrix} x_i^t \\ y_i^t \\ 1 \end{pmatrix} \tag{3}$$

In this case, the form of the eigenvector is shown in Formula (4):

$$\begin{cases} U_1 \in R^{56\times 56\times 64} \\ U_2 \in R^{28\times 28\times 128} \\ U_3 \in R^{14\times 14\times 256} \end{cases} \tag{4}$$

When no key points are labeled, the STN allows the neural network to actively transform the feature map based on input features and learn spatial transformation parameters without requiring additional training supervision or modifications to the optimization process. As illustrated in Figure 2, the STN can effectively align input images or learned features during training, thereby mitigating the impact of spatial geometric transformations such as rotation, translation, scale, and distortion on tasks like classification and localization. The STN facilitates the spatial transformation of the input data, thereby enhancing feature classification and enabling the network to achieve rotational invariance dynamically. Moreover, it intelligently selects the most salient region of the image and optimally transforms it into a suitable orientation. Figure 2 illustrates the depiction of feeding an inverted screw image into the STN module. Through a series of transformations, the input is effectively rectified to face forward.

The employed approach in Figure 2 involves the utilization of a Siamese network for feature encoding, wherein a negative cosine similarity loss is applied as per Formulas (5) and (6).

$$D(p_a, z_b) = -\frac{p_a}{\|p_a\|_2} \cdot \frac{z_b}{\|z_b\|_2} \tag{5}$$

$$D(p_b, z_a) = -\frac{p_b}{\|p_b\|_2} \cdot \frac{z_a}{\|z_a\|_2} \tag{6}$$

The negative cosine similarity loss is an appropriate metric for quantifying the similarity between two vectors. Furthermore, it possesses the capability to map similar vectors to adjacent points and dissimilar vectors to distant points. This characteristic facilitates feature clustering, as described in Section 3.2.3.

The objective is to maximize the similarity between p_a and z_b, as well as between p_b and z_a. To prevent the input data from converging to a constant value after convolution activation, resulting in identical outputs regardless of the input image, we adopt the approach described in [27] by halting the gradient operation on one of the branches to avoid model collapse. Finally, we define Formula (7) as the registration loss for symmetric features.

$$\mathcal{L} = \frac{1}{2}[(\mathcal{D}(p_a, z_b) + (\mathcal{D}(p_b, z_a)] \tag{7}$$

The STN is employed in this stage to perform feature rotation and inversion, facilitating the model's determination of image similarity. Following each training iteration, a negative cosine similarity loss is obtained.

3.2. *Memory Bank Acquisition*

3.2.1. Feature Extraction

Let $\varphi(\mathbf{h}, \mathbf{w}, \mathbf{c})$ represent the feature mapping of the second layer, where h denotes the height, w denotes the channel width, and c denotes the number of channels. Like PatchCore, the patch-level features of local features in clustered neighborhoods can be represented as Formula (8):

$$P = f_{agg}(\varphi) \tag{8}$$

Here, f_{agg} represents the aggregation function within the neighborhood. As mentioned in Figure 1, we use a combination of the first three layers of wide_resnet_50_2 and *STN* to extract features and build the memory library, with each layer followed by the *STN*. Inspired by the PatchCore method, we likewise did not adopt the last layer of the Resnet network since it lost a lot of the features' spatial information. As shown in Figure 3, the features of the second and third layers of the *STN* can retain global information while containing more local feature information. However, if the features of the first three or more layers are fused, the features in the input memory library will not contain enough information for accurate detection.

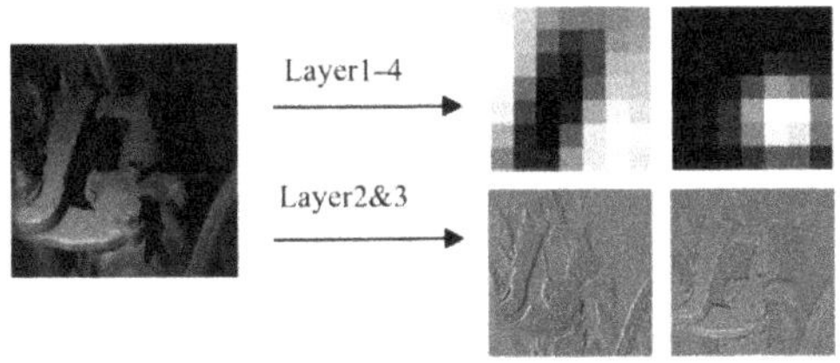

Figure 3. A visualization of features of different levels when fused. When too many levels of the feature are fused, the naked eye can no longer distinguish the outline of the original image, and the features become fine-grained.

3.2.2. Similar to Pyramid Pooling Module (SPPM)

After feature extraction, a 3-dimensional tensor with the shape (C, H, W) is obtained, where C is the sum of the dimensions of the second and third layers. We then process this three-dimensional feature vector by flattening it in three dimensions except for the channels. This results in a two-dimensional tensor with a shape of $(H \times W, C)$, which is randomly projected. Thus, we flatten the original feature vector into a column-wise phenotypic feature matrix. However, as mentioned earlier, the PatchCore method employs a pixel-by-pixel search approach, conducting nearest neighbor searches on each pixel and disregarding pixel relationships. We recognize that 2D average pooling can increase the receptive field, which is crucial for anomaly detection tasks. Hence, we employ the pooling approach illustrated in Figure 4.

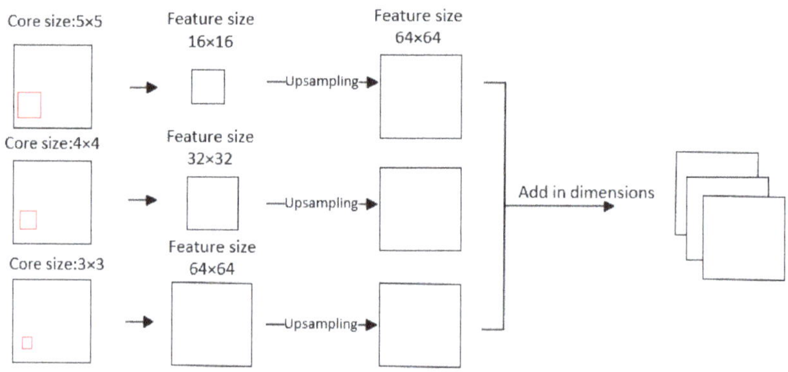

Figure 4. The structure of SPPM.

Assuming an input feature size of 64 × 64, we perform three pooling operations with pool core sizes of 3 × 3, 4 × 4, and 5 × 5. Subsequently, tensors of sizes 64 × 64, 32 × 32, and 16 × 16 are obtained. Next, the three pooled tensors are unsampled to the same size of 64 × 64, and their dimensions are concatenated. The earlier pooling regions are included within subsequent pooling regions using this approach to enlarge the receptive field and give more attention to edge information due to an overlap between pooling regions, thus establishing closer relationships between the pixels.

After the pooling step, the eigenvector of the domain $\varphi(h, w)$, as represented by Formula (9), is enacted.

$$N_p^{(h,w)} = \left\{ (a,b) \middle| a \in \left[h - \frac{p}{2}, \ldots h + \frac{p}{2} \right], b \in \left[w - \frac{p}{2}, \ldots w + \frac{p}{2} \right] \right\} \tag{9}$$

The feature aggregation operation is utilized to obtain the locally aware patch feature set p of the feature mapping tensor $\varphi_{i,j}$, enabling the successful realization of the clustering $P_{s,p}(\varphi(h, w))$ of the feature tensor, as shown in Formula (10).

$$P_{s,p}(\varphi(h,w)) = \left\{ \varphi(h,w) \left(N_p{}^{(h,w)} \right) \middle| h,\ w \ mod\ s = 0, h < h^*, w < w^*, h, w \in N \right\} \tag{10}$$

In this case, the feature repository M_1 can be described as Formula (11):

$$M_1 = \bigcup_{xi \in N} P_{s,p}(\varphi(xi)) \tag{11}$$

3.2.3. Anomaly Detection

The samples used in self-supervised learning are exclusively normal. The training process aims to identify representative features of the "normal category" and utilize them as a reference for evaluating positivity and negativity, ultimately achieving anomaly detection. During testing, we index the memory bank that stores characteristic information of positive samples and calculate the Euclidean distance between sample patches to obtain an anomaly score. In n-dimensional space, if there are two points $a(x_{11}, x_{12}, \ldots, x_{1n})$ and $b(x_{21}, x_{22}, \ldots, x_{2n})$, then the Euclidean distance d is defined as Formula (12).

$$d = \sqrt{\sum_{k=1}^{n} (x_{1k} - x_{2k})^2} \quad (12)$$

After completing the feature aggregation and pooling module in Step 2, we proceed to the feature clustering operation. The utilization of negative cosine similarity loss during training greatly facilitates feature clustering by ensuring that similar vectors are consistently mapped to proximate locations with each iteration, while dissimilar vectors are assigned to distant points. This approach effectively reduces the clustering time and enhances the operational efficiency of the model. To optimize the memory library size and improve the testing efficiency, we implement the PatchCore method and employ greedy subsampling for reducing and optimizing the memory bank based on Formula (13).

$$M_2^* = \frac{argmin}{M_2 \subset M_1} \frac{max}{m \subset M1} \frac{min}{n \in M_2} \|m - n\|_2 \quad (13)$$

The objective of our purpose is to streamline the testing process by solely conducting nearest neighbor searches for the test sample's feature in M_2, identifying its closest neighboring feature and subsequently calculating the maximum Euclidean distance from this feature to its clustering center. This approach allows us to obtain an anomaly score, facilitating effective anomaly detection.

In the memory bank, for each patch-level feature $m \in M_2$ of the training data, we select the m^* nearest neighbors from the patch-level features of the test data. The patch-level anomaly score of the test image X_{test} is estimated based on the distance between the patch-level feature m_{test} and m^*, as shown in Formulas (14) and (15).

$$m^* = arg\ min_{m \in M_2} \|m^{test} - m\|_2 \quad (14)$$

$$s^* = \left\|m^{test} - m^*\right\|_2 \quad (15)$$

3.2.4. Post Processing Method

In anomaly detection, the ground truth serves as a baseline measurement obtained from a reliable method. It is used to calibrate and improve the accuracy of new measurement methods. Many current anomaly detection methods rely on the calculated AUROC curve to determine the threshold of the entire dataset. This involves setting the threshold based on the data distribution and other characteristics. The regions above the threshold are considered abnormal, while those below are considered normal. However, if the overall threshold of an abnormal image is lower than the dataset threshold, the anomaly cannot be detected. Otherwise, false detections may occur. To address this, we propose a threshold determination method based on the image itself, which identifies the critical point between normal and anomalous regions, enabling the accurate labelling and localization of anomalies.

For each test sample, as shown in Figure 5a, we artificially construct five rectangular sampling frames in its upper left, lower left, upper right, lower right, and right center regions, as shown in Figure 5b.

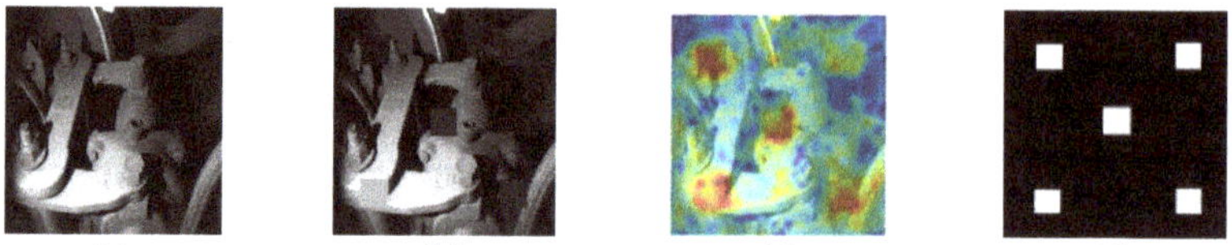

(a) (b) (c) (d)

Figure 5. (**a**) source image; (**b**) artificial construction; (**c**) heat map; (**d**) ground truth of the constructed image.

The constructed image is also fed into the network for feature extraction, and then we search for distances in the feature database, compute scores, and obtain the score matrix. Its heat map visualization is shown in Figure 5c. Obviously, as artificially constructed anomalies, these five areas will have significantly higher anomaly scores than the rest of the picture. We set the pixel values of these five areas to 0 and the other parts to 1, thus obtaining its ground truth, as depicted in Figure 5d. Then, we traverse the five regions, find the maximum and minimum fractional values ($\boldsymbol{thres_{max}}$ and $\boldsymbol{thres_{min}}$) and define the ***interval*** as outlined in Formula (16), setting ***iter_times*** to 10 to ensure the code runs efficiently:

$$\boldsymbol{interval} = (\boldsymbol{thres}_{\boldsymbol{max}} - \boldsymbol{thres_{min}}) / \boldsymbol{iter\_times} \tag{16}$$

For each iteration, we set the threshold as outlined in Formula (17):

$$\boldsymbol{th} = \boldsymbol{thres_{max}} - \boldsymbol{iter\_times} \cdot \boldsymbol{thres_{min}} \tag{17}$$

The iteration is carried out in the range of minimum and maximum thresholds, and each th can output an anomaly mask accordingly, where the white area is abnormal and the black area is normal, as shown in Figure 6a–e, which are the masks corresponding to the thresholds 9.0, 8.3, 7.6, 6.4 and 6.1, respectively. As can be seen from the five pictures shown in Figure 6, with the continuous increase in the threshold, the area of the abnormal region in the mask image corresponding to each threshold value increases first and then decreases. Finally, we calculate the intersection over union (IOU) between the prediction box and the ground truth box. The threshold is updated according to the highest IOU, and the best threshold (***best_th***) is obtained after 10 iterations. The mask graphs corresponding to different responsiveness are presented in Figure 6a–e, serving the purpose of facilitating the description of this method's principle. It is important to note that these graphs do not represent the final segmentation results but rather serve as temporary variables within the code and are subsequently deleted after calculation to optimize memory usage.

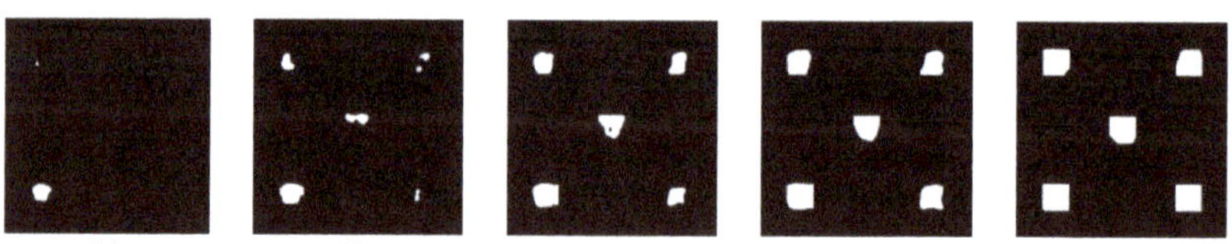

(a) (b) (c) (d) (e)

Figure 6. The corresponding anomaly mask under a certain ***th***. (**a**) ***th*** = 9; (**b**) ***th*** = 8.3; (**c**) ***th*** = 7.6; (**d**) ***th*** = 6.4; (**e**) ***th*** = 6.1.

After determining the optimal threshold value (***best_th***), based on industrial detection expertise, we establish the region of abnormal pixels (S = 25) and devise an algorithm. For the neighboring pixel region exceeding ***best_th***, if it is smaller than S, the pixel value in this region will be set to 0 (normal); otherwise, it will be set to 1 (abnormal). The implementation of this setting will effectively eliminate the occurrence of false positives in small areas, resulting in a reduction in the overall false alarm rate observed in the test results.

4. Experiment

4.1. Datasets and Metrics

We first evaluate the performance of our method on two public datasets, MVTec and MPDD, which cover both classical industrial anomaly detection and defect detection in manufacturing processes.

- MVTec dataset: The MVTec dataset is a well-known benchmark for industrial anomaly detection. It consists of 15 categories with a total of 3629 training and validation images and 1725 test images. The training set contains only normal images, while the test set includes images with various defects and normal images. The images have resolutions ranging from 700 × 700 to 1024 × 1024 pixels, and pixel-level ground truth labels are provided for each defect.
- MPDD dataset: The MPDD dataset is a newly proposed dataset focused on defect detection in the manufacturing process of painted metal parts. It comprises six types of metal parts captured under different spatial orientations, positions, distances, light intensities, and backgrounds. We utilize this dataset to evaluate the registration effect of our method and compare it with several other methods to highlight the superiority of our approach.
- Industrial scenario dataset: To demonstrate the robustness and generalization of our approach, we have selected an additional three datasets from the field of high-speed rail (HSR) component inspection for experimentation. Each of these industrial datasets consists of 460 normal samples and 115 abnormal samples, all of which exhibit visual bias in terms of their positioning. Figure 7 showcases the appearance of these normal samples.

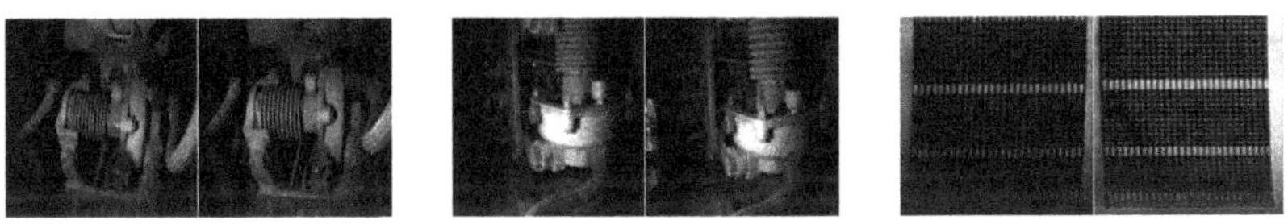

(a) (b) (c)

Figure 7. (**a**) HSR dataset 1; (**b**) HSR dataset 2; (**c**) HSR dataset 3.

We first adopt the traditional model evaluation method, randomly sampling normal data and abnormal data, and divide the dataset according to the ratio of training set to test set of 8:2. In order to avoid overfitting and other effects caused by unreasonable dataset division, we adopt the cross-validation method to evaluate the model in Section 4.4.

For the evaluation, we employ two threshold-independent metrics. The first is the area under the receiver operating characteristic curve (AUROC), which measures the true positive rate (percentage of correctly classified anomalous pixels). The image-level AUROC scores assess the model's detection accuracy, denoted as detection (Det.), while the pixel-level AUROC scores evaluate the model's positioning and segmentation accuracy, indicated as segmentation (Seg.). The second metric is the per-region-overlap score (PRO-Score) [28], which considers the mean rate of correctly classified pixels as a function of the false positive rate between 0 and 0.3 for each connected component, as shown in Formula (18). Higher scores indicate better localization of both large and small anomalies [29].

$$PRO = \frac{1}{N}\sum_{n=1}^{N}\frac{P \cap G_n}{G_n} = \frac{1}{N}\sum_{n}^{N}\frac{TP_n}{TP_n + FN_n} \tag{18}$$

where P represents the prediction result, G_n represents the true value, then the intersection of the two is the true positive sample TP_n, and FN_n represents the false negative sample.

The above two indicators are commonly employed in anomaly detection tasks and are widely acknowledged and utilized as benchmarks within the detection industry. However, due to the intricacy of industrial detection scenarios, there exists variation in data quality as well as resolution. To demonstrate the robustness of our method, we also employ three additional indicators, namely the recall rate, accuracy rate, and false alarm rate (***FAR***), for testing and cross-validation using the high-speed railway dataset. The calculation formulas for these indicators are presented in Formulas (19)–(21), respectively.

$$Recall = \frac{TP}{TP + FN} \tag{19}$$

$$Accuracy = \frac{TP + TN}{TP + TN + FP + FN} \tag{20}$$

$$FAR = \frac{FP}{FP + TN} \tag{21}$$

The term ***FP*** denotes the number of false positives, while ***TN*** represents the count of true negatives.

4.2. Implementation Details

We utilize wide_resnet_50_2 and the *STN* as the underlying architecture to construct a convolution-based encoder and predictor. These models are trained on images with a resolution of 224 × 224 using an Asus NVIDIA GTX 3090 GPU sourced from Chengdu, China. For parameter updates, we employ momentum SGD for 55 epochs with a learning rate of 0.0001 and a batch size of 32. The duration of each training round is approximately 8 min, with an average training epoch lasting for about 8.7 s.

4.3. Anomaly Detection

To assess the efficacy of FR-PatchCore, we conduct a comparative analysis with other cutting-edge anomaly detection techniques, such as SPADE, PaDiM, Patch-SVDD [30], Cutpaste [31], RegAD, and PatchCore. We meticulously replicate their reported outcomes using each method's official source code to ensure an equitable comparison. The last row displays the macro-averaged scores for all categories, highlighting the highest scores in bold.

The MVTec dataset is the most widely used benchmark for anomaly detection, and FR-PatchCore achieves a 98.81% image-level AUROC score and a 97.86% pixel-wise AUROC score when compared with the other six methods in Table 1. In the comparison of detection results of "similar categories" such as bottle, leather, cable, etc., our method is slightly inferior to PatchCore, but it still has high detection scores and accuracy, and the average detection scores of all categories are better than the other six methods. For the three "spatial transformation classes" such as hazelnut, screw, and metal nut in the dataset, we present partial representative results of their visualizations in Section 4.6.

To evaluate the robustness of our method against the "rotation class" data that we defined, we conducted experiments on the MPDD dataset, which are displayed in Table 2, from ten runs. The final row presents the macro-averaged scores across all categories, with the best scores indicated in bold.

We present partial representative results of the MPDD dataset visualizations in Section 4.5. Unlike the MVTec dataset, all six categories in the MPDD dataset have spatial differences and position shifts. This is challenging for most methods because anomaly detection models are not trained enough to deal with morphologically diverse samples. We believe that increasing the training time is not enough to make the model learn more. On the contrary, it may cause the model to overfit. Most models cannot detect the "spatial transformation class" of the sample effectually. The FR-PatchCore proposed by us is innovative in model learning. As shown in Table 2, our method leads in comparison with the other six methods. This proves that the idea of using feature-level registration as a pretext task is worthwhile. Only when the model learns the spatial information that has been ignored before can it better deal with such problems, improve the generalization and be applied to the field of high-standard industrial anomaly detection.

Table 1. The results on the MVTec dataset (AUROC—image%, AUROC—pixel%).

Category	SPADE		PaDiM		SVDD		Cutpaste		RegAD		PatchCore		Ours	
	Det.	Seg.	Det.	Seg.	Det.	Seg.	Det.	Seg.	Det.	Seg.	Det.	Seg.	Det.	Seg.
carpet	92.80	97.50	**99.40**	**98.90**	92.90	92.60	93.10	98.30	98.50	**98.90**	98.00	**98.90**	97.13	98.35
grid	47.30	93.70	95.70	94.90	94.60	96.20	**99.90**	**97.50**	91.50	88.70	97.50	**97.50**	98.65	95.14
leather	95.40	97.60	99.80	99.10	90.90	97.40	**100.0**	**99.50**	**100.0**	98.90	**100.0**	99.10	99.84	98.84
tile	96.50	87.40	97.40	91.20	97.80	91.40	93.40	90.50	97.40	95.20	**99.40**	94.90	98.46	**97.67**
wood	95.80	88.50	98.80	93.60	96.50	90.80	98.60	**95.50**	**99.40**	94.60	98.90	93.60	98.73	95.45
bottle	97.20	**98.40**	99.80	98.10	98.60	98.10	98.30	97.60	99.80	97.50	**100.0**	98.30	99.76	98.39
cable	84.80	97.20	92.20	95.80	90.30	96.80	80.60	90.00	80.60	94.90	99.30	98.90	**99.78**	**98.17**
capsule	89.70	**99.00**	91.50	98.30	76.70	95.80	96.20	97.40	76.30	98.20	97.60	98.50	**98.83**	98.82
hazelnut	88.10	**99.10**	93.30	97.70	92.00	97.50	97.30	97.30	96.50	98.50	**100.0**	98.40	99.59	97.96
metal nut	71.00	98.10	95.80	96.70	94.00	98.00	99.30	93.10	98.30	96.90	**99.70**	97.70	99.67	**98.77**
pill	80.10	96.50	94.40	94.70	86.10	95.10	92.40	95.70	80.60	97.80	95.90	97.70	**98.82**	**98.21**
screw	66.70	**98.90**	84.40	97.40	81.30	95.70	86.30	96.70	63.40	97.10	94.90	98.60	**96.86**	98.50
toothbrush	88.90	97.90	97.20	**98.70**	**100.0**	98.10	98.30	98.10	98.50	**98.70**	**100.0**	97.20	98.06	98.63
transistor	90.30	94.10	97.50	97.20	91.50	97.00	95.50	93.00	93.40	96.80	**99.90**	94.90	98.96	**97.26**
zipper	96.60	96.50	90.80	98.20	97.90	95.10	99.40	**99.30**	94.00	97.40	99.20	98.40	**99.58**	97.79
average	85.41	96.03	95.20	96.70	92.07	95.71	95.19	95.97	91.21	96.67	98.69	97.51	**98.81**	**97.86**

Table 2. The results on the MPDD dataset (AUROC—image%, AUROC—pixel%).

Category	SPADE		PaDiM		SVDD		Cutpaste		RegAD		PatchCore		Ours	
	Det.	Seg.	Det.	Seg.	Det.	Seg.	Det.	Seg.	Det.	Seg.	Det.	Seg.	Det.	Seg.
bracket black	62.80	**97.70**	75.60	94.20	60.17	89.46	77.50	85.20	67.30	72.58	68.15	90.55	**85.98**	96.74
bracket brown	88.00	**98.50**	85.40	95.20	64.56	89.69	**93.40**	93.80	69.60	93.25	74.96	86.46	83.50	95.40
bracket white	83.60	97.00	82.20	98.10	50.89	89.48	74.60	82.70	61.40	95.68	72.56	98.31	**88.25**	**99.50**
connector	90.70	98.60	91.70	97.90	94.52	90.98	97.60	96.80	94.90	92.16	80.95	96.74	**99.95**	**99.11**
metal plate	91.70	97.40	56.30	93.40	92.80	92.70	88.00	92.50	90.20	89.26	89.98	96.30	**98.48**	**98.22**
tubes	51.00	97.50	57.50	92.10	22.15	85.81	**98.20**	97.40	67.90	92.27	69.07	94.11	81.25	**98.85**
average	77.97	97.78	74.78	95.15	64.18	89.69	88.22	91.40	71.90	89.20	75.95	93.75	**89.57**	**97.97**

To evaluate the applicability of our method in real-world scenarios, we conducted experiments using data from high-speed rail (HSR) components in industrial anomaly detection. The scores are displayed from ten runs in Table 3.

Table 3. The results on the HSR dataset (AUROC—image%, AUROC—pixel%).

Category	SPADE		PaDiM		SVDD		Cutpaste		RegAD(k = 8)		PatchCore		Ours	
	Det.	Seg.	Det.	Seg.	Det.	Seg.	Det.	Seg.	Det.	Seg.	Det.	Seg.	Det.	Seg.
1	98.78	90.70	98.67	93.70	95.46	93.25	78.70	85.27	92.25	90.75	98.28	97.25	**99.32**	**96.58**
2	97.25	95.26	97.68	97.25	98.56	98.25	**99.25**	**98.75**	94.87	95.76	98.26	99.27	98.86	98.86
3	95.45	94.28	96.87	97.36	94.28	96.28	98.36	97.16	98.23	96.14	97.52	97.25	**99.28**	**99.38**
average	97.16	93.41	97.74	96.10	96.10	95.93	92.10	93.73	95.12	94.22	98.02	97.92	**99.15**	**98.27**

We present partial representative results of the HSR dataset visualizations in Section 4.5. As can be seen from Table 3, FR-PatchCore still achieves an excellent detection performance even when detecting real industrial datasets with a complex environment, noisy images, and many influencing factors, which proves that our method has good generalization and robustness.

Most anomaly detection researchers evaluate the performance of their methods using both image-level and pixel-wise AUROC curves. AUROC is a widely used metric for assessing the ability of models to discriminate between abnormal and normal samples.

However, in anomaly detection, it is not only important to identify anomalies but also to accurately localize them. The pixel-wise AUROC evaluation provides a broad assessment, where the score is significantly improved if a large region is correctly localized but has a minimal impact if a small region is incorrectly positioned.

To address this limitation, we also employed the PRO score [24] criteria, which considers the rate of correctly classified pixels as a function of the false positive rate (fpr). We calculated the average PRO score of the three datasets using an fpr of 30%, following the methodology described in [25]. In Table 4, we compare our method with SPADE, PaDiM, and PatchCore, which are all based on feature embedding. Our approach achieves the highest PRO-Score, indicating a superior localization performance.

Table 4. Average PRO scores on three datasets.

Datasets	SPADE	PADIM	PatchCore	Ours
MVTec	91.70	90.10	90.17	**92.90**
MPDD	91.20	84.70	93.20	**94.90**
HSR 1–3	90.70	94.81	94.72	**95.81**
Mean score	91.20	89.87	92.69	**94.54**

In the comparison of the PRO score, we only compare four methods based on feature embedding: SPADE, PaDiM, PatchCore and our proposed method. Since these four methods have the same basic principle, they are easier to compare, and it is more convenient to use the principle of control variables. For example, the same backbone wide_resnet_50_2 is used at the same time. As shown in Figure 8, we report the results of the PRO score comparison on three datasets of different difficulty, which indicates that FR-PatchCore has an excellent detection performance and still obtains good indicators under this more stringent score limitation.

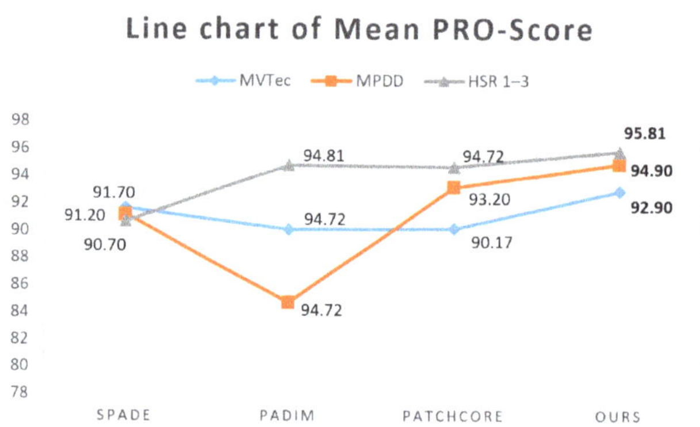

Figure 8. Average PRO score line chart.

In addition, as presented in Tables 5–7, we have documented the scores for the accuracy rate, recall rate, and false positive rate of the three high-speed rail datasets. Among the four comparative methods employed, our approach stands out as the most advanced, with the lowest occurrence of false alarms. Furthermore, apart from these metrics, industrial tests also assess efficient reasoning performance. By employing identical parameter settings across all four methods, we conducted comprehensive evaluations. The inference time considered here encompasses both inputting samples to be detected into the network for forward propagation and the processing time. The original image resolution was adjusted from 1280 $\times$ 1280 to a 224 $\times$ 224 input network size. Our method's inference speed remains comparable to that of well-known PatchCore networks while maintaining superior anomaly detection and segmentation performance at the image-level.

Table 5. Industrial inspection score on HSR dataset 1.

Metrics	SPADE	PaDiM	PatchCore	Ours
Recall%	88.00	96.00	94.00	**96.00**
Accuracy%	87.80	97.60	**97.92**	97.39
FAR%	12.00	4.00	**3.52**	4.00
TIMES(s)	0.37	0.35	**0.20**	0.22

Table 6. Industrial inspection score on HSR dataset 2.

Metrics	SPADE	PaDiM	PatchCore	Ours
Recall%	84.00	94.00	96.00	**98.00**
Accuracy%	82.60	93.04	92.17	**93.91**
FAR%	16.00	6.00	4.00	**2.00**
TIMES(s)	0.42	0.34	0.33	**0.30**

Table 7. Industrial inspection score on HSR dataset 3.

Metrics	SPADE	PaDiM	PatchCore	Ours
Recall%	86.00	92.00	96.41	**98.00**
Accuracy%	86.09	92.17	97.92	**98.45**
FAR%	14.00	0.08	4.00	**2.00**
TIMES(s)	0.31	0.26	**0.21**	**0.21**

4.4. Cross-Validation Experiment

To assess the performance of our model on limited data samples, we conducted five-fold cross-validation using the high-speed rail component dataset. The training set was divided into five folds, with four folds used for training and the remaining one for testing. Additionally, within the four-fold training data, an additional 5% of the data was extracted as a verification set. Each training and test iteration involved 442 training data, 18 validation data, and 115 test data. This approach ensures that all samples in the training set are utilized for both training and testing purposes.

Since this study is based on self-supervised learning and utilizes unlabeled positive samples for training, there is no actual labeled data available. However, it should be noted that anomaly detection inherently involves a classification task where the "normal" label is carried by the data itself during training. Therefore, in each of the five cross-validations performed, classification accuracy serves as our evaluation metric. Subsequently, we select the model with highest classification accuracy on the verification set to evaluate its performance on separate test sets through five experiments. The resulting test errors from each experiment can be obtained using Formula (22).

$$E_T(\omega) = \sum_{x\in D}^{n} \left(\hat{f}(x|\omega) - f(x)\right)^2 \tag{22}$$

Set D consists of N samples, where each sample $(x_i, y_i = f(x_i))$ represents a subset of the total set χ.

The generalization error in this case can be approximated as the average of the five test errors, as demonstrated in Formula (23).

$$E_G(\omega) = \sum_{x\in\chi} p(x)\left(\hat{f}(x|\omega) - f(x)\right)^2 \approx E_T(\omega) \tag{23}$$

where $p(x)$ represents the probability that x occurs in the total set χ.

The results of the five cross-validation iterations are presented in Tables 8–10. Taking high-speed rail parts dataset 1 as an example, the recall rate is utilized as an evaluation metric to compare the four methods in the classification task of positive and negative samples, as depicted in Table 10. The model achieving the highest score during iteration will be selected for testing purposes.

Table 8. Five train iterations on HSR dataset 1.

Iteration	SPADE	PADIM	PatchCore	Ours
1	83.33	88.89	94.44	94.44
2	94.44	94.44	88.89	100.00
3	83.33	83.33	94.44	88.89
4	100.00	94.44	100.00	94.44
5	88.89	100.00	94.44	94.44
Mean Recall	89.99	92.22	94.44	94.44

Table 9. Five cross-validation iterations on HSR datasets 1–3.

Iteration	SPADE	PADIM	PatchCore	Ours
1	94.78	98.26	95.65	98.26
2	95.65	93.04	96.52	95.65
3	92.17	93.91	94.78	93.91
4	97.39	95.65	94.78	97.39
5	90.43	94.78	99.13	96.52
Mean Accuracy	94.08	95.13	96.17	**96.35**

Table 10. Mean error calculation.

Iteration	SPADE	PADIM	PatchCore	Ours
Mean error	5.92	4.87	3.83	**3.65**

After selecting the optimal model for each of the four comparison methods, we assessed its performance by conducting tests on five separate datasets and calculated the average accuracy based on the results obtained from these five tests, as presented in Table 9.

Finally, we computed the average test error after five iterations of the four models utilized in the testing process. It can be observed from Table 10 that our method demonstrates superior generalization capability for the HSR dataset, as evidenced by its significantly lower generalization error compared to other methods employed across the three industrial datasets.

4.5. Ablation Experiment

Experiments were conducted to assess the contribution of different components in our proposed approach. The MVTec dataset was used as the subject of these experiments, where we evaluated the impact of pooling, registration, and post-processing methods individually. The experimental results clearly demonstrated that our method outperformed the other methods in terms of performance.

4.5.1. Backbone Selection Process

The CNN and transformer have emerged as the predominant backbone networks for deep learning in recent years, owing to their robust feature extraction capabilities. Initially designed for natural language processing, the transformer has also been adapted by researchers for image processing through vision transformer (VIT) networks. In this section, we randomly select three subgraphs to visualize the initial three layers of features extracted from VIT and wide_resnet_50_2 networks, using the screw class as an example (Figures 9 and 10). Notably, the features extracted by VIT exhibit enhanced recognizability, with reduced noise compared to those obtained by wide_resnet_50_2.

Figure 9. Feature visualization extracted by wide_resnet_50_2 network.

Figure 10. Feature visualization extracted by VIT network.

The disparity in the experimental outcomes arises from variations in the scale of the training data. In contrast to VIT's remarkable efficacy on extensive datasets, CNNs exhibit commendable feature extraction capabilities even with limited data, rendering them more amenable for deployment within data-scarce industrial domains.

4.5.2. SPPM

In the registration module depicted in Figure 1, we incorporate a 2D average pooling module alongside the CNN + STN for extracting image features. This addition serves to mitigate the over-sensitivity of the convolutional layers to positional information. We have opted not to use the SPPM in this module, as our experiments have shown that 2D average pooling yields better results. As shown in Table 11, the application of AvgPool2d in the registration module received the highest score because this module plays a crucial role in maximizing the absolute value of the negative cosine similarity loss, thereby significantly enhancing the performance of anomaly detection.

Table 11. Comparison of different pooling (AUROC—image%, AUROC—pixel%).

Metrics	No Pooling		SPPM		AvgPool2d	
	Det.	**Seg.**	**Det.**	**Seg.**	**Det.**	**Seg.**
scores	95.50	88.50	96.80	90.60	**98.20**	**91.70**

For evaluation purposes, we employ a pixel-wise anomaly detection AUROC score and PRO score as our chosen metrics. To illustrate the convergence of loss, we utilize the screw category as an example, which is depicted in Figure 11.

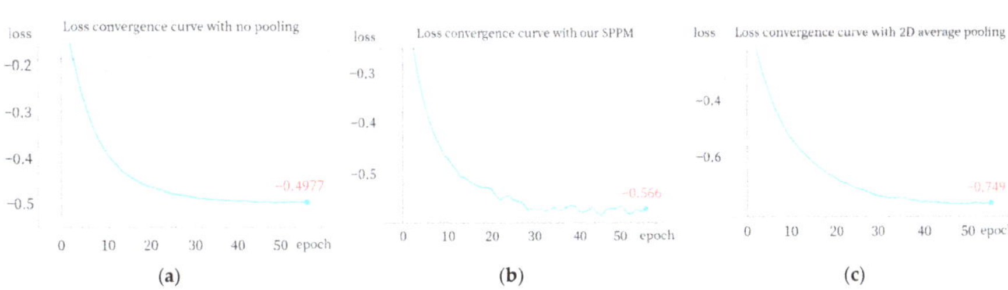

Figure 11. (**a**) Loss convergence curve with no pooling; (**b**) our SPPM; (**c**) 2D average pooling.

The convergence of the negative cosine similarity loss for the screw class is analyzed under different settings. Figure 11a demonstrates the absence of pooling results in a loss value of approximately −0.49. When our SPPM is utilized, the loss slightly increases to around −0.57, as depicted in Figure 11b. However, by incorporating 2D average pooling, the loss significantly improves and reaches approximately −0.75, as shown in Figure 11c. This observation suggests that integrating 2D average pooling enhances the model's capacity to learn features during the registration process.

4.5.3. Registration Module

For the SPPM and the registration module, we conducted ablation experiments to assess the impact of specific components in our model, namely the SPPM and the registration module. By removing and combining these components individually, we aim to understand their contributions to the overall performance. The results in Table 12 show that when neither the SPPM nor the registration module is used, the two metrics used for evaluation have the lowest score and the lowest anomaly detection result. When both are included in the model, we will get the best anomaly detection result.

Table 12. Module ablation (AUROC%).

Module	MVTec		MPDD	
	Det.	Seg.	Det.	Seg.
No SPPM	95.50	97.51	96.80	90.60
no Registration	95.50	97.64	96.80	90.60
neither	95.50	97.51	95.50	88.50
Both	**95.50**	**97.92**	**95.50**	**88.50**

4.5.4. Post-Processing Method

In Figure 12, we demonstrate the application of our proposed post-processing method to generate a visual representation of the mask. Specifically, we present an abnormal image from the high-speed rail dataset and show the original image, segmentation image, heat map, ground truth value, the mask generated without the post-processing method, and the mask generated after the post-processing method from left to right. It is evident from Figure 12e that there is a large false positive area; however, this issue is greatly reduced in Figure 12f due to our post-processing method.

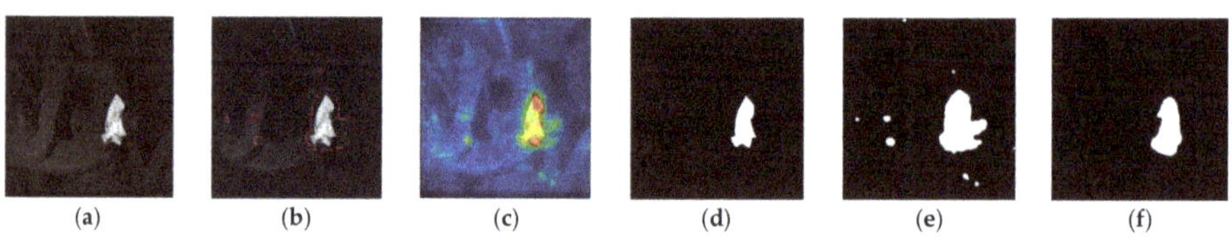

Figure 12. (**a**) Original image; (**b**) segmentation result, the red part represents the initial segmentation of abnormal areas; (**c**) predicted result; (**d**) ground truth; (**e**) the mask has been segmented without the use of any post-processing methods; (**f**) the mask has been segmented using a post-processing method.

The proposed post-processing method for segmentation exceptions employs intersection over union (IoU) as a metric to evaluate the test image, as depicted in Formula (24):

$$IoU = \frac{p_{ii}}{p_{ij} + p_{ji} - p_{ii}} \tag{24}$$

The true value is represented by $\boldsymbol{i}$, the predicted value is represented by $\boldsymbol{j}$, and p_{ij} represents the number of pixels predicting $\boldsymbol{i}$ as $\boldsymbol{j}$. Therefore, Formula (25) is equivalent to Formula (11).

$$IoU = \frac{P \cap G}{P \cup G} \tag{25}$$

The letter $\boldsymbol{P}$ denotes the prediction, while $\boldsymbol{G}$ represents the ground truth.

The intersection over union (***IoU***) algorithm is commonly utilized in deep learning for object detection and semantic segmentation tasks to compute the overlap ratio between different images. In Table 13, we present the average ***IoU*** metrics for the three datasets used in our experiment, while Section 4.5 showcases visualizations of selected test results. The mask graph generated after applying our proposed post-processing method exhibits a higher degree of proximity to the ground truth value, thereby resulting in an elevated average IOU value. This substantiates the efficacy of our proposed approach.

Table 13. The mask generated by the non-post-processing method and the mask generated by our method.

Datasets	$\overline{IoU}$ (Before)	$\overline{IoU}$
HSR dataset 1	0.42	0.68
HSR dataset 2	0.33	0.75
HSR dataset 3	0.37	0.71

As can be seen from Figure 12e,f, the mask result obtained after segmentation via post-processing method is closer to the heat map. The responsiveness of the sample in Figure 12a is depicted in a histogram format, as illustrated in Figure 13b, while the responsiveness of the corresponding class's normal sample is presented in Figure 13a.

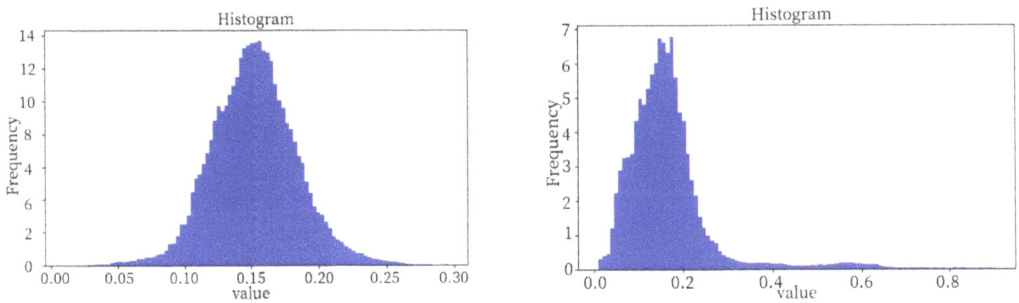

(**a**) (**b**)

Figure 13. (**a**) Responsiveness histogram (normal); (**b**) responsiveness histogram (abnormal).

The test image predominantly consists of normal areas with a high frequency of response but a low level of responsiveness, whereas abnormal areas exhibit a low frequency of response but a high level of responsiveness. The responsivity distribution in the normal region typically conforms to an approximate normal distribution, while the responsivity in the abnormal region lies outside this distribution. The process of calculating responsiveness is analogous to binary classification. Initially, the positive and negative attributes of the image are evaluated. Subsequently, threshold-based truth rules may be employed to exclude certain normal regions exhibiting higher responsiveness and identify regions demonstrating elevated levels of responsiveness. The response value of the normal region is observed to be approximately distributed within the range of (0, 0.7), while our post-processing method calculates an optimal threshold of 0.75. The proposed approach ensures that the highly responsive normal region, depicted in Figure 13b, with a responsivity range of 0.4–0.6 is not erroneously classified as an exception, and the mask prediction yields superior results. By employing this post-processing method, the occurrence of false positives is effectively minimized.

4.6. Visualization

A visualization of selected categories with "spatial transformation" characteristics from various datasets used in the experiment was employed to provide a more intuitive analysis of the detection capability of FR-PatchCore, as depicted in Figure 14.

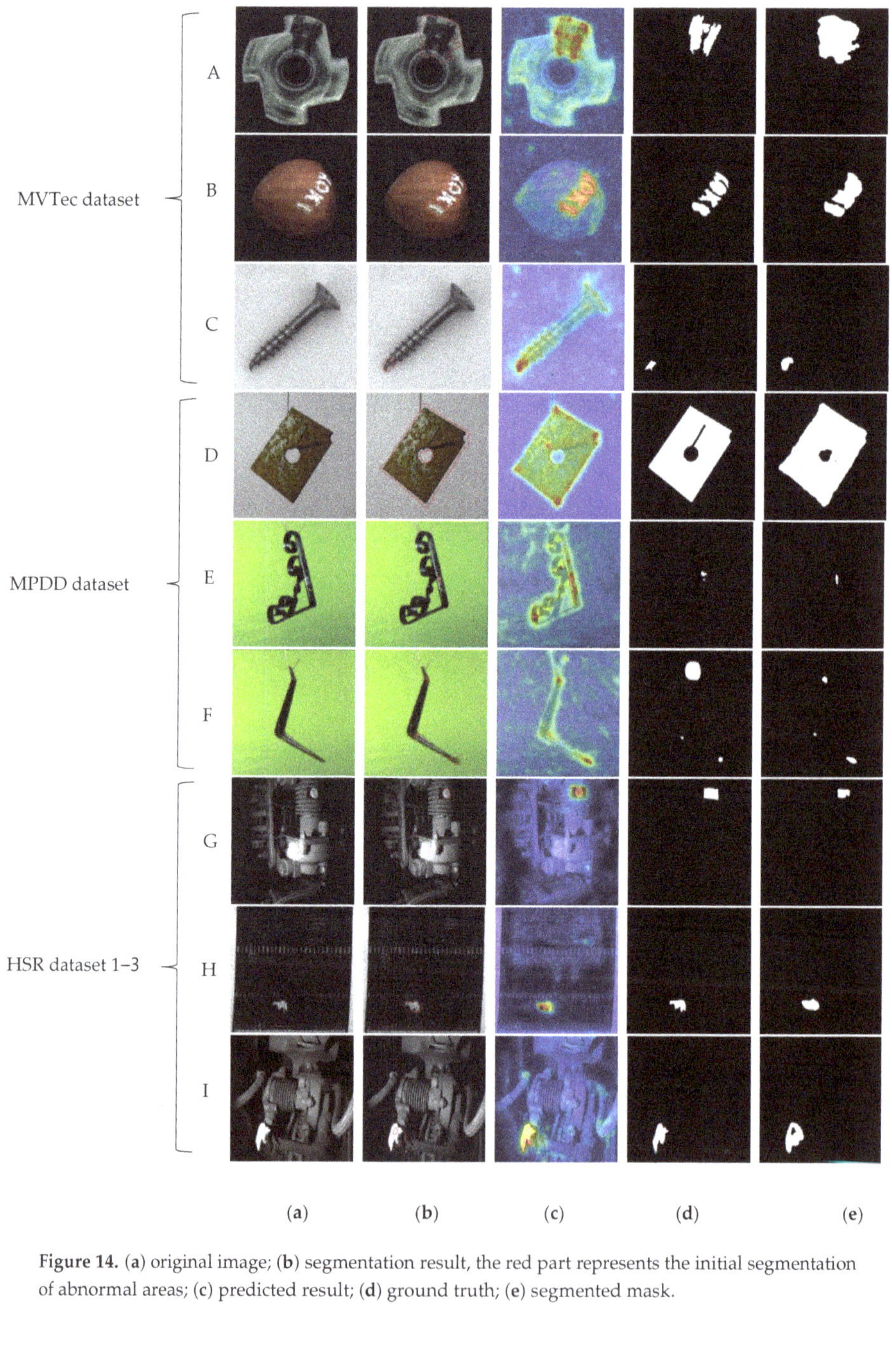

Figure 14. (**a**) original image; (**b**) segmentation result, the red part represents the initial segmentation of abnormal areas; (**c**) predicted result; (**d**) ground truth; (**e**) segmented mask.

The "spatial rotation" category of MVTec and MPDD in the two public datasets is selected for display in Figure 14, lines A–F. In Figure 14b, highly accurate anomaly segmentation results are showcased, while Figure 15c visualizes the anomaly score. Once the model successfully locates and evaluates anomalies, the threshold selection method proposed by us can be utilized to obtain mask prediction results, as depicted in Figure 14e.

(a) (b) (c) (d) (e)

Figure 15. (**a**) original image; (**b**) segmentation result, the red part represents the initial segmentation of abnormal areas; (**c**) predicted result; (**d**) ground truth; (**e**) segmented mask.

The anomaly detection results for three different categories in HSR 1–3 datasets are shown in line G–I of Figure 14. In contrast to the single simple scenes of the public dataset, the data captured in industrial scenes exhibit higher levels of noise and complexity in the background, making anomaly detection susceptible to environmental factors. To capture more comprehensive visual information of the inspected objects and prevent data overexposure caused by lighting or other factors during raw data collection for industrial inspection, an unfixed shooting point is adopted instead of continuous shooting from a fixed position. This leads to a significant presence of "spatial transformation" category data in industrial inspections, which further validates the necessity of our research. Based on the segmentation and prediction results from these three columns, FR-PatchCore demonstrates accurate localization of anomalies amidst complex backgrounds while exhibiting high sensitivity toward abnormal areas and low response toward normal areas. Moreover, it maintains a low false positive rate when ensuring reliable anomaly detection, thereby showcasing its potential applications and strong generalization capabilities.

4.7. Discussion

The AUROC and PRO scores were utilized in Tables 1–4 to perform category-by-category anomaly detection on the three datasets employed. In comparison with other notable methods, FR-PatchCore demonstrated a superior average performance by not only achieving excellent anomaly detection for conventional "similar class" images but also effectively detecting the challenging "spatial transformation class" in the dataset.

An evaluation of three challenging datasets for high-speed rail parts, which are critical in industrial testing, was conducted using recall, accuracy, and false positive rates in Tables 5–7. Additionally, the test speed of the comparison method was assessed, and the results indicate that at an image resolution of 224 $\times$ 224, the test speed is comparable to that of the PatchCore method, with a maximum speed of 0.21 s per frame. These findings demonstrate that our approach ensures a superior recall rate and accuracy while maintaining a minimal false positive rate without compromising efficiency.

Furthermore, to provide a more comprehensive discussion on FR-PatchCore's capabilities, we designed a five-fold cross-validation experiment utilizing the three industrial datasets. The average error from five tests was adopted as the final evaluation index. Our proposed method exhibited the lowest average error among all tests conducted, further highlighting its competitiveness and strong generalization abilities.

In the ablation experiment conducted in Section 4.5, we performed comprehensive analyses and experiments on the utilized model modules and network selection process. Furthermore, IoU was employed as a metric to assess the effectiveness of the proposed post-processing approach. Additionally, a subset of experimental results was visually presented in Section 4.6 to illustrate the recovery process of FR-PatchCore segmentation and anomaly positioning.

The effectiveness of self-supervised learning heavily relies on a well-designed auxiliary task. To enhance the model's understanding of local feature relationships during registration, we propose a feature-level registration task that captures and stores the maximum distance between normal features and their respective clustering centers in the feature database. In testing, precise image alignment is no longer necessary. Instead, we can simply query the memory database to calculate the maximum distance between target features and their most similar clustering centers. As a result, our approach achieves robust anomaly detection in two-dimensional images. However, it is important to acknowledge that our method also has certain limitations and disadvantages.

In the introduction, we define the "similarity category" and "spatial transformation category" based on the spatial position relations of objects in the picture, such as rotation and flip. However, this definition is limited to two-dimensional planes and does not account for variations in depth between images that may exist across different planes. Our method is therefore not suitable for capturing spatial position relationships with significant depth differences. As illustrated in Figure 15, we present detection results of data exhibiting substantial disparities in depth information within industrial detection scenarios, along with the corresponding registration loss curve for this class shown in Figure 16. The high occurrence of false positives can be attributed to our feature-grade registration method's inability to achieve accurate depth alignment. Figure 16 depicts the registration loss curve specifically for this category, revealing convergence when the loss reaches approximately 0.4. This indicates that our model struggles to effectively register features associated with this type of data, resulting in a scarcity of learned features compared to previous categories.

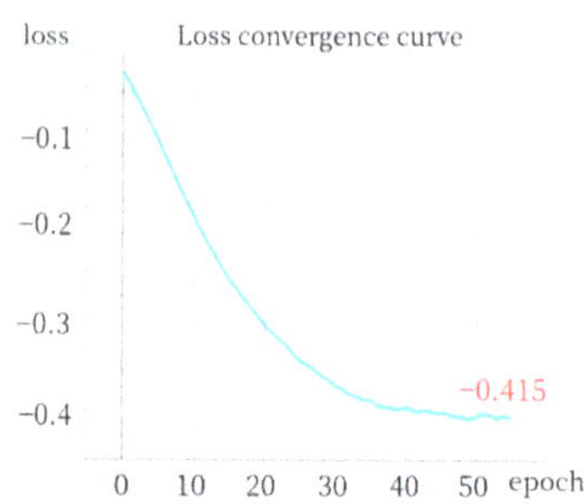

Figure 16. Registration loss function curve of Figure 15a.

The integration of two-dimensional images with depth maps in the above scenarios can potentially yield significant benefits, as it encompasses more comprehensive spatial information. By combining these two types of images, the model can acquire an enhanced feature representation, thereby enhancing its performance.

In addition, the proposed post-processing method is based on a single test image and determines the threshold by calculating the responsivity of each part of the image, thereby mitigating the occurrence of false positives to some extent. However, when dealing with large-scale test data and high-resolution images, this approach may introduce significant computational overhead and pose challenges for practical implementation. Furthermore, our exploration into adaptive threshold calculations remains limited, without incorporating a comprehensive and universally applicable algorithm. The responsivity values for each area of the image in Section 4.5.4 were visually represented, offering a novel perspective. If we consider the data distribution, assuming that responsiveness follows a normal distribution, then according to the three-sigma rule, an outlier is defined as a value in a set of result values that deviates from the mean value by more than three standard deviations. However, if the responsiveness does not follow a normal distribution and this distribution can be computed with minimal effort, then the outlier can be defined as k times the standard deviation from the mean, where k serves as a threshold. Further investigation into this issue will be conducted in future research.

5. Conclusions

In this study, we propose FR-PatchCore, a feature-level registration method that leverages PatchCore, one of the most advanced anomaly detection methods on the MVTec AD benchmark. Feature-level registration is employed as a pre-text task, and the features stored in memory are optimized through registration loss to approach an optimal feature representation. Additionally, we introduce an innovative ground-truth segmentation method to address false positives resulting from artificial threshold determination. The performance of FR-PatchCore excels in anomaly detection on MVTec datasets and achieves state-of-the-art functionality on MPDD datasets (datasets exhibiting spatial location differences). Furthermore, our method is validated using the high-speed rail dataset. The experimental results demonstrate that FR-PatchCore effectively handles two types of data, "similar category" and "spatial transformation category", thereby enhancing the generalization capability of PatchCore. Our approach showcases its potential for industrial anomaly detection applications while maintaining high accuracy for effective generalization purposes. However, during our discussion, we also analyze certain limitations of this approach. When significant depth differences exist between samples within the same category, our approach does not sufficiently structure the registration task to cope with such variations, leaving room for future improvements.

Author Contributions: Conceptualization, Y.Z. and Z.J.; methodology, Y.W.; software, Z.J.; validation, Z.J. and J.L.; formal analysis, J.L.; investigation, Y.Z.; resources, X.G.; data curation, X.G.; writing—original draft preparation, Z.J.; writing—review and editing, X.G.; visualization, Z.J.; supervision, Y.Z.; project administration, J.L.; funding acquisition, Y.Z. All authors have read and agreed to the published version of the manuscript.

Funding: This work was funded by the Funds for International Cooperation and Exchange of the National Natural Science Foundation of China (Grant No. 61960206010).

Institutional Review Board Statement: Not applicable.

Informed Consent Statement: Not applicable.

Data Availability Statement: Data underlying the result presented in this paper are available in Refs. [13,14]. Other data are not publicly available at this time but can be obtained from the corresponding author upon reasonable request.

Conflicts of Interest: The authors declare no conflicts of interest.

References

1. Wang, Z.; Zhang, Y.; Luo, L.; Wang, N.; Lim, S. AnoDFDNet: A Deep Feature Difference Network for Anomaly Detection. *J. Sens.* **2022**, *2022*, 3538541. [CrossRef]
2. Nguyen, D.T.; Lou, Z.; Klar, M.; Brox, T. Anomaly detection with multiple-hypotheses predictions. In Proceedings of the International Conference on Machine Learning, Long Beach, CA, USA, 31 May 2019; pp. 4800–4809.
3. Sakurada, M.; Yairi, T. Anomaly detection using autoencoders with nonlinear dimensionality reduction. In Proceedings of the MLSDA 2014 2nd Workshop on Machine Learning for Sensory Data Analysis, Dunedin, New Zealand, 2 December 2013; pp. 4–11.
4. Akcay, S.; Atapour-Abarghouei, A.; Breckon, T.P. Ganomaly: Semi-supervised anomaly detection via adversarial training. In Proceedings of the Computer Vision–ACCV 2018: 14th Asian Conference on Computer Vision, Perth, Australia, 2–6 December 2018; Revised Selected Papers, Part III 14. pp. 622–637.
5. Arisoy, S.; Nasrabadi, N.M.; Kayabol, K. Unsupervised pixel-wise hyperspectral anomaly detection via autoencoding adversarial networks. *IEEE Geosci. Remote Sens. Lett.* **2021**, *19*, 1–5. [CrossRef]
6. Li, D.; Chen, D.; Jin, B.; Shi, L.; Goh, J.; Ng, S.-K. MAD-GAN: Multivariate anomaly detection for time series data with generative adversarial networks. In Proceedings of the International Conference on Artificial Neural Networks, Lugano, Switzerland, 15 January 2019; pp. 703–716.

7. Carrara, F.; Amato, G.; Brombin, L.; Falchi, F.; Gennaro, C. Combining gans and autoencoders for efficient anomaly detection. In Proceedings of the 2020 25th International Conference on Pattern Recognition (ICPR), Milan, Italy, 10–15 January 2021; pp. 3939–3946.
8. Lyu, Y.; Han, Z.; Zhong, J.; Li, C.; Liu, Z. A Generic Anomaly Detection of Catenary Support Components Based on Generative Adversarial Networks. *IEEE Trans. Instrum. Meas.* **2020**, *69*, 2439–2448. [CrossRef]
9. Son, H.; Jang, Y.; Kim, S.-E.; Kim, D.; Park, J.-W. Deep Learning-Based Anomaly Detection to Classify Inaccurate Data and Damaged Condition of a Cable-Stayed Bridge. *IEEE Access* **2021**, *9*, 124549–124559. [CrossRef]
10. Mitra, J.; Qiu, J.; MacDonald, M.; Venugopal, P.; Wallace, K.; Abdou, H.; Richmond, M.; Elansary, N.; Edwards, J.; Patel, N.; et al. Automatic Hemorrhage Detection From Color Doppler Ultrasound Using a Generative Adversarial Network (GAN)-Based Anomaly Detection Method. *IEEE J. Transl. Eng. Health Med.* **2022**, *10*, 1800609. [CrossRef] [PubMed]
11. Xiao, K.; Cao, J.; Zeng, Z.; Ling, W.-K. Graph-based Active Learning with Uncertainty and Representativeness for Industrial Anomaly Detection. *IEEE Trans. Instrum. Meas.* **2023**, *72*, 5016114. [CrossRef]
12. Roth, K.; Pemula, L.; Zepeda, J.; Schölkopf, B.; Brox, T.; Gehler, P. Towards total recall in industrial anomaly detection. In Proceedings of the IEEE/CVF Conference on Computer Vision and Pattern Recognition, New Orleans, LA, USA, 18–24 June 2022; pp. 14318–14328.
13. Bergmann, P.; Fauser, M.; Sattlegger, D.; Steger, C. MVTec AD—A comprehensive real-world dataset for unsupervised anomaly detection. In Proceedings of the IEEE/CVF Conference on Computer Vision and Pattern Recognition, Long Beach, CA, USA, 15–20 June 2019; pp. 9592–9600.
14. Yan, X.; Zhang, H.; Xu, X.; Hu, X.; Heng, P.-A. Learning semantic context from normal samples for unsupervised anomaly detection. *AAAI Conf. Artif. Intell.* **2019**, *35*, 3110–3118. [CrossRef]
15. Zhou, C.; Paffenroth, R.C. Anomaly detection with robust deep autoencoders. In Proceedings of the 23rd ACM SIGKDD International Conference on Knowledge Discovery and Data Mining, Halifax, NS, Canada, 13–17 August 2017; pp. 665–674.
16. Bergmann, P.; Löwe, S.; Fauser, M.; Sattlegger, D.; Steger, C. Improving unsupervised defect segmentation by applying structural similarity to autoencoders. *arXiv* **2018**, arXiv:1807.02011. [CrossRef]
17. You, Z.; Cui, L.; Shen, Y.; Yang, K.; Lu, X.; Zheng, Y.; Le, X. A unified model for multi-class anomaly detection. *Adv. Neural Inf. Process. Syst.* **2022**, *35*, 4571–4584. [CrossRef]
18. Bergman, L.; Cohen, N.; Hoshen, Y. Deep nearest neighbor anomaly detection. *arXiv* **2020**, arXiv:2002.10445. [CrossRef]
19. Cohen, N.; Hoshen, Y. Sub-image anomaly detection with deep pyramid correspondences. *arXiv* **2020**, arXiv:2005.02357. [CrossRef]
20. Altman, N.S. An Introduction to Kernel and Nearest-Neighbor Nonparametric Regression. *Am. Stat.* **1992**, *46*, 175–185. [CrossRef]
21. Defard, T.; Setkov, A.; Loesch, A.; Audigier, R. Padim: A patch distribution modeling framework for anomaly detection and localization. In Proceedings of the International Conference on Pattern Recognition, Milan, Italy, 10–15 January 2020; pp. 475–489.
22. Zheng, Y.; Wang, X.; Deng, R.; Bao, T.; Zhao, R.; Wu, L. Focus your distribution: Coarse-to-fine non-contrastive learning for anomaly detection and localization. In Proceedings of the 2022 IEEE International Conference on Multimedia and Expo (ICME), Taipei, Taiwan, 18–22 July 2022; pp. 1–6.
23. Jaderberg, M.; Simonyan, K.; Zisserman, A. Spatial transformer networks. In Proceedings of the Advances in Neural Information Processing Systems 28 (NIPS 2015), Montreal, QC, Canada, 7–12 December 2015; Volume 28. [CrossRef]
24. Hanley, J.A.; McNeil, B.J. The meaning and use of the area under a receiver operating characteristic (ROC) curve. *Radiology* **1982**, *143*, 29–36. [CrossRef] [PubMed]
25. Huang, C.; Guan, H.; Jiang, A.; Zhang, Y.; Spratling, M.; Wang, Y.-F. Registration based few-shot anomaly detection. In Proceedings of the European Conference on Computer Vision, Tel Aviv, Israel, 23–27 October 2022; pp. 303–319.
26. Zagoruyko, S.; Komodakis, N. Wide residual networks. *arXiv* **2016**, arXiv:1605.07146. [CrossRef]
27. Chen, X.; He, K. Exploring simple siamese representation learning. In Proceedings of the IEEE/CVF Conference on Computer Vision and Pattern Recognition, Nashville, TN, USA, 20–25 June 2021; pp. 15750–15758.
28. Cai, L.Z.; Lin, J.; Starr, M.R.; Obeid, A.; Ryan, E.H.; Ryan, C.; Forbes, N.J.; Arias, D.; Ammar, M.J.; Patel, L.G.; et al. PRO score: Predictive scoring system for visual outcomes after rhegmatogenous retinal detachment repair. *Br. J. Ophthalmol* **2023**, *107*, 555–559. [CrossRef] [PubMed]
29. Yang, J.; Shi, Y.; Qi, Z. Dfr: Deep feature reconstruction for unsupervised anomaly segmentation. *arXiv* **2020**, arXiv:2012.07122. [CrossRef]
30. Yi, J.; Yoon, S. Patch svdd: Patch-level svdd for anomaly detection and segmentation. In Proceedings of the Asian Conference on Computer Vision, Kyoto, Japan, 30 November–4 December 2020.
31. Li, C.-L.; Sohn, K.; Yoon, J.; Pfister, T. Cutpaste: Self-supervised learning for anomaly detection and localization. In Proceedings of the IEEE/CVF Conference on Computer Vision and Pattern Recognition, Nashville, TN, USA, 20–25 June 2021; pp. 9664–9674.

sensors

MDPI

Article

Advancing Ultrasonic Defect Detection in High-Speed Wheels via UT-YOLO

Qian Zhang [1], Jianping Peng [1,*], Kang Tian [1], Ai Wang [2], Jinlong Li [1] and Xiaorong Gao [1]

[1] School of Physical Science and Technology, Southwest Jiaotong University, Chengdu 610036, China; zhangqian94@my.swjtu.edu.cn (Q.Z.); tiankang@my.swjtu.edu.cn (K.T.); jinlong_lee@swjtu.edu.cn (J.L.); gxrr@vip.163.com (X.G.)

[2] Chengdu Lead Science & Technology Co., Ltd., Chengdu 610073, China; ai.wang.ms@gmail.com

* Correspondence: adams.peng@swjtu.edu.cn

Abstract: In the context of defect detection in high-speed railway train wheels, particularly in ultrasonic-testing B-scan images characterized by their small size and complexity, the need for a robust solution is paramount. The proposed algorithm, UT-YOLO, was meticulously designed to address the specific challenges presented by these images. UT-YOLO enhances its learning capacity, accuracy in detecting small targets, and overall processing speed by adopting optimized convolutional layers, a special layer design, and an attention mechanism. This algorithm exhibits superior performance on high-speed railway wheel UT datasets, indicating its potential. Crucially, UT-YOLO meets real-time processing requirements, positioning it as a practical solution for the dynamic and high-speed environment of railway inspections. In experimental evaluations, UT-YOLO exhibited good performance in best recall, mAP@0.5 and mAP@0.5:0.95 increased by 37%, 36%, and 43%, respectively; and its speed also met the needs of real-time performance. Moreover, an ultrasonic defect detection data set based on real wheels was created, and this research has been applied in actual scenarios and has helped to greatly improve manual detection efficiency.

Keywords: ultrasonic testing; B-scan images; defect detection; UT-YOLO

1. Introduction

Wheels play a pivotal role as integral components of trains, directly influencing the safety and reliability of train operations [1]. With the rapid development and increasing speed of high-speed trains, corresponding safety risks are also escalating. The quality and structural condition of high-speed rail wheels have a direct impact on the safety and overall performance of an entire train. Operating at high speeds exposes the wheels to various external factors, including wear, impacts, and fatigue cracks, which will lead to deformation, damage, or even detachment, particularly under high-load and -frequency conditions. Wheel defects can lead to severe traffic accidents.

Non-destructive testing is a method that preserves the integrity of the object being examined [2], primarily being used for detecting defects in materials, components, and equipment. Traditional non-destructive testing (NDT) methods encompass ultrasonic testing (UT) [3], magnetic particle testing (MT) [4], and penetrant testing (PT) [5]. In the railway industry, owing to the unique environment, nearly all inspections rely on non-destructive testing. X-ray computed tomography has been employed to address the reconstruction of rolling contact fatigue cracks in rails [6], and various ultrasonic testing methods have been applied in axle defect detection [7,8]. Given the complex structures of wheels and the multitude of components, there are a variety of detection methods [9–11].

Traditional wheel defect detection methods predominantly hinge on visual inspection and ultrasonic testing. Visual inspection, however, is inefficient and prone to inaccuracies. Ultrasonic testing, relying on the principles of ultrasonic wave propagation in materials, involves emitting ultrasonic waves and capturing reflected signals. This approach enables the detection of various defects within wheels, such as cracks, inclusions, and porosities. Characterized by high sensitivity and resolution, this technology proves effective in identifying minute defects, offering early warnings of potential issues and contributing to the prevention of safety hazards arising from wheel failures.

The application of ultrasonic testing to train wheels goes beyond routine periodic inspections; it is also applicable in emergencies, such as post-collision or other external impact scenarios. Continuous monitoring of wheel health empowers operators to implement timely maintenance and repair measures, ensuring the safety and stability of trains during high-speed operations. However, due to the huge quantity of data, relying solely on manual detection is too inefficient. Traditional detection methods rely more on fixed methods. Matching and efficiency are difficult to guarantee. At the same time, subject to the limitations of expert knowledge, the detection process is relatively complicated. Therefore, a set of novel methods that can meet the above needs continues to be applied and promoted.

Deep learning technology represents a significant advancement in enhancing the efficiency and accuracy of computer vision as well as NDT. Its application extends to the automated analysis of various NDT data, including ultrasonic signals, magnetic particle images, and penetrant images, thereby substantially improving defect detection accuracy. Specifically, in ultrasonic wheel defect detection, deep learning technology offers several advantages. Firstly, it facilitates the automatic analysis of ultrasonic signals, thereby significantly enhancing detection efficiency. Secondly, it can extract intricate features from ultrasonic signals that may be challenging for human recognition, leading to a substantial improvement in detection accuracy. Additionally, deep learning technology contributes to cost reduction and minimizes manual intervention in the testing process. One notable strength of deep learning models is their ability to classify defects into different types, accurately locate defect positions, and measure defect sizes. These affordances not only streamline the analysis process but also provide a comprehensive understanding of the detected defects. By annotating ultrasound B-scan images, we can annotate a defect, which will give us a UT dataset. By learning from a large quantity of effective data, a model can efficiently detect data.

Despite these advantages, the application of deep learning technology in ultrasonic wheel defect detection still encounters certain challenges. One of the primary challenges is the creation of high-quality datasets. The performance of deep learning models heavily relies on the quality and diversity of the training data. Ensuring that the datasets used for training are representative of real-world scenarios is essential for achieving robust and reliable defect detection. Another challenge lies in the high complexity of deep learning models. While these models exhibit remarkable capabilities, their complexity can lead to challenges in interpretability and the need for significant computational resources. Striking a balance between model complexity and interpretability is crucial for practical implementation and widespread adoption. While deep learning technology has immense potential for revolutionizing ultrasonic wheel defect detection by significantly improving efficiency and accuracy, addressing challenges related to dataset creation and model complexity is essential for realizing its full benefits in real-world applications. To this end, we need to explain the categories of defects. In this work, the defect data are divided into four different forms according to the depth of scanning, namely, surface, near-surface, internal, and wheel plate cracks.

Given that natural images seldom exhibit exaggerated aspect ratios, standard deep-learning approaches may fall short when applied directly to ultrasonic images. Hence, this article explores specialized preprocessing techniques aimed at enhancing the compatibility between deep learning models and the intricate features inherent in ultrasonic wheel defect B-scan images. By addressing these specific challenges through dedicated preprocessing methods, the goal is to optimize the learning capacity of deep neural networks, fostering improved accuracy in detecting defects. This approach is crucial in bridging the gap between conventional image-processing methodologies and the distinctive characteristics of ultrasonic datasets, ultimately enhancing the efficacy of defect detection in high-speed railway train wheels. The design and training of deep learning models need to be tailored to accommodate these distinctive characteristics, ensuring optimal performance in detecting ultrasound defects. The main contributions of this paper are as follows:

- An ultrasonic defect detection data set based on real wheels was created, and the defects were subdivided according to different categories;
- An advanced UT-YOLO network was proposed to increase a network's ability to perceive small defects and detection;
- This research has been applied in actual scenarios and has helped to greatly improve manual detection efficiency.

2. Related Work

2.1. Traditional Defect Detection

Traditional defect detection methods are based on the characteristics of an object, such as color, size, shape, etc., and use image-processing methods and traditional machine learning methods for detection. Among them, the representative target detection methods include Scale-Invariant Feature Transform (SIFT) [12], Histogram of Oriented Gradients (HOG) [13], Oriented FAST and Rotated BRIEF (ORB) [14], and Harr [15]. It is necessary to use a pre-designed template to slide the window on the target image, and the steps are cumbersome. The extracted object features are all low-level, manually selected features, so they are not conducive to accurate detection and offer poor real-time detection. Moreover, these methods are mostly used for obvious features in natural images which have rich information. For these simple tasks, generally, Prewitt, Sobel, and Canny can be used to find solutions. However, in practice, especially in complicated wheel defect detection situations, noise is very serious, and much depends on the accuracy of the equipment and the skilled operators. All in all, due to the limitations of many factors, ultrasonic detection images are often not used in this type of detection method.

While conventional methods exhibit proficiency in straightforward industrial scenarios governed by clear rules, they encounter challenges when confronted with the intricacies and diversities inherent in real-world defects. Preceding the era of deep learning, object detection heavily relied on meticulously crafted features. Although this approach attained some degree of success, it grappled with notable limitations. For instance, the manual design of features often necessitated subjective decisions, resulting in inconsistencies and a lack of generalizability. The manual approach struggled to accommodate the vast diversity and complexity of real world objects, and as the number of object categories and image variations expanded, the manual design process became increasingly cumbersome and impractical. Moreover, the broad spectrum of potential defects demanded multiple sets of pre-defined templates, constraining these methods' capacity to identify novel and unexpected issues. Consequently, when confronted with intricate data, traditional methods frequently falter, necessitating laborious post-processing efforts.

2.2. Deep-Learning-Based Defect Detection

The rise of deep learning has transformed object detection, offering notable advantages over traditional methods. Convolutional neural networks (CNNs) excel at extracting high-level features from images, overcoming the limitations of manual feature crafting. Their seamless integration of feature extraction, selection, and classification enhances overall efficiency and effectiveness. Scholarly work consistently demonstrates the prowess of deep learning in defect detection, particularly in industrial settings.

The R-CNN network was proposed by Girshick et al. [16] in 2013. It is a landmark effort in the field of object detection. He et al. proposed SPP-Net [17], which no longer extracts features from all candidate areas and only extracts features of an entire image once, which greatly reduces the amount of calculation. Faster-RCNN [18] was the most representative model in the R-CNN series that was proposed by Ren et al. Because the Region Proposal Network (RPN) realized end-to-end training and testing of a network, detection accuracy was highly improved. However, Faster-RCNN is generally more complex, and its performance on small objects is not very impressive. He et al. proposed Mask RCNN [19], which has high accuracy in detection, but at the same time, instance segmentation will increase label price. The biggest problem for practical use is that the speed of the algorithm still cannot meet the industrial real-time detection requirements. In the industrial field, some researchers have adjusted the common detection frameworks to suit the detection of industrial data. Liu et al. [20] improved Faster-RCNN by utilizing cascade heads to detect metal defects. Cha et al. [21] changed the backbone to improve the detection time for bridge defects based on Faster-RCNN. He et al. [22] improved detection accuracy using multi-scale feature diffusion. The pursuit of greater detection speed has favored a large number of one-stage algorithms.

Redmon et al. [23] proposed the first one-stage detector, YOLO v1, which uses a multi-scale region centered on a grid to replace the RPN, greatly improving detection speed and meeting the needs of real-time detection but not in terms of accuracy. A novel object detection method called SSD [24] was proposed in 2016, combining the advantages of both Faster-RCNN and YOLO v1. It ensures high-precision detection while taking into account detection speed by using multi-scale regional features for regression, particularly via combining high-level and low-level feature maps. YOLO v2 [25] employs Darknet-19 as a backbone network of a model to extract features and reduce the computational complexity of the model. In addition, it further improves detection speed and accuracy by adding batch normalization, multi-scale training, and K-means dimension clustering after each convolutional layer. YOLO v3 [26] introduced several key advancements over its predecessors, leading to improved performance in both speed and accuracy. One of its most significant contributions is the Darknet-50 residual network, a newly designed architecture that utilizes residual connections for deeper feature extraction. This network was combined with the feature pyramid network (FPN) [27], which enabled multi-scale fusion prediction and detection on three different-scale feature maps. YOLO v4 [28] uses CSPDarkNet53 as a backbone network and optimizes to varying degrees using data processing, training methods, activation functions, loss functions, etc., to achieve the best detection effect at a given time. Li et al. [29] provided an end-to-end solution for the surface defect detection of steel strips based on improved YOLO. Zhang et al. [30] improved YOLO v3 by introducing a novel transfer learning method with fully pre-trained weights from a geometrically similar dataset to detect bridge surface damage.

However, all algorithms are based on the premise of sufficient data. However, in actual situations, especially in industrial environments, due to the particularity of application scenarios, sufficient and balanced labeled data sets are difficult to obtain, even if ordinary data augmentation cannot completely solve this problem. A novel automated defect synthesis network called Defect-GAN was designed by Zhang et al. [31] to generate realistic and diverse defect samples for training accurate and robust defect inspection networks. Valente et al. [32] used synthetic training data by simulating two types of print effects with image-processing and computer graphic techniques.

Building upon YOLO v5s as the baseline, this work introduces UT-YOLO, a novel ultrasonic testing defect detection method. UT-YOLO incorporates several key advancements over the baseline model, namely, enhanced backbone architectures with Swin-Transformer [33] and ResNet [34] variants, improved neck modules like BiFPN [35] for better feature aggregation, a dedicated multi-detection head for the accurate localization of small defects, a novel attention module for enhanced feature representation, and a practical pre-processing pipeline specifically tailored for real-world ultrasonic B-scan images, ensuring robust and effective defect detection in industrial applications.

3. Methodology

The basic architecture of YOLO v5 is shown in Figure 1. The whole operation process is as follows: Firstly, the images are readied for feature extracted by the backbone network, named CSPDarkNet53 [28]; then, the features in the neck section are separated into different sizes for feature fusion to obtain more feature information. After that, the different detection heads are designed, which allows the algorithm to fulfill the specific requirements of object detection.

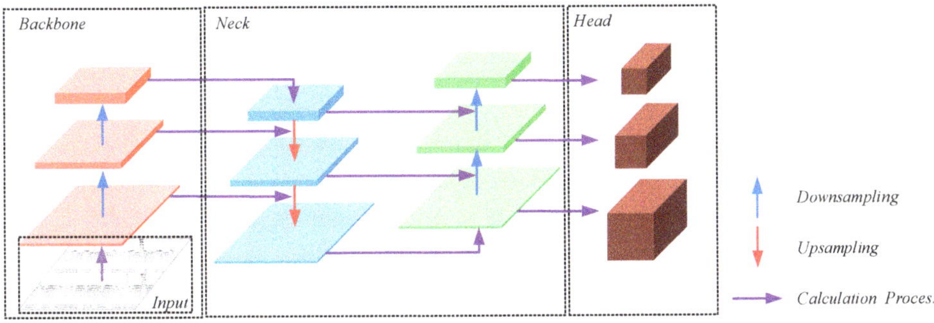

Figure 1. The architecture of YOLO v5.

3.1. Backbone

The architecture of the backbone is designed for feature extraction, but its efficiency should be given more attention. Due to this consideration, some novel backbone structures were applied in this task. The defects from ultrasonic testing B-scan images appear in different regions, so a useful feature extraction backbone should be designed to solve this problem. Regarding the wheels of railway trains, the defects are always small and noise-like, making them hard to identify easily. Figure 2 shows a typical residual connection, which is widely used in the design of backbone improvements. By introducing residual connections, the neural network is allowed to skip certain levels of learning, thereby alleviating the vanishing and exploding gradient problems and making training easier. This network also supports deeper networks. Traditional deep neural networks suffer from the problem wherein the training difficulty increases as the depth of the network increases. The residual structure allows building deeper networks by adding more layers without degradation issues.

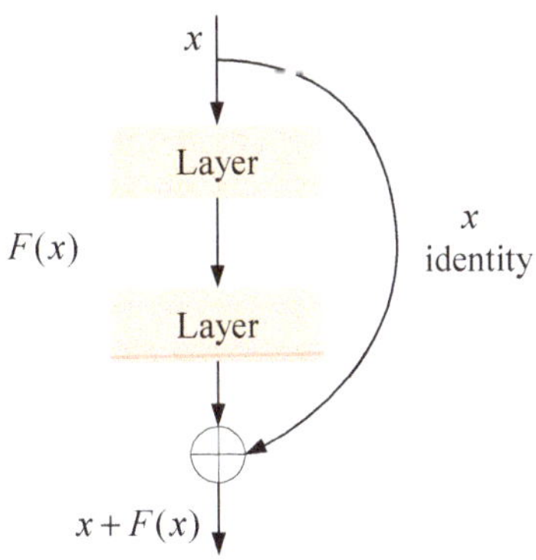

Figure 2. The structure of a residual connection.

Figure 3 shows the structure of the Swin-Transformer block. It allows a model to better capture global relationships among pixels in an image and avoids full connection to reduce calculation costs, affordances that are of vital importance in object detection, especially in ultrasonic defect testing. At the same time, by stacking multiple transformer blocks, the network is able to learn more complex and abstract feature representations.

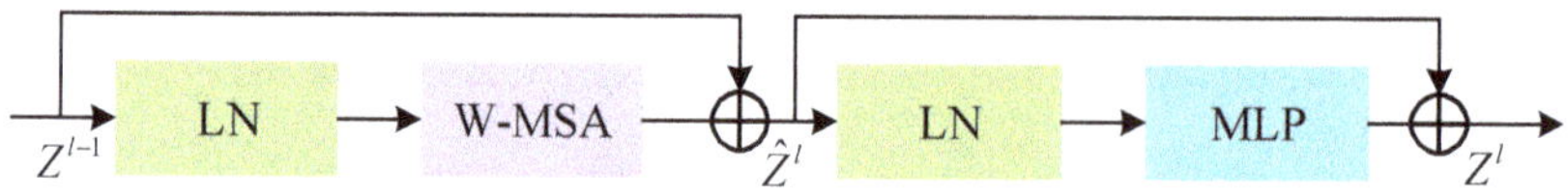

Figure 3. The structure of Swin-Transformer.

Here, Z^{l-1} represents the output of layer $l-1$ or the input of layer l. The information passes from one layer to the next layer sequentially.

Firstly, the input is linearly transformed and normalized; then, it is passed through Weighted Multi-Head Self Attention (W-MSA), which focuses on a different area of the input rather than the global average. Secondly, the output will be added by the residual connection. The new output will repeat another Layer Normalization (LN) step and then connect with a Multi-Layer Perceptron (MLP) to fit the complicated input data. Finally, the residual connection will be used again to improve model stability. It is usually used as a repeated block, which helps alleviate vanishing or exploding gradient problems during training, making the network easier to train and optimize. In wheel defect feature extraction, after convolution and residual connection, the feature extraction module of Swin-Transformer can be superimposed multiple times to obtain more semantic information, which plays an important role in distinguishing background noise and defects.

3.2. Neck

BiFPN, a bidirectional feature propagation network, facilitates the bidirectional flow of information, allowing features to traverse from high-level layers to low-level layers, as well as enabling the propagation of low-level features to high-level layers. This dual-directional information exchange enhances the model's ability to capture semantic information across various feature levels, thereby augmenting the model's perceptual capabilities towards the target. In contrast, PANet adopts unidirectional propagation, limiting the flow of information to a single direction. The corresponding structure is shown in Figure 4.

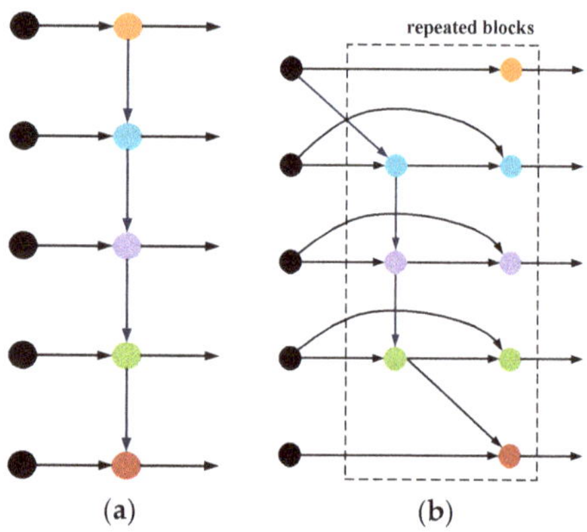

Figure 4. (**a**) Schematic diagram of PANet; (**b**) schematic diagram of BiFPN.

Another noteworthy aspect is that BiFPN was meticulously designed to mitigate computational complexity without compromising model efficacy. It achieves this by judiciously fusing features at distinct levels, thereby sidestepping the computational overhead associated with full connections. In contrast, PANet exhibits a more intricate design that could entail a higher computational burden. Nevertheless, a novel enhancement of BiFPN made by incorporating the Simple, Parameter-Free Attention Module (SimAM) [36] was proposed. This addition enables the network to selectively concentrate on varying levels of features, thus refining its capacity for nuanced feature extraction. A diagram of SimAM is shown in Figure 5.

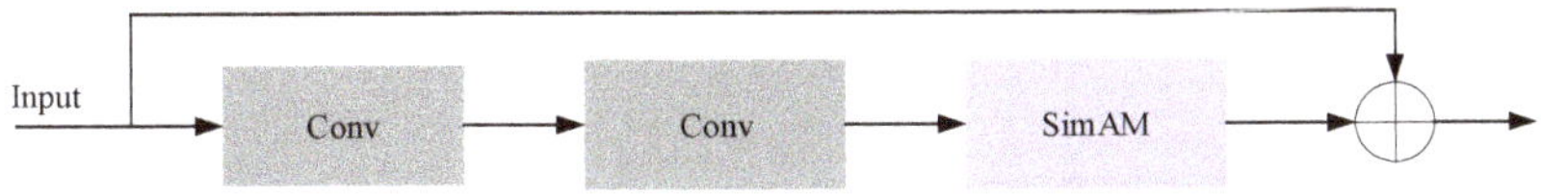

Figure 5. The diagram of SimAM.

3.3. Head

Given that ultrasonic testing typically involves small targets, the existing detection heads may struggle to meet the required standards. Consequently, this study enhances the number of detection heads and simultaneously augments the depth of the detection heads within the target detection model. The objective is to enhance the model's discernment of intricate scenes and nuanced features in pursuit of more effective performance. The new structure is shown in Figure 6.

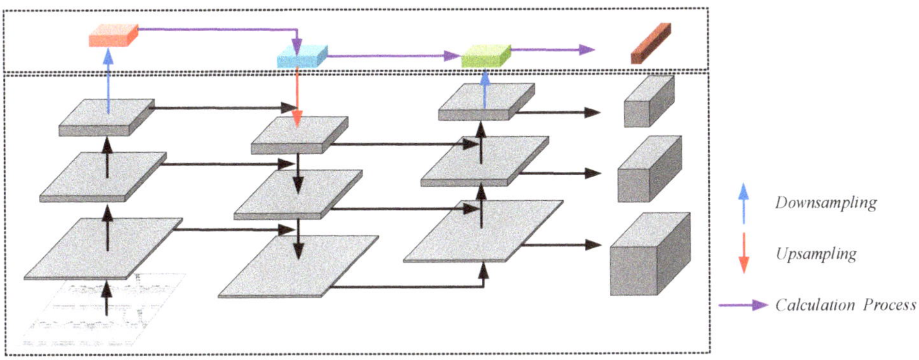

Figure 6. The new small-object detection head.

The added detection head is shown in Figure 6. The structure in the dotted box in the lower half is consistent with that in Figure 1 but is rendered in gray. On this basis, in the upper block diagram of the network, a small target detection layer was added. Using color marking, the structure is used for the detection of small targets.

3.4. Loss Function

The original YOLOv5 loss function generally consists of three components, as represented by Equation (1). The first term quantifies localization loss, evaluating the accuracy of the model's predictions for the position of the target bounding box. The second term corresponds to confidence loss, measuring how accurately the model predicts the presence of a target. The final term assesses the accuracy of the classification

$$\mathcal{L}_{\text{ori}}\left(t_{\text{p}}, t_{\text{gt}}\right) = \sum_{k=0}^{K}\left[\alpha_k^{\text{W}}\alpha_{\text{box}}\sum_{i=0}^{S^2}\sum_{j=0}^{B}\mathbb{I}_{k_{ij}}^{\text{obj}}\mathcal{L}_{\text{CIoU}}+\alpha_{\text{obj}}\sum_{i=0}^{S^2}\sum_{j=0}^{B}\mathbb{I}_{k_{ij}}^{obj}\mathcal{L}_{\text{obj}}+\alpha_{\text{cls}}\sum_{i=0}^{S^2}\sum_{j=0}^{B}\mathbb{I}_{k_{ij}}^{obj}\mathcal{L}_{\text{cls}}\right] \quad (1)$$

where K, S^2, and B are the number of output features, the number of grids, and the number of anchors, respectively. α_* is the weight for different sections; α_{box}, α_{obj}, and α_{cls} are set as 0.05, 0.3, and 0.7, respectively. $\mathbb{I}_{k_{ij}}^{\text{obj}}$ represents whether the $i-$ th cell, the $j-$ th anchor box in the $k-$ th feature map, is a positive sample. If it is positive, it is defined as 1; otherwise, it is set as 0. t_{p} and t_{gt} are the prediction vector and ground truth (GT), and α_k^{W} is the weight for different sizes of the output.

Equation (2) is the new loss function called $\mathcal{L}_{\mathrm{UT}}$, which is designed for UT defect detection. The localization loss is replaced by $\mathcal{L}_{\mathrm{EIoU}}$.

$$\mathcal{L}_{\mathrm{UT}}(t_p, t_{gt}) = \sum_{k=0}^{K}\left[\alpha_k^{w}\alpha_{\mathrm{box}}\sum_{i=0}^{S^2}\sum_{j=0}^{B}\mathbb{I}_{k_{ij}}^{\mathrm{obj}}\mathcal{L}_{\mathrm{EIoU}}+\alpha_{\mathrm{obj}}\sum_{i=0}^{S^2}\sum_{j=0}^{B}\mathbb{I}_{k_{ij}}^{obj}\mathcal{L}_{\mathrm{obj}}+\alpha_{\mathrm{cls}}\sum_{i=0}^{S^2}\sum_{j=0}^{B}\mathbb{I}_{k_{ij}}^{obj}\mathcal{L}_{\mathrm{cls}}\right] \quad (2)$$

Intersection over Union (IoU) is used to evaluate the quality of object detection models. A schematic diagram of IoU is shown in Figure 7.

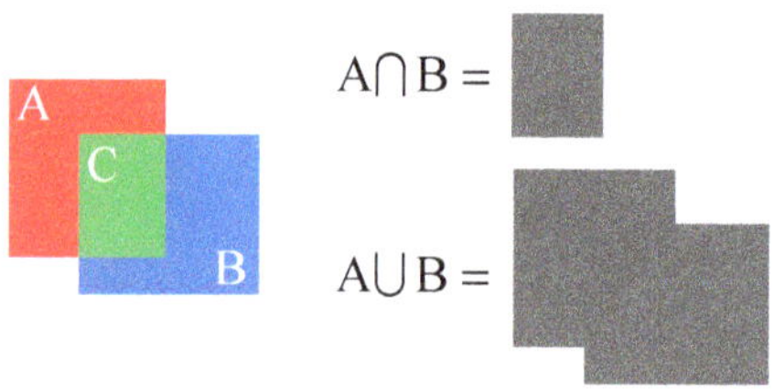

Figure 7. The schematic diagram of IoU.

Here, the value of IoU should be described as in Equation (3).

$$\mathrm{IoU} = \frac{\mathrm{A}\cap\mathrm{B}}{\mathrm{A}\cup\mathrm{B}} \quad (3)$$

Simply put, the overlapping region of A and B is C, which is denoted as $\mathrm{A}\cap\mathrm{B}$. The area of $\mathrm{A}+\mathrm{B}-\mathrm{C}$ is denoted as $\mathrm{A}\cup\mathrm{B}$. The ratio of $\mathrm{A}\cap\mathrm{B}$ and $\mathrm{A}\cup\mathrm{B}$ is IoU, which is an indicator of prediction accuracy. In the original loss function, a complete IoU (CIoU) was used to solve the special cases with the same central point but different ratios of height to width. Equations (4)–(6) define the loss of CIoU

$$\mathcal{L}_{\mathrm{CIoU}} = 1 - \mathrm{CIoU} \quad (4)$$

$$\mathrm{CIoU} = \mathrm{IoU} - \left(\frac{\rho^2(b, b^{gt})}{c^2} + \alpha\nu\right) \quad (5)$$

$$\nu = \frac{4}{\pi^2}\left(arctan\frac{w^{gt}}{h^{gt}} - arctan\frac{w}{h}\right)^2 \quad (6)$$

where α is weight and the ν is the parameter for measuring the ratio of height to width. A schematic diagram is shown in Figure 8: the green line (AB) is the parameter c, which signifies the diagonal distance of the minimum bounding box (the gray, dotted one) capable of encompassing both the predicted box and the GT box. Meanwhile, the parameter ρ is the Euclidean distance between two center points b and b^{gt}, which are indicated by the red line.

At the same time, the definition of ν is shown in Equation (7):

$$\nu = \frac{4}{\pi^2}\left(arctan\frac{w^{gt}}{h^{gt}} - arctan\frac{w}{h}\right)^2 \quad (7)$$

In Equation (2), the penalty term of EIoU consists of separating the influence factors of the aspect ratio based on the penalty term of CIoU to calculate the length and width of the target frame and anchor frame, respectively. So, $\mathcal{L}_{\mathrm{EIoU}}$ is changed according to Equation (8).

$$\mathcal{L}_{\mathrm{EIoU}} = 1 - \mathrm{IoU} + \frac{\rho^2(b, b^{gt})}{c^2} + \frac{\rho^2(w, w^{gt})}{c_w^2} + \frac{\rho^2(h, h^{gt})}{c_h^2} \quad (8)$$

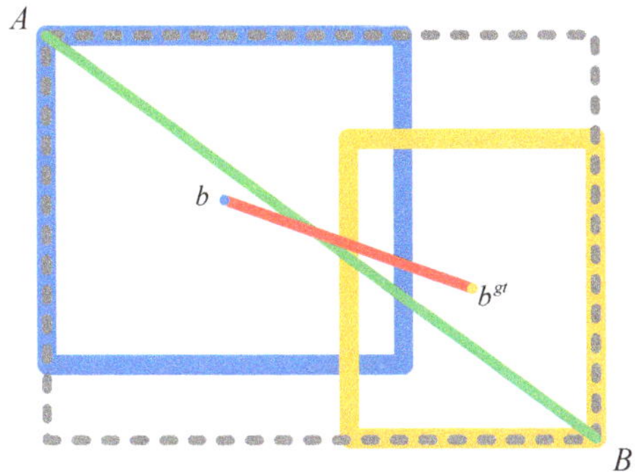

Figure 8. A schematic diagram of prediction box (blue) and GT (yellow).

Therefore, $\mathcal{L}_{\mathrm{EIoU}}$ can provide a more accurate assessment when an object's position is imprecise or slightly offset. However, to use $\mathcal{L}_{\mathrm{EIoU}}$ as the loss function, the dataset should be more closely scrutinized, especially regarding the balance between various types of data.

To strike a balance between real-time requirements and computational complexity, the original YOLOv5 algorithm was chosen as the baseline model. This work enhanced the baseline model by (1) utilizing dataset-preprocessing methods that incorporate the characteristics of ultrasonic B-scan data, enhancing the model's compatibility with neural networks; (2) employing a combination of ResNet, Swin-Transformer, and BiFPN structures for feature extraction to capture richer feature information, thereby increasing the capability of feature extraction; (3) introducing SimAM and a dedicated small-object detection head to further optimize the detection accuracy for small objects; and (4) enhancing the accuracy of bounding box regression through the refinement of the loss function's design.

4. Experiments and Results

Diverging from natural images, ultrasonic detection images primarily convey intensity information, rendering them particularly susceptible to noise interference. Moreover, as the focus of this study is on detecting defects in wheels, the acquired data hold exceptional value. The experiments were conducted alongside an exploration of datasets and various experiments.

4.1. Dataset

This work was evaluated with reference to a real dataset, which was collected with phased array UT (PAUT) and transmit receive (TR) probes to attain B-scan images about four different defects according to different depths, surfaces near other surfaces, and internal and rim cracks in different service depots using the LU system [37]. The entire dataset of UT defects consisted of around 15,000 B-scan images, in which peeling, scratches, and cracks in different positions and depths were recorded. It is hard to find different real images with which to show the different types of defects, especially some inner defects, so a diagram of typical wheel defects is shown in Figure 9.

UT is a very useful solution for detecting inner defects, but analyzing surface and sub-surface areas is not its strength because of the large amount of initial wave interference at the contact surface, as well as the noise; last but not the least, due to the particularity of the wheel structure, the collection is generally completed in one cycle, so the aspect ratio of the collected images will be very large. Examples of the original collection data are shown in Figure 10. Image (a) below depicts a surface defect, and the periodic signal represents the wheel plate hole; there are six holes in a wheel.

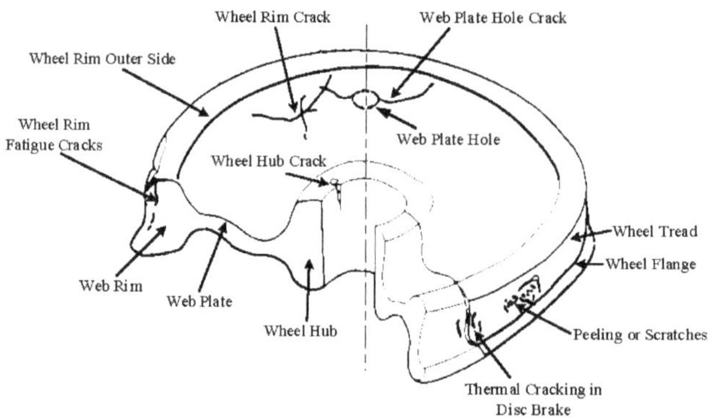

Figure 9. Diagram of typical wheel defects.

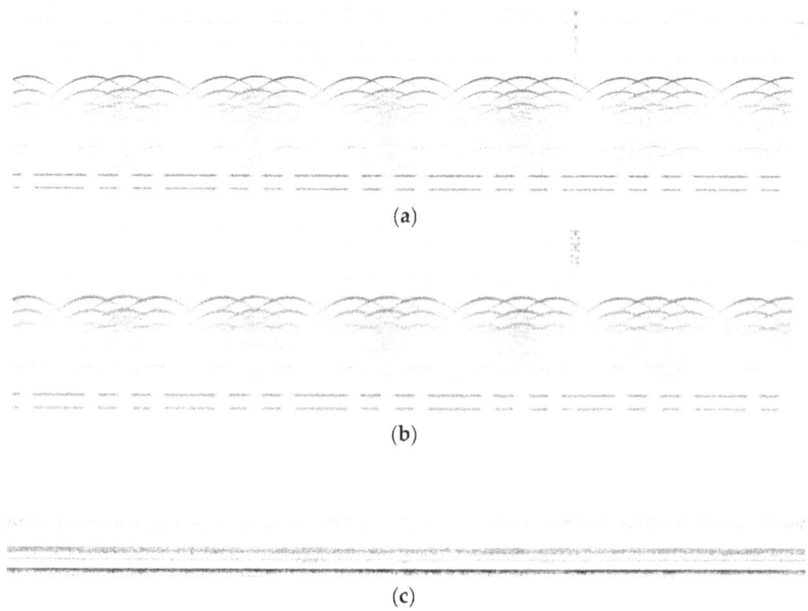

Figure 10. Examples of (**a**) original UT B-scan with a defect; (**b**) original UT B-scan with a defect and annotation; and (**c**) original UT B-scan data without defects.

The environment of the experimental data acquisition is shown in Figure 11. In the service depots, the wheel runs in the designated area, and the robotic arm will carry a phased array probe to collect the testing information. The collected signals will be stored in the industrial computer for preliminary analysis like different gains to give a short judgement.

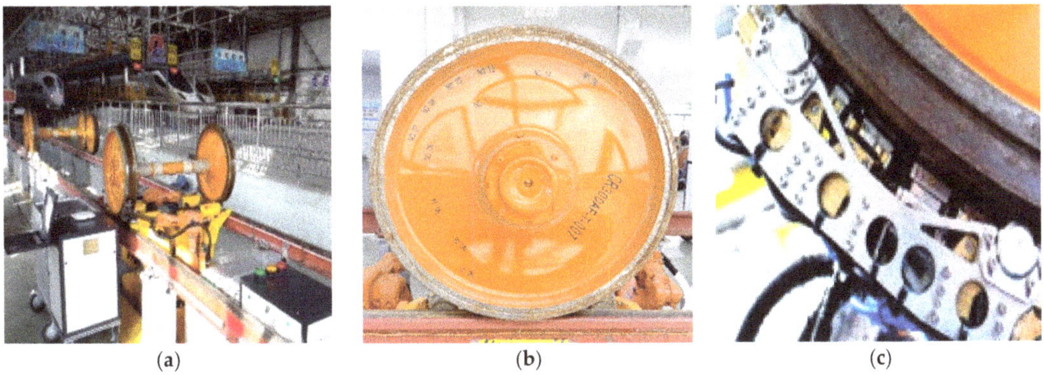

Figure 11. Typical scenarios: (**a**) wheel detection in service depots; (**b**) wheel; (**c**) PA probes.

To illustrate the practicability of the proposed method, this work briefly displays the typical real defects, including wheel rim cracks and some surface defects like scratches, with their B-scan images given below in Figure 12.

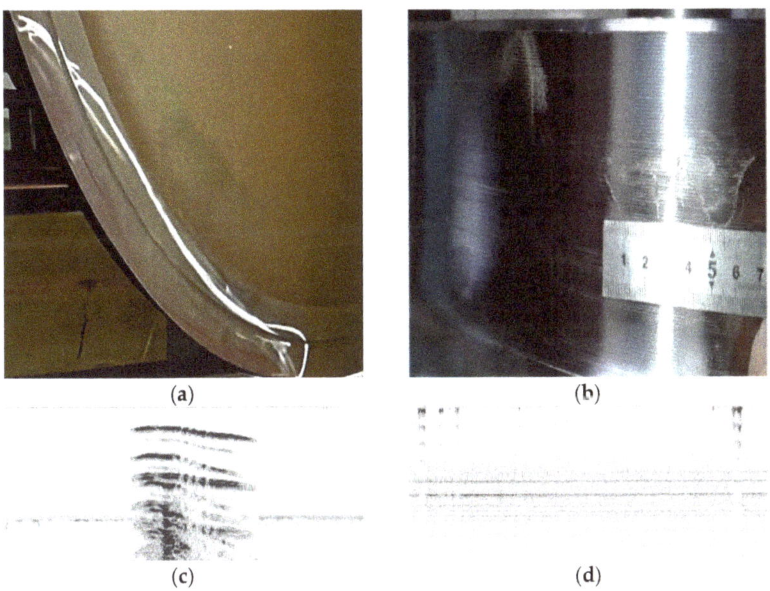

Figure 12. Typical defects: (**a**) wheel rim crack; (**b**) scratch; (**c**) ultrasonic b-scan defect image of rim crack; (**d**) surface and near-surface b-scan defect images.

Another thing that should be taken into consideration is the variation in the depths of the scan with the different ratios of width to height. It was necessary to perform a data-preprocessing procedure. First, the images were cropped and reorganized into rectangles with an aspect ratio of 1 to the greatest extent possible, as in the algorithm for data preprocessing shown below (Algorithm 1).

Then, the processed images needed to be labeled with different classes. Finally, the images were divided into three sets: a training set, a validation set, and a testing set with a ratio of 7:2:1, containing 10,395, 2970, and 1485 images, respectively.

Algorithm 1: Data Preprocessing

Input: Original Images.
Output: Processed Image after Data Preprocessing.

```
// Obtain the width (w) and height (h) of the image
w, h = getImageDimensions(originalImage)
// Check if the width is greater than 3 times the height
if w > 3 * h:
// Crop the image into two halves and stack them vertically
processedImage = cropAndStackImage(originalImage)
else:
// If width is not greater than 3 times height, no processing is done
processedImage = originalImage
end if
// Return the processed image
return processedImage
```

4.2. *Evaluated Index*

In this work, two sets of metrics from academia and industry, respectively, are used to evaluate the model. Generally, the academic indexes that should be used are average precision (AP), recall, mAP, and frames per second (FPS); the industrial indexes always focus on the true alarm rate (TAR) and the false alarm rate (FAR) of detection.

4.2.1. Academic Indexes

A confusion matrix is often used to summarize the prediction effect of classification models, as shown in Table 1.

Table 1. Confusion matrix.

	Actual Positive	**Actual Negative**
Predicted Positive	TP	FP
Predicted Negative	FN	TN

The calculations are shown in Equations (9)–(13):

$$precision = \frac{TP}{TP + FP} \tag{9}$$

$$Recall = \frac{TP}{TP + FN} \tag{10}$$

$$AP = \int_0^1 p(r)dr \tag{11}$$

$$mAP = \frac{1}{n}\sum_{k=1}^{n} AP_k \tag{12}$$

$$FPS = \frac{frameNum}{elapsedTime} = \frac{N}{T} \tag{13}$$

4.2.2. Industrial Indicators

An example describing the true alarm rate (TAR) and false alarm rate (FAR) is provided here. Assuming there are T samples in total, and D is the number of defect samples, the rest of the samples, called N, are normal. Using an algorithm of detection, in the problem section, D − 1 out of D samples were found to be defects, and in the normal section, 2 out of N were identified as defects; an explanatory diagram is shown in Figure 13, and the calculation formulae should be like those in Equations (14) and (15).

$$TAR = \frac{D-1}{D} \quad (14)$$

$$FAR = \frac{2}{N} \quad (15)$$

Figure 13. Example explaining the industrial indicators.

In this example, there is one more definition to provide, that is, the miss alarm rate (MAR), which in this case is 1 out of D, as represented in Equation (16).

$$MAR = 1 - \frac{D-1}{D} = \frac{1}{D} \quad (16)$$

In practice, there is a big gap between academia and industry. The requirements for statistical results are often stringent in the industrial setting, making the utilization of industrial indexes more challenging. In this work, academic metrics were adopted primarily to validate the feasibility of algorithmic improvements, serving as an assessment tool for evaluating the effectiveness of the algorithm.

4.3. Implementation Details

All the experiments were carried out on a computer running Windows 10. Two Nvidia GeForce RTX3090 GPUs were used, and the CPU version was Intel Core i9-1098XE. The deep learning framework was Pytorch 1.7.1, and the python version was 3.7.11. The image input was resized to 640 × 640, the batch size was 128, the optimizer was SGD, the initial learning rate was set to 0.001, the total number of iterations was set as 500 epochs, and the patience was set to 50, signifying that if there are 100 consecutive rounds without improvement, the system defaults to convergence.

4.4. Experiment Results

To evaluate the performance of the algorithm in question in terms of accuracy and speed in terms of both academic and industrial applications, UT-YOLO was compared with different algorithms. Table 2 presents the best recall, mAP@0.5, and mAP@0.5:0.95 results for each algorithm, derived from the experimental observations of the proposed model with respect to the test set. Figures 14 and 15 depict the mAP@0.5 and mAP@0.5:0.95.

Table 2. Results regarding detection for the UT-B scan dataset.

Model	Best Recall	mAp@0.5	mAp@0.5:0.95	FPS
YOLOv5s	0.57	0.53	0.21	55
YOLOv5s + BiFPN	0.63	0.58	0.21	58
YOLOv5s + BiFPN + SimAM	0.67	0.63	0.25	57
YOLOv5s + Swin-Transformer	0.85	0.82	0.61	64
UT-YOLO (Ours)	**0.94**	**0.89**	**0.64**	**69**

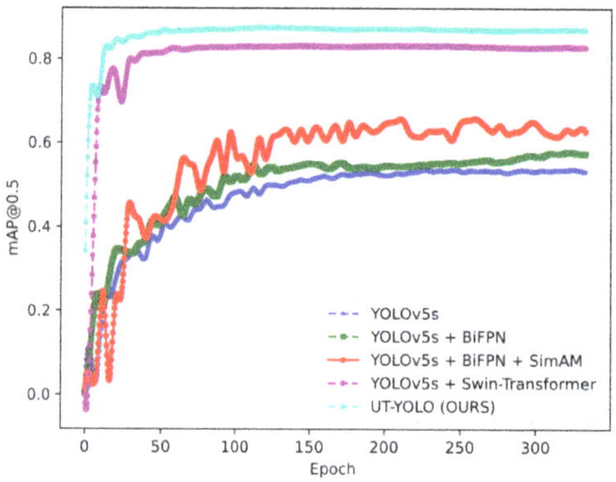

Figure 14. mAP@0.5 for the ablation experiment.

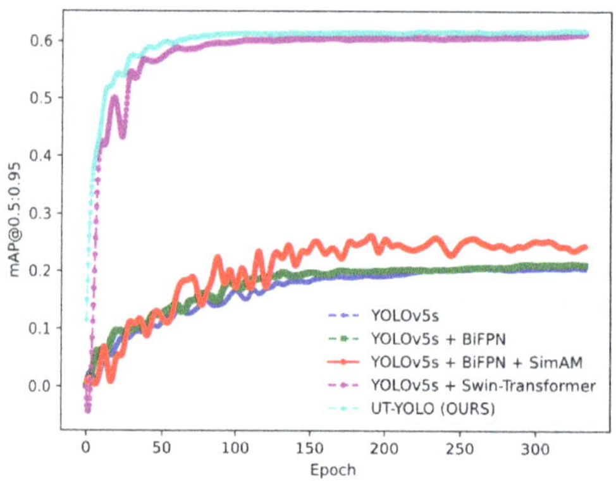

Figure 15. mAP@0.5:0.95 for the ablation experiment.

The results depicted in Figures 14 and 15 show the effectiveness of UT-YOLO. Compared with the original baseline, the IoU0.5 indicator and the 0.5:0.95 indicator have been greatly improved, which is a very exciting result. In Figure 16, a comparative analysis of the detection results across different models is presented, highlighting the performance variations for a common object. Notably, the examination distinctly underscores the superior efficacy of the UT-YOLO model.

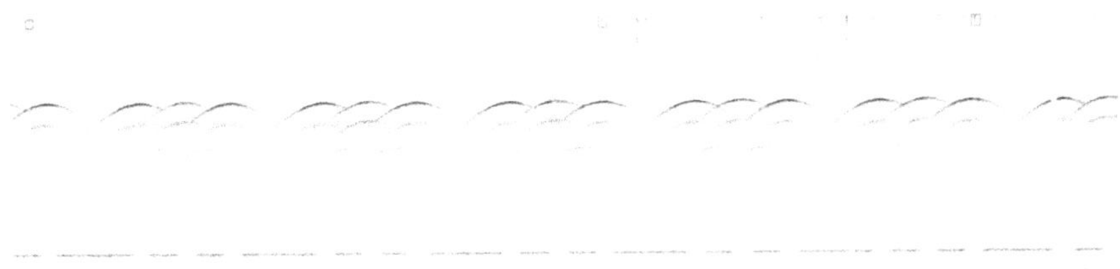

(**a**) YOLOv5s

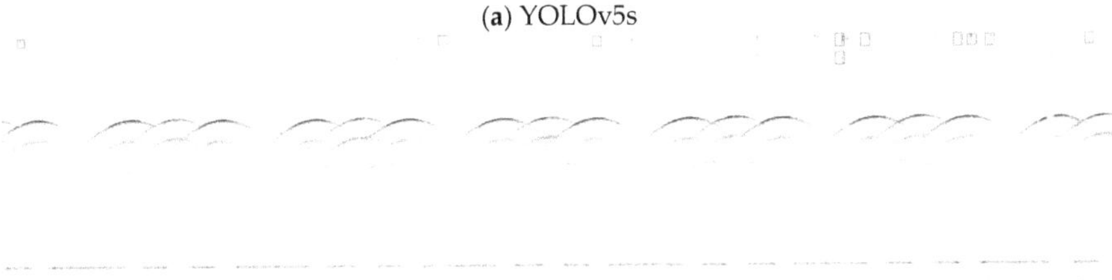

(**b**) YOLOv5s + BiFPN

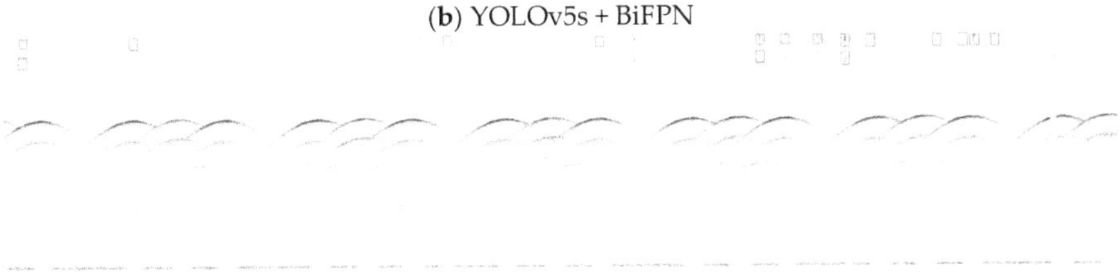

(**c**) YOLOv5s + BiFPN + SimAM

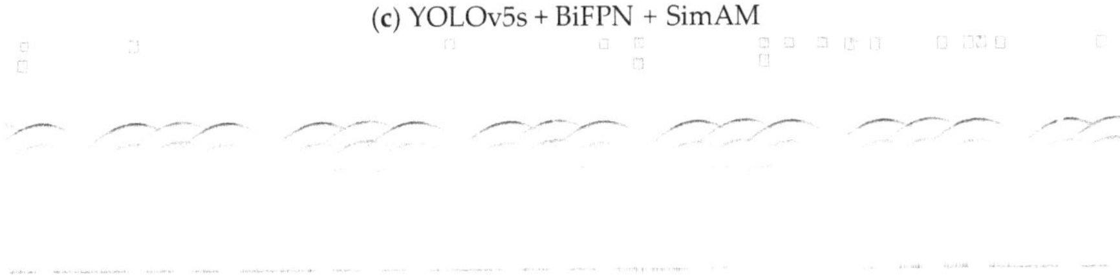

(**d**) YOLOv5s + Swin-Transformer

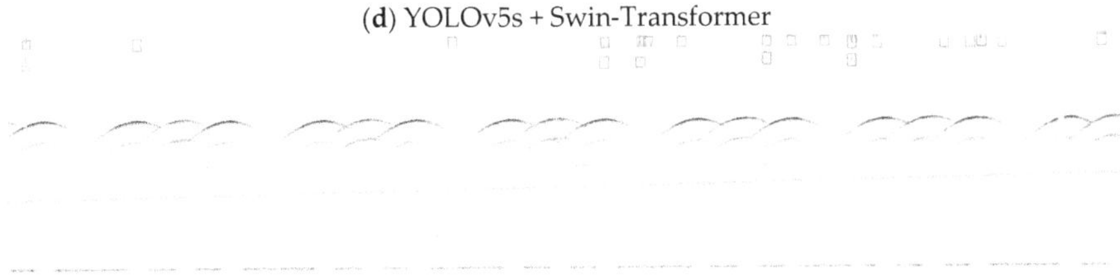

(**e**) UT-YOLO (Ours)

Figure 16. Comparison of detection results achieved using different models.

It can be observed in Figure 17 that the UT-YOLO model demonstrated precise localization of the defects, especially the small-sized defects. Furthermore, for defects near the wheel plate hole, the model exhibited commendable feature extraction capabilities, as evident in the outcomes presented in Figure 17. Despite the presence of some minor defects associated with background blur, the model displays robustness and can effectively detect defects, even in challenging conditions like strong background noise and electrical interference. The left side of each image is the original detection result, and the blue box on the right corresponds to the enlarged local features.

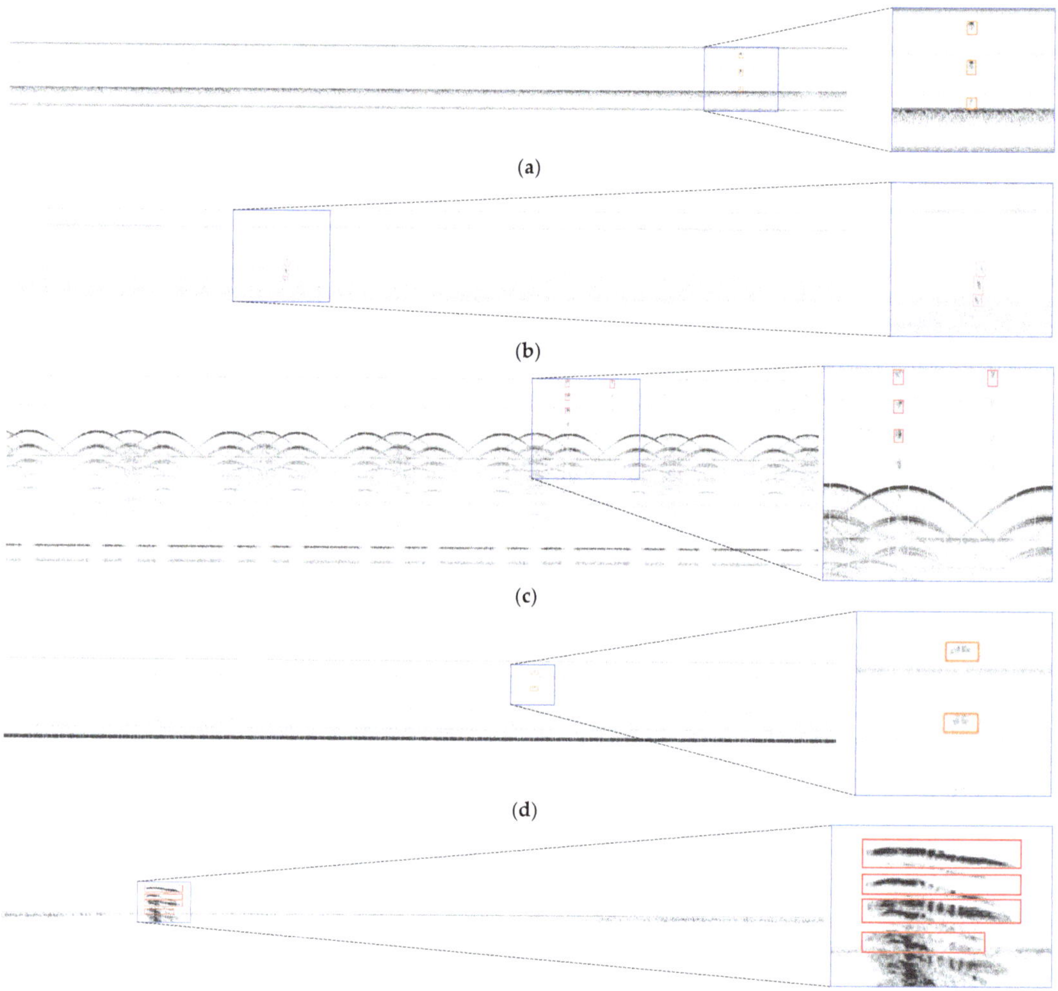

Figure 17. Detection results and magnified details. (**a**) surface and near-surface crack; (**b**) inner defect; (**c**) the defect near the web plate hole region; (**d**) Surface and near-surface hole-like defects; (**e**) wheel rim crack.

For real-world industrial inspection, additional sets of 50 defective samples and 50 non-defective samples were carefully chosen to assess the detection rate and false alarm rate. Among the 50 defective samples, consisting of a total of 6000 images, 210 were identified as defects by the UT inspectors. Using UT-YOLO, a total of 206 defect instances were successfully detected, resulting in a TAR of 98.10% (206 out of 210). Notably, the four undetected defects existed in three distinct wheelsets. In practical applications, the achieved TAR was 94% (47 out of 50), underscoring the heightened challenges faced in real-world scenarios.

Furthermore, a thorough examination of 6000 defect-free images from 50 wheelsets revealed 83 instances of false alarms, originating from 6 different wheelsets. According to statistical analysis based on the images, the calculated FAR was 1.38% (83 out of 6000). However, in actual application scenarios, the observed FAR was 12% (6 out of 50). These results emphasize the complexity and intricacies encountered in practical industrial applications, where achieving optimal detection indicators proves to be a more demanding task.

Last but not least, the efficiency of this method aligns with the demands of real-time collection and detection. In typical scenarios, a wheel completes one full rotation in about one minute. With a phased array probe utilizing 120 channels for data collection, the algorithm achieves a detection speed of approximately 2 s per wheel. Considering factors such as software response and data loading time, on-site application can accomplish the detection of a wheel and provide defect location information within 10 s, significantly diminishing the expenses associated with manual inspection.

5. Discussion

In this study, the application of the UT-YOLO method to ultrasonic inspection data of railway wheels was inspected, with a focus on detecting various types of defects. The evaluation of results underscores the superiority of this method, emphasizing its reliability both in field applications and on site. Among the models employed for object detection, the proposed UT-YOLO model exhibits significant advancements compared to several benchmark models. The outcomes indicate that UT-YOLO achieved a best recall of 0.94, an mAP@0.5 of 0.89, and an mAP@0.5:0.9 of 0.64, surpassing baseline YOLOv5s by 37%, 36%, and 42%, respectively. Particularly noteworthy is UT-YOLO's substantial superiority over other comparative algorithms in terms of mAP@0.5:0.9. Moreover, UT-YOLO demonstrated the fastest speed among the evaluated models. This is attributed to its unique features, such as an added small object detection layer, an attention mechanism module, data enhancement, and data preprocessing, specially designed for ultrasonic wheel b-scan image defect detection.

While the current study has effectively addressed challenges in ultrasonic wheel detection, especially in regard to speed and accuracy, achieving promising outcomes in terms of efficiency and practical applicability, it is crucial to acknowledge the diverse nature of real-world defects. Despite this success, certain limitations may arise in addressing specific prevalent issues. In particular, some false alarms caused by external factors such as electrical interference are still relatively serious, so the data collection process is very dependent on the stability of the equipment because the surface conditions of the wheels are not the same, which will cause the data to be easily invalidated, and a large amount of noise, especially electrical noise, is generated at the same time. However, to avoid excessive investment, more targeted distinctions will be made on the algorithm side to eliminate some special interferences. This will also be reflected in future work. Therefore, future endeavors will prioritize an in-depth exploration of actual wheel inspection equipment and the quality of inspection data. These tasks will involve a meticulous classification of various defects, aiming to enhance this methodology's performance across a broader spectrum of real-life scenarios.

6. Conclusions

The study presents a noteworthy breakthrough in defect detection methods utilizing ultrasound B-scan images obtained through the application of UT-YOLO. The UT-YOLO model incorporates specific modifications that significantly enhance the accuracy of defect detection, particularly in the context of B-scan images of ultrasonic defects. The notable enhancements include the integration of small-object detection layers, residual structures for feature fusion, attention mechanisms, and specialized processing methods. Collectively, these adjustments culminate in a robust model poised to address the challenges inherent in wheel defect detection in real-world ultrasonic B-scan images. When compared with the baseline YOLOv5s model, UT-YOLO demonstrated marked performance improvements. In the wheel defect dataset test results, it achieves noteworthy enhancements of 37%, 36%, and 43% in best recall, mAP@0.5, and mAP@0.5:0.95, respectively. Simultaneously, the speed of UT-YOLO reached an impressive 69FPS, indicating its capacity to operate effectively in real-time applications and making it well suited for deployment in practical scenarios.

Despite the UT-YOLO algorithm's notable successes in UT defect detection, there are still some areas that may be further investigated and refined. To further lower the false alarm rates and missed alarm rates, future work will concentrate on distinguishing the noise and the noise-like defects as well as the whole automatic process in train wheel defect detection to achieve truly intelligent detection.

Author Contributions: Conceptualization, Q.Z.; methodology, Q.Z.; software, Q.Z. and K.T.; validation, Q.Z., K.T. and J.P.; formal analysis, Q.Z.; investigation, Q.Z. and K.T.; resources, A.W. and J.P.; data curation, Q.Z., K.T. and J.P.; writing—original draft preparation, Q.Z.; writing—review and editing, J.P.; visualization, Q.Z. and K.T.; supervision, J.P., J.L. and X.G.; project administration, J.P., J.L. and X.G.; funding acquisition, J.P. All authors have read and agreed to the published version of the manuscript.

Funding: This work was supported in part by the National Nature Science Foundation of China (No. 61771409) and Science and Technology Project of Sichuan (No. 2021YJ0080).

Institutional Review Board Statement: Not applicable.

Informed Consent Statement: Not applicable.

Data Availability Statement: Data are contained within the article.

Acknowledgments: We appreciate the hard work of every author, the valuable suggestions of every reviewer, and the patience and meticulousness of the academic editors.

Conflicts of Interest: Author Ai wang was employed by the company Chengdu Lead Science & Technology. The remaining authors declare that the research was conducted in the absence of any commercial or financial relationships that could be construed as a potential conflict of interest.

References

1. Zerbst, U.; Mädler, K.; Hintze, H. Fracture mechanics in railway applications-An overview. *Eng. Fract. Mech.* **2005**, *72*, 163–194. [CrossRef]
2. Dwivedi, S.K.; Vishwakarma, M.; Soni, A. Advances and researches on non destructive testing: A review. *Mater. Today Proc.* **2018**, *5*, 3690–3698. [CrossRef]
3. Drinkwater, B.W.; Wilcox, P.D. Ultrasonic arrays for non-destructive evaluation: A review. *NDT E Int.* **2006**, *39*, 525–541. [CrossRef]
4. Kasai, N.; Takada, A.; Fukuoka, K.; Aiyama, H.; Hashimoto, M. Quantitative investigation of a standard test shim for magnetic particle testing. *NDT E Int.* **2011**, *44*, 421–426. [CrossRef]
5. Roshan, C.C.; Raghul, C.; Ram, H.V.; Suraj, K.; Solomon, J. Non-destructive testing by liquid penetrant testing and ultrasonic testing—A review. *Int. J. Adv. Res. Ideas Innov. Technol.* **2019**, *5*, 694–697.
6. Naeimi, M.; Li, Z.; Qian, Z.; Zhou, Y.; Wu, J.; Petrov, R.H.; Sietsma, J.; Dollevoet, R. Reconstruction of the rolling contact fatigue cracks in rails using X-ray computed tomography. *NDT E Int.* **2017**, *92*, 199–212. [CrossRef]
7. Kappes, W.; Hentschel, D.; Oelschlägel, T. Potential improvements of the presently applied in-service inspection of wheelset axles. *Int. J. Fatigue* **2016**, *86*, 64–76. [CrossRef]
8. Mineo, C.; Cerniglia, D.; Pantano, A. Numerical study for a new methodology of flaws detection in train axles. *Ultrasonics* **2014**, *54*, 841–849. [CrossRef] [PubMed]
9. Le, M.; Jun, J.; Kim, J.; Lee, J. Nondestructive testing of train wheels using differential-type integrated Hall sensor matrixes embedded in train rails. *NDT E Int.* **2013**, *55*, 28–35. [CrossRef]
10. Montinaro, N.; Epasto, G.; Cerniglia, D.; Guglielmino, E. Laser ultrasonics inspection for defect evaluation on train wheel. *NDT E Int.* **2019**, *107*, 102145. [CrossRef]
11. Cavuto, A.; Martarelli, M.; Pandarese, G.; Revel, G.M.; Tomasini, E.P. Train wheel diagnostics by laser ultrasonics. *Measurement* **2016**, *80*, 99–107. [CrossRef]
12. Lindeberg, T. Scale invariant feature transform. *Scholarpedia* **2012**, *7*, 10491. [CrossRef]
13. Dalal, N.; Triggs, B. Histograms of oriented gradients for human detection. In Proceedings of the IEEE Conference on Computer Vision and Pattern Recognition, San Diego, CA, USA, 20–25 June 2005; pp. 886–893.
14. Rublee, E.; Rabaud, V.; Konolige, K.; Bradski, G. ORB: An efficient alternative to SIFT or SURF. In Proceedings of the 2011 International Conference on Computer Vision, Barcelona, Spain, 6–13 November 2011; pp. 2564–2571.

15. Rainer, L.; Maydt, J. An extended set of Haar-like features for rapid object detection. In Proceedings of the International Conference on Image Processing, Rochester, NY, USA, 22–25 September 2002; Volume 1.
16. Girshick, R.; Donahue, J.; Darrell, T.; Malik, J. Rich feature hierarchies for accurate object detection and semantic segmentation. In Proceedings of the IEEE Conference on Computer Vision and Pattern Recognition, Columbus, OH, USA, 23–28 June 2014; pp. 580–587.
17. He, K.; Zhang, X.; Ren, S.; Sun, J. Spatial pyramid pooling in deep convolutional networks for visual recognition. *IEEE Trans. Pattern Anal. Mach. Intell.* **2015**, *37*, 1904–1916. [CrossRef] [PubMed]
18. Ren, S.; He, K.; Girshick, R.; Sun, J. Faster R-CNN: Towards real-time object detection with region proposal networks. In Proceedings of the Advances in Neural Information Processing Systems 28 (NIPS 2015), Montreal, QC, Canada, 7–12 December 2015; Volume 28.
19. He, K.; Gkioxari, G.; Dollár, P.; Girshick, R. Mask R-CNN. In Proceedings of the IEEE International Conference on Computer Vision, Venice, Italy, 22–29 October 2017; pp. 2961–2969.
20. Liu, Z.; Tang, R.; Duan, G.; Tan, J. TruingDet: Towards high-quality visual automatic defect inspection for mental surface. *Opt. Lasers Eng.* **2021**, *138*, 106423. [CrossRef]
21. Cha, Y.J.; Choi, W.; Suh, G.; Mahmoudkhani, S.; Büyüköztürk, O. Autonomous structural visual inspection using region-based deep learning for detecting multiple damage types. *Comput.-Aided Civ. Infrastruct. Eng.* **2018**, *33*, 731–747. [CrossRef]
22. He, Y.; Song, K.; Meng, Q.; Yan, Y. An end-to-end steel surface defect detection approach via fusing multiple hierarchical features. *IEEE Trans. Instrum. Meas.* **2019**, *69*, 1493–1504. [CrossRef]
23. Redmon, J.; Divvala, S.; Girshick, R.; Farhadi, A. You only look once: Unified, real-time object detection. In Proceedings of the IEEE Conference on Computer Vision and Pattern Recognition, Las Vegas, NV, USA, 27–30 June 2016; pp. 779–788.
24. Liu, W.; Anguelov, D.; Erhan, D.; Szegedy, C.; Reed, S.; Fu, C.Y.; Berg, A.C. SSD: Single shot multibox detector. In Proceedings of the Computer Vision–ECCV 2016: 14th European Conference, Amsterdam, The Netherlands, 11–14 October 2016; Proceedings, Part I 14. Springer International Publishing: Berlin/Heidelberg, Germany, 2016; pp. 21–37.
25. Redmon, J.; Farhadi, A. YOLO9000: Better, faster, stronger. In Proceedings of the IEEE Conference on Computer Vision and Pattern Recognition, Honolulu, HI, USA, 21–26 July 2017; pp. 7263–7271.
26. Redmon, J.; Farhadi, A. Yolov3: An incremental improvement. *arXiv* **2018**, arXiv:1804.02767.
27. Lin, T.Y.; Dollár, P.; Girshick, R.; He, K.; Hariharan, B.; Belongie, S. Feature pyramid networks for object detection. In Proceedings of the IEEE Conference on Computer Vision and Pattern Recognition, Honolulu, HI, USA, 21–26 July 2017; pp. 2117–2125.
28. Bochkovskiy, A.; Wang, C.Y.; Liao, H.Y.M. Yolov4: Optimal speed and accuracy of object detection. *arXiv* **2020**, arXiv:2004.10934.
29. Li, J.; Su, Z.; Geng, J.; Yin, Y. Real-time detection of steel strip surface defects based on improved yolo detection network. *IFAC-PapersOnLine* **2018**, *51*, 76–81. [CrossRef]
30. Zhang, C.; Chang, C.C.; Jamshidi, M. Concrete bridge surface damage detection using a single-stage detector. *Comput.-Aided Civ. Infrastruct. Eng.* **2020**, *35*, 389–409. [CrossRef]
31. Zhang, G.; Cui, K.; Hung, T.-Y.; Lu, S. Defect-GAN: High-fidelity defect synthesis for automated defect inspection. In Proceedings of the IEEE/CVF Winter Conference on Applications of Computer Vision, Waikoloa, HI, USA, 3–8 January 2021; pp. 2524–2534.
32. Valente, A.; Wada, C.; Neves, D.; Neves, D.; Pere, F.V.M.; Megeto, G.A.S.; Cascone, M.H.; Gomes, O.; Lin, Q. Print defect mapping with semantic segmentation. In Proceedings of the IEEE/CVF Winter Conference on Applications of Computer Vision, Snowmass, CO, USA, 1–5 March 2020; pp. 3551–3559.
33. Liu, Z.; Lin, Y.; Cao, Y.; Hu, H.; Wei, Y.; Zhang, Z.; Lin, S.; Guo, B. Swin transformer: Hierarchical vision transformer using shifted windows. In Proceedings of the IEEE/CVF International Conference on Computer Vision, Montreal, QC, Canada, 10–17 October 2021; pp. 10012–10022.
34. He, K.; Zhang, X.; Ren, S.; Sun, J. Deep residual learning for image recognition. In Proceedings of the IEEE Conference on Computer Vision and Pattern Recognition, Las Vegas, NV, USA, 27–30 June 2016; pp. 770–778.
35. Tan, M.; Pang, R.; Le, Q.V. Efficientdet: Scalable and efficient object detection. In Proceedings of the IEEE/CVF Conference on Computer Vision and Pattern Recognition, Seattle, WA, USA, 13–19 June 2020; pp. 10781–10790.
36. Yang, L.; Zhang, R.Y.; Li, L.; Xie, X. Simam: A simple, parameter-free attention module for convolutional neural networks. In Proceedings of the International Conference on Machine Learning, Virtual Event, 18–24 July 2021; pp. 11863–11874.
37. Zhang, Y.; Tan, Y.; Peng, J.; Peng, C.; Yang, K.; Gao, X. LU automatic ultrasonic inspection system for in-service wheelset and its application. In Proceedings of the 2016 18th International Wheelset Congress (IWC), Chengdu, China, 7–10 November 2016; pp. 100–104.

sensors

MDPI

Article

Context-Enhanced Network with Spatial-Aware Graph for Smartphone Screen Defect Detection

Aili Liang [1], Qishan Wang [2] and Xiaofeng Wu [1,*]

[1] School of Information Science and Technology, Fudan University, Shanghai 200433, China; 21210720031@m.fudan.edu.cn
[2] Academy for Engineering & Technology, Fudan University, Shanghai 200433, China; qswang20@fudan.edu.cn
* Correspondence: xiaofengwu@fudan.edu.cn; Tel.: +86-021-6564-3757

Abstract: Interactive devices such as touch screens have gained widespread usage in daily life; this has directed the attention of researchers to the quality of screen glass. Consequently, defect detection in screen glass is essential for improving the quality of smartphone screens. In recent years, defect detection methods based on deep learning have played a crucial role in improving detection accuracy and robustness. However, challenges have arisen in achieving high-performance detection due to the small size, irregular shapes and low contrast of defects. To address these challenges, this paper proposes CE-SGNet, a Context-Enhanced Network with a Spatial-aware Graph, for smartphone screen defect detection. It consists of two novel components: the Adaptive Receptive Field Attention Module (ARFAM) and the Spatial-aware Graph Reasoning Module (SGRM). The ARFAM enhances defect features by adaptively extracting contextual information to capture the most relevant contextual region of defect features. The SGRM constructs a region-to-region graph and encodes region features with spatial relationships. The connections among defect regions are enhanced during the propagation process through a graph attention network. By enriching the feature representations of defect regions, the CE-SGNet can accurately identify and locate defects of various shapes and scales. Experimental results demonstrate that the CE-SGNet achieves outstanding performance on two public datasets.

Keywords: defect detection; smartphone screen; attention mechanism; graph reasoning

1. Introduction

The development of modern smart devices has brought immense convenience to people's lives. As a fundamental component of smartphones, the touchscreen has attracted considerable attention regarding its quality. The quality of the touchscreen plays a crucial role in determining the lifespan, user experience and display performance of the smartphone [1]. The smartphone screen is manufactured from glass material, exhibiting favorable mechanical and optical properties. The glass screen for smartphones is produced through a series of specialized processes, including precision Computer Numerical Control (CNC) glass cutting, shaping, polishing, hardening, ultrasonic cleaning, vacuum coating, silk-screen printing, anti-fingerprint oil application, drying, cleaning and inspection [2]. The entire process imposes strict requirements on the production environment. However, during the processes of production and transportation, various defects invariably arise, including scratches, cracks, breakages, spots and so on. Such defects affect the aesthetic appeal of smartphone screens and lead to adverse effects on their production quality. Therefore, defect detection on smartphone screens is a pivotal step in the industrial process, aiming to avoid process waste, reduce production costs and ensure product quality.

Currently, smartphone screen defects are primarily detected through traditional manual inspection or machine vision methods, both of which have certain limitations. Traditional manual inspection, conducted by experienced inspectors under strong lighting, consumes significant time and energy, and may lead to vision impairment. This approach struggles to meet the demands of large-scale industrial production, hindering automation and intelligence in the production process. On the other hand, machine vision methods rely on handcrafted feature-extraction techniques. Subsequently, traditional machine learning algorithms are employed for defect detection [3,4]. The machine vision methods overcome the drawbacks of traditional manual inspection, but their reliance on manual rules or thresholds often results in an elevated occurrence of false negatives, particularly for small defects. Additionally, they have limitations in stability, accuracy and adaptability to varying environments.

With the rapid progress in computer vision technologies, the emergence of deep learning, notably convolutional neural networks (CNNs), has opened up new avenues for smartphone screen defect detection. CNNs extract features from original data without handcrafted feature extractors, offering expressive feature representation. The classical object detectors based on CNNs can be broadly categorized into two types: two-stage detectors like Faster RCNN [5] and Cascade RCNN [6], and one-stage detectors such as YOLO [7], RetinaNet [8] and FCOS [9]. The primary distinction lies in the methodology: two-stage detectors generate region proposals (RPs) for further classification and regression, while one-stage detectors predict classification and position directly from the features. In contrast, two-stage detectors typically achieve higher detection accuracy.

General detectors are often designed for natural images and may not achieve optimal performance when applied to the specific task of smartphone screen defect detection. Compared to natural images, smartphone screen defect images exhibit unique characteristics. Firstly, screen defects tend to be small and irregular, such as spots and scratches, which leads to limited feature information and a high false-negative rate. Secondly, defects like scratches and cracks exhibit minimal color contrast against the screen background. Smartphone screen images are typically captured by high-resolution line-scan cameras, with a black background to minimize interference from ambient light. Despite these efforts, the features of defects still do not stand out prominently, presenting a significant challenge to detection. Some examples are shown in Figure 1. These unique characteristics emphasize the necessity for tailored methods to improve the performance of defect detection on smartphone screens.

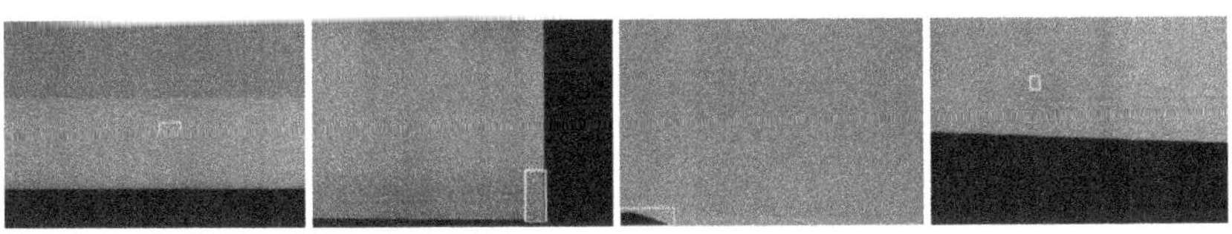

Figure 1. Some examples of smartphone screen defect images. The categories of defects include scratch, crack, breakage and spot.

Several detection methods for screen defects have been proposed to enhance defect detection performance. Wang et al. [10] applied Faster RCNN to detect smartphone screen defects. Faced with the challenge of a limited number of samples, they employed Boundary Equilibrium Generative Adversarial Networks (BEGAN) to generate and enhance defect data, achieving better performance compared to machine vision methods. Chen et al. [11] proposed a defect detection network based on the fusion of Faster RCNN, and they employed U-Net [12] and ResNet50 [13] for feature extraction. U-Net is employed to capture detailed information about defects, while ResNet50 extracts high-level semantic information. The combination of both can obtain more distinctive feature information, effectively detecting three common surface defects on screens. However, the fusion of two feature-extraction networks led to higher complexity of the model. Some works have introduced attention mechanisms to enhance the network's focus on defect features. Zhang et al. [14] proposed a category-aware network for detecting edge cracks on smartphone screens. They employed multiple parallel feature-extraction branches to capture features at different scales and adjusted channel weights through a category-aware attention mechanism. Zhu et al. [15] introduced a channel attention mechanism for detecting scratches on smartphone screens. They used high-frequency information and local cross-channel interactions to represent weighted defect features, reducing feature loss during global average pooling and enhancing the feature of tiny defects. Additionally, they employed Region of Interest (RoI) Align to handle scratches with extreme aspect ratios. However, these methods [14,15] only focus on channel attention, neglecting spatial attention. Attention to spatial positions is crucial because it determines the accuracy of localization in defect detection.

In the task of smartphone screen defect detection, previous defect detection methods face two prevalent challenges: (1) In defect images, small defects lead to an imbalance in the distribution of normal and defective pixels, which causes the network to excessively focus on the background, resulting in insufficient attention to information of defect features. (2) Most defect detection methods focus on individual regions, neglecting the relationship among defects and the implicit prior knowledge embedded in the image.

To overcome the two challenges, this paper proposes a defect detection method including an attention mechanism and spatial-aware graph reasoning, named CE-SGNet. CE-SGNet introduces two pivotal modules: the Adaptive Receptive Field Attention Module (ARFAM) and the Spatial-aware Graph Reasoning Module (SGRM). The ARFAM is applied to perform attention-weighting on the fused multi-scale features by employing channel attention and spatial selection attention. It dynamically captures diverse contextual information necessary for various defects, enhancing the expressive capability of defect features. In the SGRM, proposals are employed to construct a region-to-region relation graph under category semantic supervision. The CE-SGNet learns adjacency relationships among proposals autonomously, and propagates the updated features globally. Finally, the enhanced feature of each region is concatenated with the original RoI feature for classification and localization.

In summary, the major contributions of this paper can be summarized as follows:

1. A novel attention module called ARFAM is proposed, which integrates channel attention and spatial selection attention. The channel attention focuses on the importance of different channel features, while the spatial selection attention dynamically adjusts the requirements of diverse contextual information for different defect types, thereby enhancing the model's ability to express features.
2. The SGRM introduces the Spatial-aware Graph Attention Network (SGAT) to model the relationships among regions. A learnable relation graph is constructed to facilitate adaptive reasoning for global features, effectively propagating spatial and contextual information.
3. Experiments conducted on a real-world dataset of smartphone screen defects illustrate the effectiveness of the proposed method in defect detection. Furthermore, our method is evaluated on a public defect dataset to demonstrate its generalization capability.

The remainder of this paper is organized as follows. Section 2 describes the related works. Section 3 presents the proposed method. Section 4 shows the experimental results, analysis of every module, visualization study, generalization study, model complexity and time-efficiency evaluation. Section 5 discusses the limitations of the proposed method and the future research directions. Section 6 presents the conclusion.

2. Related Works

2.1. General Object Detectors

Object detection is a crucial task in computer vision. In recent years, there has been significant progress in object detection methods based on deep learning. Generally, these methods can be divided into two categories: two-stage detectors and one-stage detectors.

Two-stage detectors, represented by the RCNN series, such as Faster RCNN [5] and Cascade RCNN [6], first propose candidate regions and then perform classification/regression. Taking Faster RCNN [5] as an example, it generates RPs through the Region Proposal Network (RPN) firstly and subsequently employs RoI pooling to map these proposals of various sizes to a fixed-size feature map for accurate classification and localization. Cascade RCNN [6] improves detection performance by cascading multiple detection heads to refine proposals. Compared to two-stage detectors, one-stage detectors directly predict classification and location without RoI proposals. For example, YOLO [7] directly predicts classification and location by generating dense anchor boxes on the input image. RetinaNet [8] introduces a loss function called Focal Loss to address class imbalance in object detection. FCOS [9] is an anchor-free detector that predicts the points of objects to form bounding boxes. In comparison, two-stage detectors achieve higher detection accuracy through refined design, which makes them more suitable for high-accuracy scenarios.

In recent years, there has been an increase in methods utilizing Transformer for detection. Carion et al. [16] introduced DETR, a pioneering method that applies Transformer structure to object detection tasks. DETR formulates object detection as a set prediction problem, achieving end-to-end detection. To solve the slow convergence and poor performance in detecting small objects in DETR, Deformable DETR [17] introduced a deformable attention mechanism to flexibly adapt to the deformation and changes in the object shapes. However, these detectors often require significant computational resources and are easy to overfit, particularly when dealing with small datasets such as defect detection datasets.

2.2. Attention Mechanisms

Attention mechanisms simulate the function of the human visual system, enabling neural networks to focus on crucial parts of an image, thus improving the model's performance and generalization capability. Generally, attention mechanisms include channel attention, spatial attention and self-attention. SENet [18] is a classic attention mechanism network that learns channel weights through global pooling layers, enhancing the importance of different channels in feature maps. Expanding on the foundation of channel attention, CBAM [19] introduces spatial attention, dynamically adjusting weights for both channels and spatial positions in feature maps to enhance overall model performance. SKNet [20] introduces a selective kernel module responsible for selectively applying convolutional kernels of different sizes on each feature map, better capturing multi-scale information in the image. Additionally, the self-attention mechanism in Transformer [21] allows each input position to interact with others, generating corresponding attention weights, which enables the model to capture long-range dependencies in sequential data. However, this method operates on the entire feature map, demanding significant computational resources. Our attention module draws inspiration from the CBAM structure, but we refine the spatial attention module through extracting adaptive contextual information for various defects. This makes it more suitable for smartphone screen defects.

2.3. Graph Neural Networks

Extensive research has been carried out in diverse visual domains to explore relation reasoning through objects. In recent works, objects and their relationships are frequently represented as graph structures, leveraging Graph Neural Networks (GNNs) [22] for their flexibility in modeling dependencies between nodes, particularly in scenarios involving irregular data structures. Graph Convolutional Networks (GCNs) [23] extend CNNs to graph-structured data, enabling nodes to focus on semantic or spatial neighborhood features. Several works have applied graph structures to defect detection. In work [24], structured knowledge graphs are integrated into a learning system to enhance the accuracy and reliability of additive manufacturing defect diagnosis. Wang et al. [25] addressed the issues of inter-class similarity and intra-class variability in defect detection by constructing a class-balanced graph. They applied this approach in domains such as steel and textiles. Zhai et al. [26] focus on insulator defect detection, constructing both a multi-geometry reasoning network and an appearance-geometry reasoning network to extract appearance and geometry features of defect samples, respectively. However, these works ignore the spatial relationships between defect regions and implicit information in images. Therefore, we construct a region-to-region graph and employ Graph Attention Networks (GATs) [27] for the propagation and updating of region features.

3. Method

3.1. Overview

The proposed architecture is illustrated in Figure 2. The two-stage detector Faster RCNN [5] with FPN [28] is employed as the baseline. Initially, the feature maps of defect image are extracted through the backbone, and then passed through the FPN for feature fusion. The fused feature maps are processed by the ARFAM for adaptive receptive field attention weighting. After extracting RPs through the RPN and RoI Align, the SGRM utilizes these proposals to construct a region-to-region graph guided by the previous classification results. The graph node embeddings of different regions are evolved and propagated by SGAT. Finally, the enhanced feature of each region is concatenated with the original feature for classification and localization.

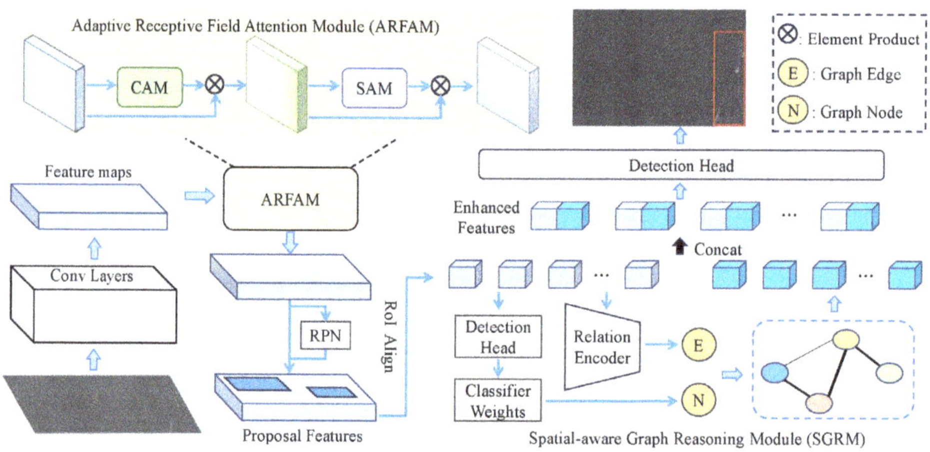

Figure 2. The overall architecture of proposed CE-SG defect detection network. It consists of two innovative modules: the Adaptive Receptive Field Attention Module and the Spatial-aware Graph Reasoning Module.

3.2. Adaptive Receptive Field Attention Module

In the task of smartphone screen defect detection, there are numerous defects with low contrast and poor saliency. During the detection process, these defects may be considered background noise, leading to missed detection. Moreover, the pixel area of defects in the image is much smaller than that of the background, leading to the network paying excessive attention to the background instead of foreground. In computer vision, attention mechanisms are widely applied, effectively improving the performance of networks. Therefore, we propose the ARFAM to enhance defects' features and obtain wide contextual information, as illustrated in Figure 3.

The ARFAM consists of a Channel Attention Module (CAM) and a Spatial Attention Module (SAM). The CAM learns channel feature weights, enhancing the features of relevant channels based on these weights while suppressing irrelevant channel features. Firstly, the input feature map F_{in} undergoes separate weighting through both global average pooling (GAP) and global maximum pooling (GMP). Afterward, the features generated by the MLP layers are subjected to element-wise summation, followed by sigmoid activation to produce the channel attention map $M_c(F_{in})$. The formula is as follows:

$$M_c(F_{in}) = \sigma(MLP(GAP(F_{in})) + MLP(GMP(F_{in}))), \quad (1)$$

where σ is the sigmoid activation function. MLP is a two-layer neural network with the first layer having N_c/r neurons, activated by ReLU, and the second layer having N_c neurons, where r is the reduction ratio and N_c is the number of neurons. The channel attention feature map F_c is combined with F_{in} as $F_c = F_{in} \times M_c(F_{in})$, generating the input features for the SAM.

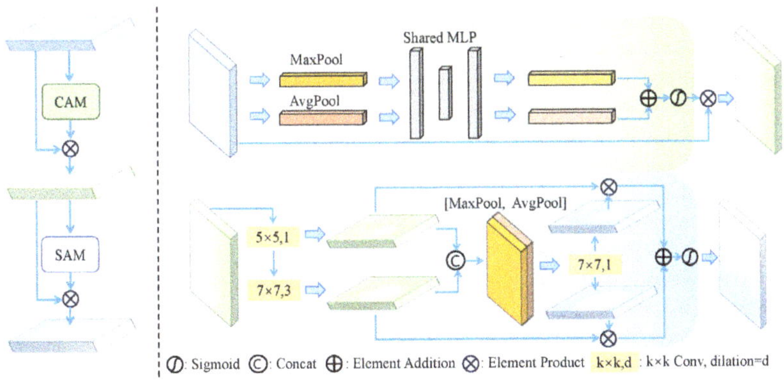

Figure 3. Illustration of the Adaptive Receptive Field Attention Module.

The SAM adaptively adjusts the receptive field of the channel-weighted feature map, effectively handling diverse contextual information for different defects. In defect detection, extensive contextual information can help the network localize defects effectively, and the range of contextual information varies for different types of defects. Therefore, it is necessary for the network to dynamically adapt the range of defect contextual information. According to [29], large kernel convolution can increase the receptive field of the image, thereby obtaining more extensive contextual information. Therefore, a series of depth-wise convolutions with different kernels are applied to construct larger kernel convolutions, which can control the range of contextual information for different types of defects. The receptive field of the i-th layer RF_i can be represented as

$$RF_i = RF_{i-1} + d_i(k_i - 1), \quad (2)$$

where d_i is the dilation rate and k_i is the kernel size. The feature maps from different convolutional kernels can be represented as:

$$F_{s1} = f_1^{dw}(F_c), F_{s2} = f_2^{dw}(F_{s1}), \quad (3)$$

where $f_i^{dw}(\cdot)$ are depth-wise convolutions. Then they are concatenated as $F' = [F_{s1}; F_{s2}]$, and F' is spatially selected through channel-based maximum pooling and average pooling. Finally, a convolution layer is applied to generate spatial selection weighting coefficients from the pooled feature maps:

$$C_s = f^{7\times7}([AvePool(F'); MaxPool(F')]). \quad (4)$$

The spatial attention map is generated by weighting each receptive field feature map, followed by sigmoid activation:

$$M_s(F) = \sigma(\sum_i C_{si} \times F_{si}). \tag{5}$$

The SAM enables adaptive selection of contextual information for different defects, enabling the network to capture spatial contextual regions most relevant to the defect features.

The ARFAM dynamically captures the long-range dependency of defect features through adjusting the range of the receptive field, enhances the network's attention to defect features and enables more accurate localization. Generally, attention mechanisms are employed in the convolution layers for extracting features. However, we extrapolate that it may still introduce noise during feature fusion due to the inconsistency of features across different scales, thereby affecting detection results. Therefore, the ARFAM is inserted after FPN in this paper and adaptive weighting is conducted individually for feature maps at different scales. The experiments in Section 4.3.3 validate its effectiveness.

3.3. *Spatial-Aware Graph Reasoning Module*

Smartphone screen defects, caused by mechanical reasons, often tend to occur in specific locations with distinct types or textures. For instance, cracks often occur in edge regions, and some defects are typically incomplete. In addition, downsampling in CNNs may lead to the loss of features in small defects, posing challenges for detection. Previous works usually relied on every RP for classification and regression. However, independent RPs contain limited feature information, which poses challenges for detection. We hope to leverage information within the images to enhance low-quality features. In order to enhance the representation of various defects and their interrelations, a graph network is proposed to model the spatial relationships between defect regions and enhance the RoI features. Figure 4 shows the flowchart of the SGRM.

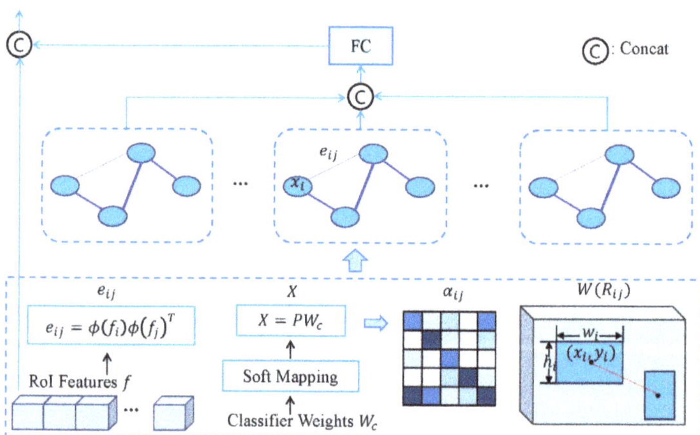

Figure 4. Flowchart of the Spatial-aware Graph Reasoning Module.

3.3.1. Spatial-Aware Graph Building

In the initial graph representation as $G = < \mathcal{V}, \mathcal{E} >$, where $\mathcal{V}$ represents node features, and each node represents each RP, each edge $e_{i,j} \in \mathcal{E}$ encodes the relationship between two regions. Some works use handcraft linguistic knowledge to build graphs, which may ignore the spatial relationships. Due to the gap between visual and language domains, we refrain from introducing explicit linguistic knowledge. Instead, the relationship matrix between regions is learned through RoI features F, i.e., $e_{ij} = \phi(f_i)\phi(f_j)^T$, where $\phi(\cdot)$ denotes a non-linear function. In this paper, $\phi(\cdot)$ is obtained through two fully connected layers with ReLU activation, and $f_i \in F$ represents the feature of RP i.

The regions proposed by RoI Align are independent of each other and exhibit low correlation between regions. Furthermore, these proposals may not accurately depict defect information. When a region proposal is a negative sample, it can propagate incorrect information. Hence, utilizing proposal features directly as graph node feature embeddings not only affects detection results but also introduces noise. By guiding the regions with advanced semantic information, the initial feature embeddings of graph node can be aligned more closely with the actual region features. This facilitates a more accurate modeling of relationships among defects and enhances the network's learning capability. Inspired by the work in [30], the previous classifier weights $\mathbf{W}_c \in \mathbb{R}^{C\times(D+1)}$ are employed, which contain advanced semantic information in the graph node embedding. The weights can be obtained by copying the parameters with the bias from the previous classification layer, updated throughout network training. To build a region-to-region graph, it is necessary to map category embeddings to region feature representations of nodes. While the previous classifier scores could achieve this mapping, incorrect classification results could lead to erroneous node feature representations. Therefore, we propose a soft mapping denoted by

$$p_i = \frac{\exp(s_{ij})}{\sum_j \exp(s_{ij})}, \tag{6}$$

where s_{ij} represents the score of RP i being classified into category j, $p_i \in \mathbf{P} \in \mathbb{R}^{N\times C}$ is the mapping weight coefficient. The node feature embedding can be computed as $\mathbf{X} = \mathbf{P}\mathbf{W}_c$, where $\mathbf{X} \in R^{N\times(D+1)}$, N is the number of nodes, C is the number of categories, and D is the dimension of the graph node features. Employing classification weights as supervision can improve the correlation among similar defects and accentuate the distinctions between different defect categories. This facilitates more accurate propagation of node feature information.

3.3.2. Spatial-Aware Graph Reasoning

To construct a graph with rich spatial information, the relative positional relationship between defects is crucial. Therefore, the spatial relationship between region i and region j is formulated as

$$R_{ij} = \left(log(\frac{x_i - x_j}{w_i}), log(\frac{y_i - y_j}{h_i}), log(\frac{w_j}{w_i}), log(\frac{h_j}{h_i}) \right)^T, \tag{7}$$

where (x_i, y_i) represents the central coordinates of region i, and (w_i, h_i) denotes the width and height of region i, which are normalized. The feature will be mapped to a high-dimensional space to encode spatial information.

The SGAT is employed for updating the graph, which allows the selection of more relevant nodes based on their importance, implicitly assigning different weights to nodes in the neighborhood. The attention mechanism is used to autonomously learn and optimize the connection relationships between graph nodes. The update formula for nodes is as follows:

$$\mathbf{X}^{l+1} = \sigma\left(\sum_{j\in N_i} \alpha_{ij}\mathbf{W}^l(R_{ij})\mathbf{X}^l \right), \tag{8}$$

where $\mathbf{X}^l$ represents the node feature embedding for the l-th layer of the SGAT, $\mathbf{W}^l(R_{ij})$ is a learnable weight matrix encoding spatial position information of the nodes. α_{ij} represents the normalized spatial relation coefficient, computed as the weighted sum of node feature representations:

$$\alpha_{ij} = \frac{\exp(\text{LeakyReLU}(\mathbf{a}^T[\mathbf{W}X_i||\mathbf{W}X_j]))}{\sum_{l\in\mathcal{N}_i} \exp(\text{LeakyReLU}(\mathbf{a}^T[\mathbf{W}X_i||\mathbf{W}X_l]))}, \tag{9}$$

where $\mathbf{X} = \{X_1, X_2, \cdots, X_N\}$, N is the number of regions, $||$ denotes the concatenation operator and $\mathbf{a}^T$ represents the attention coefficients. Multi-head attention layers are used to propagate spatial relations among nodes, and the propagation process of the network is defined as:

$$F_{enc} = \mathop{||}_{k=1}^{K} \sigma\left(\sum_{j \in \mathcal{N}_i} \alpha_{ij}^k \mathbf{W}^l (R_{ij}) \mathbf{X}^l \right), \tag{10}$$

where K is the number of multi-head attention layers. Finally, the enhanced features F_{enc} generated by SGAT are dimensionally mapped and concatenated with the original features F for classification and localization in the detection head.

In the SGRN, advanced semantic features and implicit information in the images guide the construction of the graph and the propagation of node features. In this process, proposal features with semantic similarities strengthen each other, while those with significant differences are suppressed, reducing the positional offset between bounding boxes and ground truth. It is beneficial to capture complex relationships between regions globally, aiding the network in understanding more intricate scenes, such as dense distributions.

4. Experimental Results and Analysis

4.1. Experimental Setup

4.1.1. Datasets

The smartphone screen glass defect (SSGD) dataset [31] is a popular public dataset designed for smartphone screen glass defect detection tasks. It comprises 2504 images, each with dimensions of 1500 × 1000 pixels. It includes seven common defect categories: crack, broken, spot, scratch, light-leakage, blot and broken membrane. All images are captured from real-world scenarios, providing a comprehensive representation of potential screen defects with a certain level of complexity. The dataset is challenging, and some examples are shown in Figure 1. The dataset comprises two sub-datasets, namely LB101 and LB201, collected from different working platforms. These sub-datasets exhibit variations in defect sample distribution, types of screens and screen background colors. We validate the effectiveness of the proposed method on each of the two subsets separately. The dataset is randomly separated into a training and a testing set with a ratio of 4:1.

To prove the generalization of the proposed method, the printed circuit board (PCB) dataset released by Peking University is introduced. It consists of 693 images with 6 types of defects: missing hole, mouse bite, open circuit, short, spur and spurious copper. The image resolutions range from 3056 × 2464 to 2904 × 1521. The dataset is randomly separated into a training and a testing set with a ratio of 4:1.

4.1.2. Evaluation Metrics

In the defect detection task, precision and recall are crucial metrics. The Precision–Recall (PR) curve illustrates the relationship between precision and recall, and the Average Precision (AP) is calculated as the area under this curve. AP combines precision and recall to evaluate the overall detection performance of the model across different thresholds. The mean Average Precision (mAP) is the average of AP values across all categories. We employ evaluation metrics following the COCO standard, including AP, AP_{50}, AP_{75}, AP_s, AP_m and AP_l. Here, AP represents the average precision at IoU thresholds ranging from 0.5 to 0.95 with intervals of 0.05. AP_{50} and AP_{75} represents average precision at IoU thresholds of 0.5 and 0.75, respectively. AP_s, AP_m and AP_l represent average detection precision for small (area $< 32^2$), medium ($32^2 <$ area $< 96^2$) and large (area $> 96^2$) objects, respectively. These metrics comprehensively evaluate the defect detection performance in general scenes, high-precision localization scenes and multi-scale detection capability.

4.1.3. Implementation Details

The CE-SGNet is constructed of the PyTorch framework and the experiments are performed on 8 NVIDIA GeForce RTX 3090 GPUs. Stochastic Gradient Descent (SGD) with a momentum of 0.9 and weight decay of 0.05 serves as the optimizer. The initial learning rate is set to 0.02, with a batch size of 8. The training strategy follows a 1× schedule (12 epochs), with learning rate reductions by a factor of 10 at the 8th and 11th epochs to prevent overfitting due to excessively large learning rates in the later iteration of training. All models are initialized with weights pre-trained on the COCO dataset, ensuring a solid starting point for defect detection. Random flips and multi-scale training are applied to images during training. To avoid potential experimental bias, the final evaluation metrics are obtained by averaging the results from five experiments.

4.2. Comparative Experiments

To assess the performance of the proposed method, the CE-SGNet is compared with nine representative detectors. These detectors include: two-stage typical detectors Faster RCNN [5] and Cascade RCNN [6], one-stage anchor-based detector RetinaNet [8], GFL [32] and YOLOv5 [33], anchor-free detector FCOS [9] and ATSS [34], transformer-based detectors DETR [16] and Deformable DETR [17]. YOLOv5 stands out as one of the most widely used one-stage detectors in the field of defect detection. It employs CSPDarkNet53 as the backbone and is trained for 100 epochs. To avoid parameter redundancy, the lightest model YOLOv5-s is selected for the experiment. DETR and Deformable DETR represent typical transformer-based detectors, utilized for performance comparison with CNN-based detectors. They are trained for 50 epochs using the Adam optimizer. All the other models use ResNet50 as the backbone and FPN as the neck. All methods are implemented based on MMDetection with official hyperparameter settings. The same preprocessing operations are applied, including multi-scale data augmentation and dataset splitting, to ensure a fair comparison.

The detailed results on the SSGD dataset are presented in Table 1. It is observed that the proposed method outperforms other methods in most metrics, demonstrating superior performance. Compared to the baseline model, the CE-SGNet achieves better performance in all metrics, which indicates the effectiveness of both ARFAM and SGRM. ARFAM enables the model to focus more on defects and their contextual information, enhancing the model's ability to learn defect features. Subsequently, due to the improved localization accuracy achieved by SGRM, CE-SGNet demonstrates robust performance across various IoU thresholds, making it suitable for both general scenarios and high-precision localization scenarios. In terms of multi-scale detection capability, it also achieves excellent performance. The performance improvement in both small and medium defects are higher than that in large defects, as CE-SGNet enriches the limited features in small and medium defects, thereby enhancing performance. In summary, the CE-SGNet improves detection performance across the board and does not sacrifice the performance of one metric to improve another metric.

Table 1. Comparison results of different methods on the SSGD dataset. Bold data denote the best performance. Underlined data denote the second best performance.

Model	Backbone	LB101						LB201					
		AP	AP_{50}	AP_{75}	AP_s	AP_m	AP_l	AP	AP_{50}	AP_{75}	AP_s	AP_m	AP_l
Faster RCNN [5]	ResNet50	20.3	<u>47.3</u>	<u>17.0</u>	13.4	<u>28.2</u>	34.4	20.4	43.8	16.4	20.7	18.7	22.1
Cascade RCNN [6]	ResNet50	<u>21.0</u>	44.2	16.8	15.1	25.3	<u>41.0</u>	<u>22.2</u>	<u>47.1</u>	<u>18.4</u>	22.3	19.3	**23.8**
RetinaNet [8]	ResNet50	16.5	35.5	13.8	12.2	24.6	29.4	17.8	39.3	13.0	<u>22.5</u>	16.7	17.3
FCOS [9]	ResNet50	18.2	43.1	13.2	13.4	24.1	26.4	19.9	44.1	15.5	19.3	20.2	16.9
ATSS [34]	ResNet50	18.3	42.0	13.9	<u>15.8</u>	25.2	26.2	21.5	44.6	17.5	21.9	19.5	23.3
GFL [32]	ResNet50	18.1	40.6	14.6	14.1	25.1	28.0	21.7	44.7	<u>18.4</u>	20.9	<u>20.4</u>	<u>23.5</u>
YOLOv5-s [33]	CSPDarkNet53	16.4	42.7	9.3	8.3	23.3	25.1	15.2	34.1	12.4	8.7	17.0	14.7
DETR [16]	ResNet50	11.5	23.0	10.1	3.5	12.3	27.4	11.4	25.2	9.0	3.3	7.3	13.1
Deformable DETR [17]	ResNet50	18.5	40.0	15.7	13.7	25.2	29.3	19.7	42.8	16.8	14.5	18.4	20.3
CE-SGNet (Ours)	ResNet50	**24.0**	**51.2**	**20.3**	**17.5**	**30.3**	**41.9**	**22.7**	**48.1**	**18.8**	**22.9**	**21.6**	23.4

Furthermore, it is observed that two-stage methods generally outperform one-stage detectors in AP. This is attributed to the effectiveness of RPN in distinguishing foreground and background, reducing false positives. One-stage detectors, without rough filtering by RPN, exhibit lower accuracy compared to two-stage detectors. Furthermore, there exists a notable performance disparity between transformer-based models and CNN-based methods. This indicates that transformer-based models encounter challenges in demonstrating their advantages, especially when handling small datasets. The comprehensive analysis of results highlights the efficacy of the CE-SGNet in smartphone screen defect detection.

4.3. Performance Analysis of Modules

In order to validate the effectiveness of the proposed modules, we conduct comprehensive experiments on the SSGD-LB101 dataset, utilizing metrics such as AP and AP_{50} to evaluate the detection performance of different modules.

4.3.1. Analysis of Each Module

The results of the effectiveness of each module are shown in Table 2. As observed, each module played an effective role, improving performance independently. The ARFAM enhances the sensitivity of the detection model, improving the model's perception and recognition capabilities for detecting defects. Although the SGRM partially improves performance through constructing relationships between regions, the RoI features become more accurate after ARFAM extracting extensive contextual information, thereby resulting in higher performance improvement. The combined action of the two modules results in an increase of AP to 24.0 and AP_{50} to 51.2, showing improvements of 3.7 AP and 3.9 AP_{50}, respectively. This indicates the effectiveness of each proposed module and the combined action of both does not have a negative impact on performance.

Table 2. Ablation study of different models. Bold data denote the best performance.

Methods	ARFAM	SGRM	AP	AP_{50}
Baseline			20.3	47.3
	✓[1]		22.6	50.0
CE-SGNet (ours)		✓	22.3	48.1
	✓	✓	**24.0**	**51.2**

[1] ✓ signifies the utilization of this algorithm.

4.3.2. Analysis of the Attention Mechanisms

The compared attention mechanisms include SENet [18], CBAM [19], ARFAM with only channel attention (CAM), ARFAM with only spatial attention (SAM) and the proposed ARFAM. The computational complexity (Giga Floating Point Operations per Second, GFLOPs) and the total number of trainable parameters in the model (Params) are also employed for evaluating the efficiency of models. For a fair comparison, all other parameters and experimental settings remain consistent, with the only difference being the type of attention mechanism. As shown in Table 3, the results demonstrate that the combination of channel attention and spatial attention can effectively improve the detection performance. CAM, which adds a global maximum pooling layer compared to SENet [18], can effectively extract the most prominent features of defects, aligning with conclusions from CBAM [19]. In comparison to CBAM [19], the ARFAM increases 0.7 AP and 1.3 AP_{50} respectively, without a significant increase in computational resources or model parameters. This indicates that the ARFAM can fully extract the most relevant contextual information of defects by adaptively selecting receptive fields.

Table 3. Analysis of different attention mechanisms. Bold data denote the best performance.

Methods	GFLOPs	Params	AP	AP_{50}
FPN	332.01	60.68 M	22.3	48.1
SENet	332.04	60.72 M	22.9	48.6
CAM	332.08	60.72 M	23.1	48.4
SAM	347.20	61.27 M	23.2	48.8
CBAM	332.09	60.72 M	23.3	49.9
ARFAM	347.27	61.31 M	**24.0**	**51.2**

4.3.3. Analysis of Different Positions of the ARFAM

We also conduct an analysis of the different positions of ARFAM. "ARFAM+FPN" represents ARFAM inserted before FPN, "FPN+ARFAM" represents ARFAM inserted after FPN without shared parameters and "FPN+ARFAM-shared" represents ARFAM inserted after FPN with shared parameters. According to the experimental results in Table 4, "ARFAM+FPN" reduces AP by 0.8 and AP_{50} by 2.4 compared to "FPN+ARFAM". If the ARFAM is inserted before FPN, it may introduce noise during the feature fusion process. The defect features on different feature maps may be inconsistent, which affects the stability of model learning. Additionally, "ARFAM+FPN" leads to an excessive increase in model parameters, causing parameter redundancy and decreasing accuracy. Shared parameters tend to promote consistent attention across different scales, but the focus positions often vary across scales in practice. By inserting ARFAM after FPN, irrelevant feature interference before fusion is reduced, resulting in more accurate candidate boxes generated by RPN.

Table 4. Analysis of different positions of the ARFAM. Bold data denote the best performance.

Methods	GFLOPs	Params	AP	AP_{50}
FPN	332.01	60.68 M	22.3	48.1
ARFAM+FPN	373.58	70.03 M	23.2	48.8
FPN+ARFAM-shared	347.27	60.80 M	23.1	49.0
FPN+ARFAM	347.27	61.31 M	**24.0**	**51.2**

4.3.4. Analysis of the SGRM

Finally, we analyze different mapping ways for node embedding. According to the results shown in Table 5, the soft mapping increases 0.6 AP and 0.8 AP_{50} compared to the hard mapping. Moreover, it does not require additional computational costs or an increase in model parameters. However, hard mapping may increase the error rate if the initial classification results are incorrect. In addition, We also conduct an analysis with different numbers of the head in SCAT, setting k to 1, 2, 4, 6, 8, 10, as shown in Figure 5. Based on the results, the detection performance continuously improves with an increase in the number of heads until a sudden drop is observed when there are 10 heads in the multi-head attention mechanism. Considering both AP and AP_{50}, the number of heads is set to 8 in this paper.

Table 5. Analysis of different node embedding methods. Bold data denote the best performance.

Methods	GFLOPs	Params	AP	AP_{50}
SGRM-hard	347.27	61.31 M	23.4	50.4
SGRM-soft	347.27	61.31 M	**24.0**	**51.2**

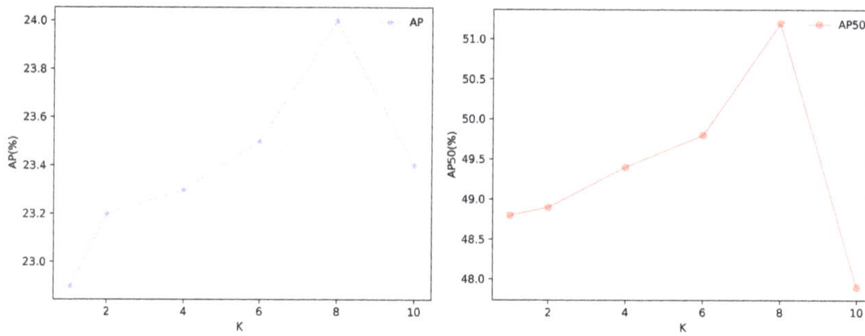

Figure 5. Comparison results of different numbers of attention heads.

4.4. Visualization of Results

To visually show the effectiveness of the CE-SGNet, Figure 6 presents qualitative comparison results with the baseline model. It is observed that the CE-SGNet is more effective than the baseline model in detecting difficult defects, such as small spots. Moreover, the defect regions detected via CE-SGNet are more accurately aligned with the ground truth. The results indicate that the CE-SGNet has the capability to reduce both false positives and false negatives, improving overall detection performance. The ARFAM enhances contextual information of defects, which is beneficial for identifying defects accurately. Moreover, the SGRM strengthens the interdependence between defects, improving localization accuracy.

We also visualize the feature maps to provide an intuitive understanding of the defect regions attended by ARFAM. The results in Figure 7 demonstrate that the ARFAM effectively suppresses background feature information and enhances contextual information. For instance, in the first and second columns of Figure 7, it successfully suppresses spots or blots in the background which are not defects. It is beneficial for RPN to perform more accurate classification between foreground and background. Moreover, our CE-SGNet adapts contextual information based on defect size, as shown in the third and fourth columns of Figure 7. Compared to the baseline model, the CE-SGNet identifies regions closer to the ground truth, demonstrating its effectiveness in improving detection accuracy.

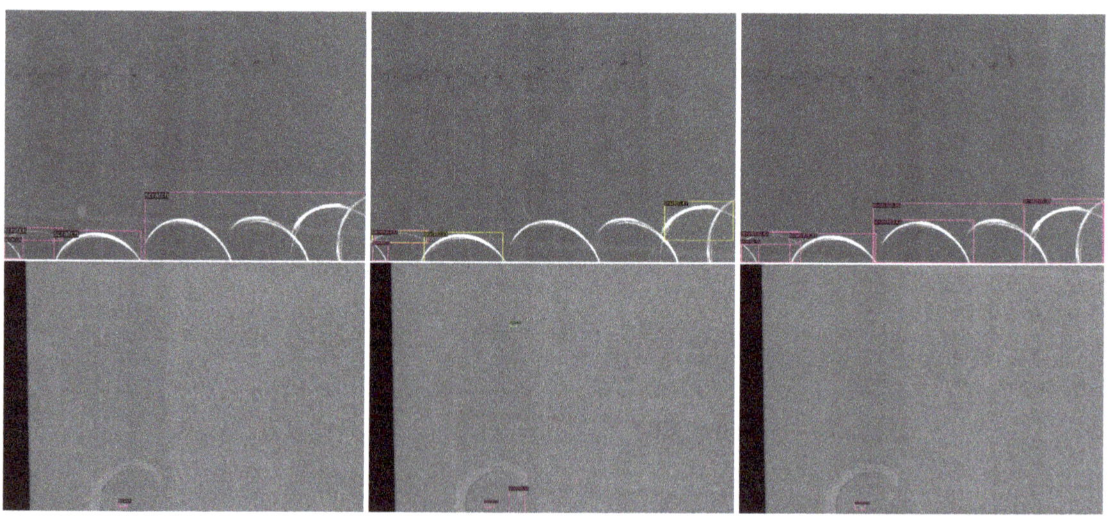

Figure 6. *Cont.*

Figure 6. Examples of visualization results on the SSGD dataset. The first column is the ground truth, the second column is the results of the baseline model and the third column is the results of our model.

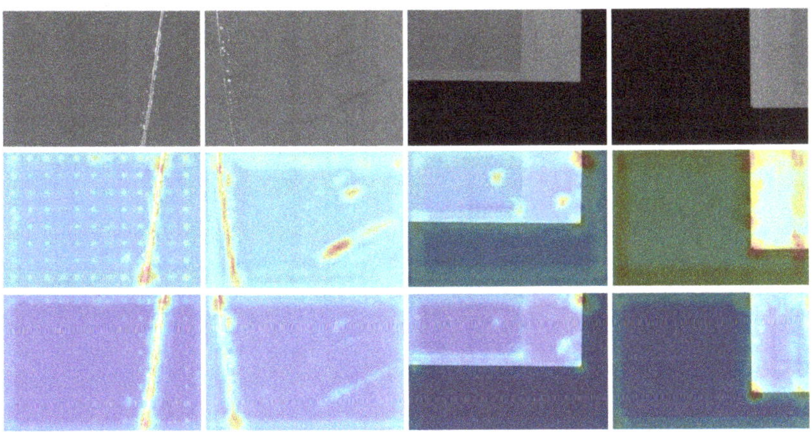

Figure 7. Examples of feature map visualization results on the SSGD dataset. The first row is original images, the second row is feature maps of baseline and the third row is feature maps of CE-SGNet.

4.5. Generalization Study

To verify the generalization capability of the CE-SGNet, we conduct generalization experiments on the PCB dataset. The CE-SGNet is compared with Faster RCNN [5], Cascade RCNN [6], RetinaNet [8], FCOS [9], ATSS [34], GFL [32], YOLOv5-s [33], DETR [16] and Deformable DETR [17]. The results are shown in Table 6. It can be observed that the proposed method achieves 53.8 AP, achieving optimal performance. In terms of multi-scale detection capability, the proposed method achieves the highest accuracy in AP_s, AP_m and AP_l compared to other methods. This shows that the CE-SGNet can be applied to scenarios with different defect sizes, especially for detecting small defects. The experimental results demonstrate that the CE-SGNet exhibits excellent generalization performance.

Table 6. Comparison results of different methods on the PCB dataset. Bold data denote the best performance. Underlined data denote second best performance.

Model	Backbone	AP	AP_{50}	AP_{75}	AP_s	AP_m	AP_l
Faster RCNN [5]	ResNet50	52.8	96.3	50.3	35.5	52.7	47.1
Cascade RCNN [6]	ResNet50	53.1	96.7	<u>53.0</u>	34.2	53.2	49.3
RetinaNet [8]	ResNet50	52.8	95.8	52.4	20.0	53.0	47.8
FCOS [9]	ResNet50	53.0	96.3	52.3	26.6	<u>53.6</u>	44.5
ATSS [34]	ResNet50	<u>53.4</u>	**97.2**	**53.3**	26.7	<u>53.6</u>	<u>50.2</u>
GFL [32]	ResNet50	53.1	**97.2**	52.6	<u>36.7</u>	53.2	46.3
YOLOv5-s [33]	CSPDarkNet53	49.9	96.6	43.7	30.3	50.3	47.2
DETR [16]	ResNet50	36.9	87.6	22.4	4.5	37.4	36.8
Deformable DETR [17]	ResNet50	40.3	89.4	25.2	13.3	40.8	40.6
CE-SGNet (Ours)	ResNet50	**53.8**	<u>97.0</u>	52.7	**40.0**	**53.7**	**50.8**

4.6. Model Complexity and Time-Efficiency Evaluation

To evaluate the complexity and efficiency of the CE-SGNet, GFLOPs, Params and frames per second (FPS) are employed as the evaluation metrics. GFLOPs is used to evaluate the computational complexity, Params is used to evaluate the model size and memory usage, while FPS is used to evaluate the real-time efficiency. The comparative results are shown in Table 7. It is observed that CE-SGNet exhibits higher GFLOPs at 347.37 compared to the baseline model, due to the inclusion of two novel components. In terms of model parameters, Cascade RCNN consumes the highest memory resources. In contrast, our CE-SGNet, while requiring additional computational resources and a slightly larger model size, obtains a significant improvement in detection performance. However, the introduction of SGAT slows down CE-SGNet's inference speed, impacting its real-time performance.

Table 7. Comparsion results of different models on the complexity and time efficiency.

Model	GFLOPs	Params	FPS
Faster RCNN [5]	303.84	41.15 M	16.1
Cascade RCNN [6]	331.63	68.94 M	14.1
RetinaNet [8]	311.23	36.23 M	16.3
FCOS [9]	296.20	31.85 M	21.3
ATSS [34]	311.23	36.23 M	16.3
GFL [32]	307.94	32.05 M	18.4
YOLOv5-s [33]	43.87	7.04 M	54.2
DETR [16]	137.44	41.28 M	14.3
Deformable DETR [17]	292.81	39.82 M	9.6
CE-SGNet (Ours)	347.27	61.31 M	7.9

5. Discussion

In this paper, we propose the CE-SGNet based on two-stage detectors for smartphone screen defect detection. While the CE-SGNet achieves excellent performance, there are still some limitations. Firstly, as shown in Table 7, our CE-SGNet consumes more computational resources and has a complex model structure. In terms of inference speed, it may face challenges in real-time detection scenarios. Future research can focus on accelerating the structure of CE-SGNet through appropriate light-weighting methods to improve the real-time detection capability of our model. Secondly, due to the unique optical properties of screen glass, the color of defects is similar to that of the background. This means that not all objects identified as defects based on visual perception are real defects. Most detection methods improve detection performance by enhancing model sensitivity, which also leads to an increase in false positives. In the future, exploring the fusion of multimodal knowledge, such as language and image, could provide a more comprehensive understanding of industrial scenarios.

6. Conclusions

In our research, a context-enhanced spatially aware defect detection network that effectively detects defects on smartphone screens is proposed. We introduce two modules: the ARFAM and the SGRM. The ARFAM, composed of a channel attention module and a spatial attention module, obtains wide contextual information and suppresses background features, effectively detecting low-contrast tiny defects. The SGRM, constructs a region-to-region graph to strengthen the spatial relationship between defects and reduce false negatives. Experimental results both on the SSGD dataset and the PCB dataset illustrate that the proposed method demonstrates significant improvement in detection accuracy, highlighting its considerable potential for practical applications.

Author Contributions: Conceptualization, A.L. and X.W.; methodology, A.L. and X.W.; software, A.L.; validation, A.L.; formal analysis, A.L.; investigation, A.L.; resources, A.L.; data curation, A.L.; writing—original draft preparation, A.L.; writing—review and editing, A.L., Q.W. and X.W.; visualization, A.L.; supervision, X.W. All authors have read and agreed to the published version of the manuscript.

Funding: This research received no external funding.

Institutional Review Board Statement: Not applicable.

Informed Consent Statement: Not applicable.

Data Availability Statement: The data presented in this study are available on request from the corresponding author.

Conflicts of Interest: The authors declare no conflicts of interest.

References

1. Wei, X.; Feng, W.; Lei, Q.; Gui, G.; Le, S.; Zhong, J.; Wang, W. Defect Detection of Using Variant CNN in the Processing of Cover Glass, Touch Screen and Display under Parallel Light. In Proceedings of the 2020 IEEE 6th International Conference on Computer and Communications (ICCC), Chengdu, China, 11–14 December 2020; pp. 1349–1355.
2. Chuanxia, J.; Jian, G. Visual Detection Method for Surface Defect of Mobile Phone Screen Glass. *Packag. Eng.* **2018**, *39*, 16–21.
3. Li, C.; Zhang, X.; Huang, Y.; Tang, C.; Fatikow, S. A novel algorithm for defect extraction and classification of mobile phone screen based on machine vision. *Comput. Ind. Eng.* **2020**, *146*, 106530. [CrossRef]
4. Jian, C.; Gao, J.; Ao, Y. Automatic surface defect detection for mobile phone screen glass based on machine vision. *Appl. Soft Comput.* **2017**, *52*, 348–358. [CrossRef]
5. Ren, S.; He, K.; Girshick, R.; Sun, J. Faster r-cnn: Towards real-time object detection with region proposal networks. *Adv. Neural Inf. Process. Syst.* **2015**, *28*. [CrossRef] [PubMed]
6. Cai, Z.; Vasconcelos, N. Cascade r-cnn: Delving into high quality object detection. In Proceedings of the IEEE Conference on Computer Vision and Pattern Recognition, Salt Lake City, UT, USA, 18–23 June 2018; pp. 6154–6162.
7. Redmon, J.; Divvala, S.; Girshick, R.; Farhadi, A. You only look once: Unified, real time object detection. In Proceedings of the IEEE Conference on Computer Vision and Pattern Recognition, Las Vegas, NV, USA, 27–30 June 2016; pp. 779–788.
8. Lin, T.Y.; Goyal, P.; Girshick, R.; He, K.; Dollár, P. Focal loss for dense object detection. In Proceedings of the IEEE International Conference on Computer Vision, Venice, Italy, 22–29 October 2017; pp. 2980–2988.
9. Tian, Z.; Shen, C.; Chen, H.; He, T. FCOS: Fully convolutional one-stage object detection. *arXiv* **2019**, arXiv:1904.01355.
10. Wang, T.; Zhang, C.; Ding, R.; Yang, G. Mobile phone surface defect detection based on improved faster r-cnn. In Proceedings of the 2020 25th International Conference on Pattern Recognition (ICPR), Milan, Italy, 10–15 January 2021; pp. 9371–9377.
11. Chen, Z.; Zha, Y.; Wu, Z.; Zeng, D. Detection of Mobile Phone Screen Defect Based on Faster R-CNN Fusion Model. In Proceedings of the 2021 China Automation Congress (CAC), Beijing, China, 22–24 October 2021; pp. 6601–6606.
12. Ronneberger, O.; Fischer, P.; Brox, T. U-net: Convolutional networks for biomedical image segmentation. In Proceedings of the Medical Image Computing and Computer-Assisted Intervention–MICCAI 2015: 18th International Conference, Munich, Germany, 5–9 October 2015; Proceedings, Part III 18; Springer: Berlin/Heidelberg, Germany, 2015; pp. 234–241.
13. He, K.; Zhang, X.; Ren, S.; Sun, J. Deep residual learning for image recognition. In Proceedings of the IEEE Conference on Computer Vision and Pattern Recognition, Las Vegas, NV, USA, 27–30 June 2016; pp. 770–778.
14. Zhang, J.; Su, H.; Zou, W.; Gong, X.; Zhang, Z.; Shen, F. CADN: A weakly supervised learning-based category-aware object detection network for surface defect detection. *Pattern Recognit.* **2021**, *109*, 107571. [CrossRef]
15. Zhu, Y.; Ding, R.; Huang, W.; Wei, P.; Yang, G.; Wang, Y. HMFCA-Net: Hierarchical multi-frequency based Channel attention net for mobile phone surface defect detection. *Pattern Recognit. Lett.* **2022**, *153*, 118–125. [CrossRef]

16. Carion, N.; Massa, F.; Synnaeve, G.; Usunier, N.; Kirillov, A.; Zagoruyko, S. End-to-end object detection with transformers. In Proceedings of the European Conference on Computer Vision, Glasgow, UK, 23–28 August 2020; Springer: Berlin/Heidelberg, Germany, 2020; pp. 213–229.
17. Zhu, X.; Su, W.; Lu, L.; Li, B.; Wang, X.; Dai, J. Deformable detr: Deformable transformers for end-to-end object detection. *arXiv* **2020**, arXiv:2010.04159.
18. Hu, J.; Shen, L.; Sun, G. Squeeze-and-excitation networks. In Proceedings of the IEEE Conference on Computer Vision and Pattern Recognition, Salt Lake City, UT, USA, 18–23 June 2018; pp. 7132–7141.
19. Woo, S.; Park, J.; Lee, J.Y.; Kweon, I.S. Cbam: Convolutional block attention module. In Proceedings of the European Conference on Computer Vision (ECCV), Munich, Germany, 8–14 September 2018; pp. 3–19.
20. Li, X.; Wang, W.; Hu, X.; Yang, J. Selective kernel networks. In Proceedings of the IEEE/CVF Conference on Computer Vision and Pattern Recognition, Long Beach, CA, USA, 15–20 June 2019; pp. 510–519.
21. Vaswani, A.; Shazeer, N.; Parmar, N.; Uszkoreit, J.; Jones, L.; Gomez, A.N.; Kaiser, Ł.; Polosukhin, I. Attention is all you need. *Adv. Neural Inf. Process. Syst.* **2017**, *30, 1–11* .
22. Scarselli, F.; Gori, M.; Tsoi, A.C.; Hagenbuchner, M.; Monfardini, G. The graph neural network model. *IEEE Trans. Neural Netw.* **2008**, *20*, 61–80. [CrossRef] [PubMed]
23. Kipf, T.N.; Welling, M. Semi-supervised classification with graph convolutional networks. *arXiv* **2016**, arXiv:1609.02907.
24. Wang, R.; Cheung, C.F. Knowledge graph embedding learning system for defect diagnosis in additive manufacturing. *Comput. Ind.* **2023**, *149*, 103912. [CrossRef]
25. Wang, Y.; Gao, L.; Li, X.; Gao, Y.; Xie, X. A new graph-based method for class imbalance in surface defect recognition. *IEEE Trans. Instrum. Meas.* **2021**, *70*, 5007816. [CrossRef]
26. Zhai, Y.; Hu, Z.; Wang, Q.; Yang, Q.; Yang, K. Multi-Geometric Reasoning Network for Insulator Defect Detection of Electric Transmission Lines. *Sensors* **2022**, *22*, 6102. [CrossRef] [PubMed]
27. Velickovic, P.; Cucurull, G.; Casanova, A.; Romero, A.; Lio, P.; Bengio, Y. Graph attention networks. *Stat* **2017**, *1050*, 10-48550.
28. Lin, T.Y.; Dollár, P.; Girshick, R.; He, K.; Hariharan, B.; Belongie, S. Feature pyramid networks for object detection. In Proceedings of the IEEE Conference on Computer Vision and Pattern Recognition, Honolulu, HI, USA, 21–26 July 2017; pp. 2117–2125.
29. Li, Y.; Hou, Q.; Zheng, Z.; Cheng, M.M.; Yang, J.; Li, X. Large selective kernel network for remote sensing object detection. In Proceedings of the IEEE/CVF International Conference on Computer Vision, Paris, France, 2–6 October 2023; pp. 16794–16805.
30. Xu, H.; Jiang, C.; Liang, X.; Lin, L.; Li, Z. Reasoning-rcnn: Unifying adaptive global reasoning into large-scale object detection. In Proceedings of the IEEE/CVF Conference on Computer Vision and Pattern Recognition, Long Beach, CA, USA, 15–20 June 2019; pp. 6419–6428.
31. Han, H.; Yang, R.; Li, S.; Hu, R.; Li, X. SSGD: A smartphone screen glass dataset for defect detection. In Proceedings of the ICASSP 2023—2023 IEEE International Conference on Acoustics, Speech and Signal Processing (ICASSP), Rhodes Island, Greece, 4–10 June 2023; pp. 1–5.
32. Li, X.; Wang, W.; Wu, L.; Chen, S.; Hu, X.; Li, J.; Tang, J.; Yang, J. Generalized focal loss: Learning qualified and distributed bounding boxes for dense object detection. *Adv. Neural Inf. Process. Syst.* **2020**, *33*, 21002–21012.
33. Ultralytics. YOLOv5: A State-of-the-Art Real-Time Object Detection System. 2021. Available online: https://docs.ultralytics.com (accessed on 12 March 2024).
34. Zhang, S.; Chi, C.; Yao, Y.; Lei, Z.; Li, S.Z. Bridging the gap between anchor-based and anchor-free detection via adaptive training sample selection. In Proceedings of the IEEE/CVF Conference on Computer Vision and Pattern Recognition, Seattle, WA, USA, 13–19 June 2020; pp. 9759–9768.

sensors MDPI

Communication

Improved YOLOv8-Based Target Precision Detection Algorithm for Train Wheel Tread Defects

Yu Wen, Xiaorong Gao *, Lin Luo and Jinlong Li

School of Physical Science and Technology, Southwest Jiaotong University, Chengdu 610031, China; 2022201057@my.swjtu.edu.cn (Y.W.); happyluolin@vip.163.com (L.L.); jinlong_lee@126.com (J.L.)
* Correspondence: gxrr@vip.163.com

Abstract: Train wheels are crucial components for ensuring the safety of trains. The accurate and fast identification of wheel tread defects is necessary for the timely maintenance of wheels, which is essential for achieving the premise of conditional repair. Image-based detection methods are commonly used for detecting tread defects, but they still have issues with the misdetection of water stains and the leaking of small defects. In this paper, we address the challenges posed by the detection of wheel tread defects by proposing improvements to the YOLOv8 model. Firstly, the impact of water stains on tread defect detection is avoided by optimising the structure of the detection layer. Secondly, an improved SPPCSPC module is introduced to enhance the detection of small targets. Finally, the SIoU loss function is used to accelerate the convergence speed of the network, which ensures defect recognition accuracy with high operational efficiency. Validation was performed on the constructed tread defect dataset. The results demonstrate that the enhanced YOLOv8 model in this paper outperforms the original network and significantly improves the tread defect detection indexes. The average precision, accuracy, and recall reached 96.95%, 96.30%, and 95.31%.

Keywords: defect detection; neural network; target recognition; YOLOv8

1. Introduction

When it comes to train travel, safety is of utmost importance [1]. These defects can even pose a threat to the safety of high-speed train operations. External factors such as inertia, rail edge stress, and thermal damage can affect train wheels, leading to defects like bruises, peeling, and scratches on the tread surface. Ensuring the safety of high-speed train operations relies heavily on detecting wheel tread defects, but in the actual detection process, defects of various shapes and water stains often present challenges.

Conventional image detection methods include image localisation and segmentation, feature extraction, classifier classification, and so on. However, such methods are complex and have poor real-time performance, low efficiency, and difficulty meeting the needs of industrial production. Consequently, they have been gradually replaced by the use of deep learning to address these issues. The deep learning-based target detection method offers the advantages of fast detection speed and high accuracy, resulting in significant cost savings.

Deep learning-based target detection methods are typically classified into two types: region-based, two-stage detection algorithms, such as regions with CNNs (R-CNNs) [2], and faster region-based convolutional neural networks (Faster R-CNNs) [3]. The two-stage target detection network is transformed into a region proposal, followed by the utilisation of a neural network for target classification prediction and localisation implementation. Mengqi Chen et al. [4] utilised the Gabor filter in frequency analysis and integrated the Gabor kernel into the Faster R-NN model to address fabric detection with background texture interference. Xiangyang Xu et al. [5] conducted a comparative analysis of Faster R-CNN and Mask R-CNN for road crack detection.

The you-only-look-once (YOLO) [6] series of one-step detection algorithms represents another category. The one-stage target detection network directly extracts features and classifies and locates targets through the neural network, resulting in improved efficiency compared to the two-stage target detection network. Abhiroop Bhattacharya et al. [7] used a combination of current techniques in deep learning-based transformers, multilevel feature fusion, data augmentation, and object detection to quickly detect PCB board defects. Stefano Frizzo Stefenon et al. [8] used a genetic algorithm to optimise the hyperparameters of the YOLO model for grid fault localisation, successfully improving performance metrics and reducing computational requirements. Fenglong Ding et al. [9] employed transfer learning to apply a single-shot multibox detector (SSD) for wooden panel defect detection, utilising DenseNet as the backbone network instead of VGG16 and incorporating residual learning to mitigate feature information loss, thereby achieving wood defect detection. Qiqi Zhou et al. [10] proposed CABF-YOLO, an accurate and efficient strip surface defect detection model that incorporates a novel Bidirectional Fusion (BF) strategy to acquire detailed information and cater to the requirements of industrial strip surface defect detection.

YOLO series algorithms are widely employed for the purpose of defect detection in various applications [11]. However, the detection performance of YOLOv8 may not always meet the requirements for detecting special scenes. Defect detection in tread defect images is susceptible to variations in lighting conditions and water stains, while the intricate background of the tread image poses interference to accurate defect detection. To address this issue, this paper proposes an enhanced version of YOLOv8 that improves upon the original model by enhancing wheel tread defect detection.

1. In order to solve the problem of the false detection of water stains in defects, this paper optimises the network detection structure by adding a shallow small-target detection layer and deleting the large-target detection layer, which not only attenuates the impact of water stains but also pays more attention to the detection of small target defects.
2. To prevent the location information of small targets from being overlooked during the learning process, this paper uses an improved SPPCSPC (Spatial Pyramid Pooling Concurrent Spatial Pyramid Convolution). Firstly, the CBS (Conv BatchNormal Silu) is trimmed to address the issue of filtering the edge information of small target defects, such as peeling. Then, the SimAM (simple attention mechanism) is introduced to minimise loss in the feature extraction of small target defects.
3. This paper replaces the loss function with SIOU (Soft IoU), which encompasses four key components: angle loss, distance loss, shape loss, and IoU loss. This modified approach offers a more comprehensive solution compared to the original CIOU (Complete IoU), as it takes into account multiple factors, resulting in improved convergence speed and higher accuracy.

2. YOLOv8 Network Structure

The YOLOv8 algorithm provides a new SOTA (state of the art) model that can be used for tasks such as target detection [12], image segmentation [13], etc., and its main structure consists of three parts: the backbone, the neck, and the head. The complete network structure of YOLOv8 is illustrated in Figure 1.

The C2f module based on the CSP is used in the main part instead of the C3 module. C2f utilises the concept of ELAN in YOLOv7 and combines C3 with ELAN. This enables YOLOv8 to maintain high computational efficiency and obtain more information about the gradient flow simultaneously. The last pyramid space pooling is replaced with the SPPF module, which can extract spatial information more effectively.

In the neck section, YOLOv8 continues to use the PAN-FPN (Path Aggregation Network with Feature Pyramid Network) feature fusion technique, which effectively combines information from feature layers at various scales. It employs multiple up-sampling procedures and C2f modules, as well as a final decoupled head structure, to create the neck module. The confidence and regression frames are combined to achieve greater accuracy.

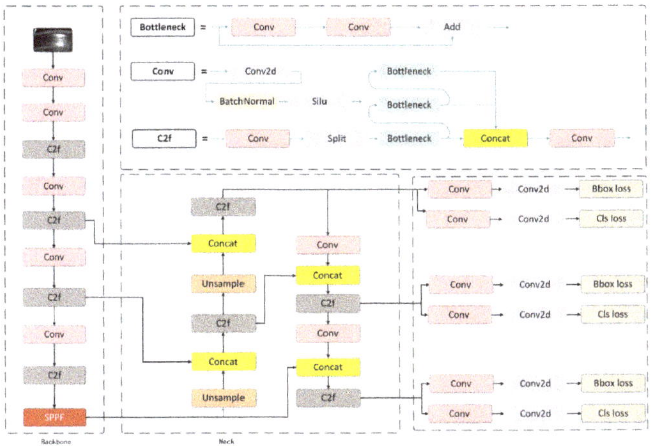

Figure 1. The structure of the YOLOv8 module.

In the head section, YOLOv8 employs an anchorless split detection head, thereby enhancing both detection accuracy and efficiency in comparison to the anchor frame-based method. The head structure is replaced with the mainstream decoupled head structure to alleviate the conflict between classification and localisation tasks. The reference is changed from anchor-based to anchor-free for YOLOX. Furthermore, YOLOX has been updated from an anchor-based to an anchor-free approach, which is particularly beneficial for detecting faces with irregular dimensions.

3. Improved YOLOv8 Structure

Although YOLOv8 has performed well in all aspects of the detection task, it still faces difficulties in recognising confusing objects. During feature extraction for defects on the tread surface, the neural network may mistake water stains for defects due to their similarity in image information. Additionally, YOLOv8 may miss small defects on the tread surface.

To address the aforementioned issues, this paper proposes a defect detection algorithm that can effectively identify small defects on the tread surface while avoiding the misdetection of water stains. Firstly, we optimised the large detection head in the original network to increase the structure of the small detection head, thereby reducing misdetection caused by water stains. Secondly, we replaced the original SPPF with the SPPCSPC module, which was optimised and introduced into the SimAM to prevent the misdetection of small defects. Throughout the learning process, the algorithm does not ignore the target's position information. Finally, the loss function is replaced with SIOU, which considers more geometric factors and improves the accuracy of detecting tread defects.

3.1. Optimisation of Network Structure

As water stains and defects share similar visual features, it is easy to mistake water stains for defects. In order to show the response of water stains and defects in different network layers for subsequent reduction of false detection of water stains. In this paper, GradCAM [14] is used to generate heat maps, based on which we can better analyse the regions of interest of the neural network.

Figure 2 integrates a heat map of water stains and defects. Upon analysis of the figure, it becomes apparent that the deep layer of the network employed for large-target detection exhibits a relatively diminished response disparity between defects and water stains. Conversely, the shallow layer of the network utilised for detecting small targets demonstrates enhanced efficacy in filtering out water stains on the tread surface. Consequently, by eliminating the large-target detection layer and incorporating a small-target detection layer, the final detection outcomes are capable of mitigating the impact of water stains, thereby enabling improved focus on identifying small target defects.

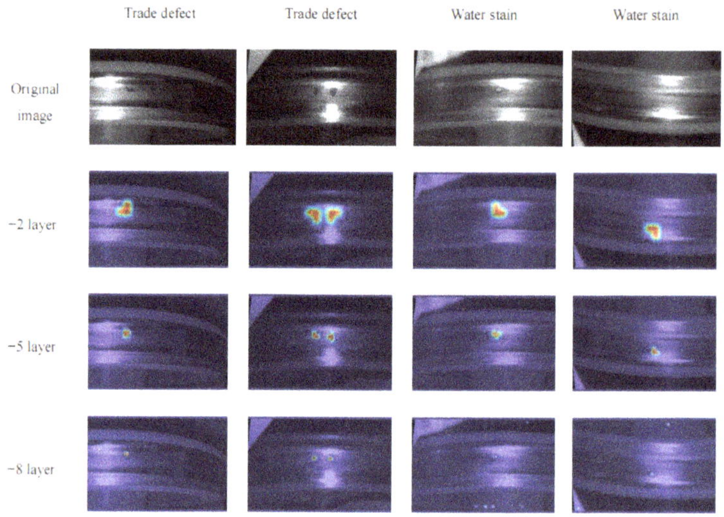

Figure 2. Tread thermal response diagram.

To address the problem of the misidentification of water stains in the original algorithm, the original YOLOv8 generates three feature maps of different sizes. In this study, we removed the 20 pixel × 20 pixel large-target detection layer and added a 160 pixel × 160 pixel small-target detection layer to improve defect detection. The architecture of the improved YOLOv8s network detection layer is shown in Figure 3. Although YOLOv8 achieves a larger perceptual domain through deep sampling, deep sampling results in the loss of some feature information. By incorporating a shallow network, we can better extract detailed features of the target, preserve defect-related information, reduce the interference of water spots on defect detection, mitigate the loss of small target defect features, and improve the small-target detection capability. The improved network exhibits reduced sensitivity to water stains and provides more accurate defect detection results.

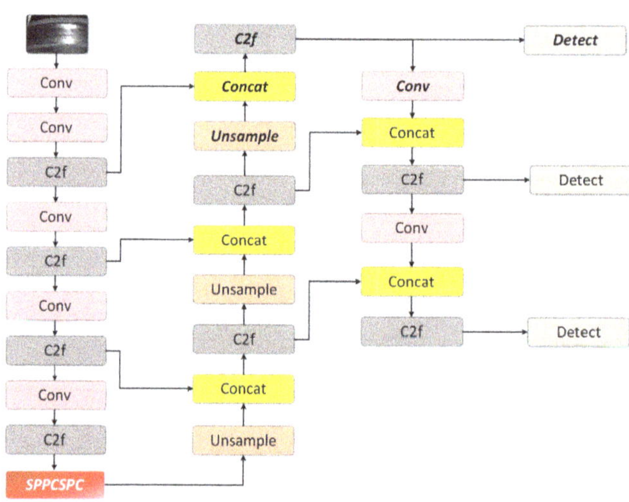

Figure 3. The structure of the improved YOLOv8 module.

3.2. Reconstructed SPPCSPC

The original network exhibits limited capability for detecting small defects in the wheel tread. To enhance its defect detection ability and accurately locate defects, we replace the initial SPPF module with an improved SPPCSPC module. This novel SPPCSPC module integrates spatial pyramid pooling (SPP) [15] and a cross-stage Partial Channel (CSPC) [16], surpassing the performance of the SPPC alone. By employing serial computation, the SPPCSPC module adapts to images of varying resolutions while accelerating the convergence speed. To better suit the spatial pyramid pooling of tread features, we reconstruct the SPPCSPC module as depicted in Figure 4a,b.

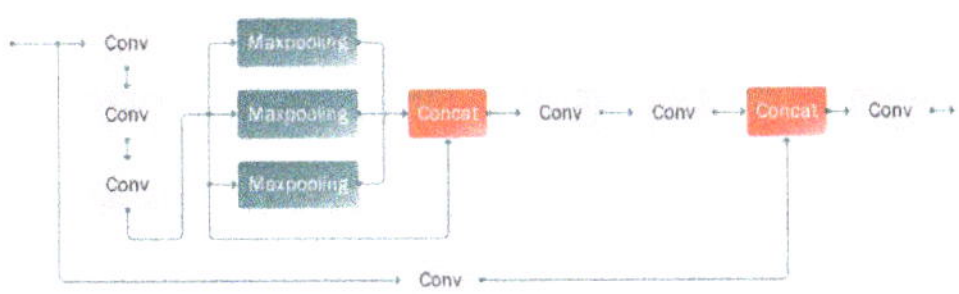

(a)

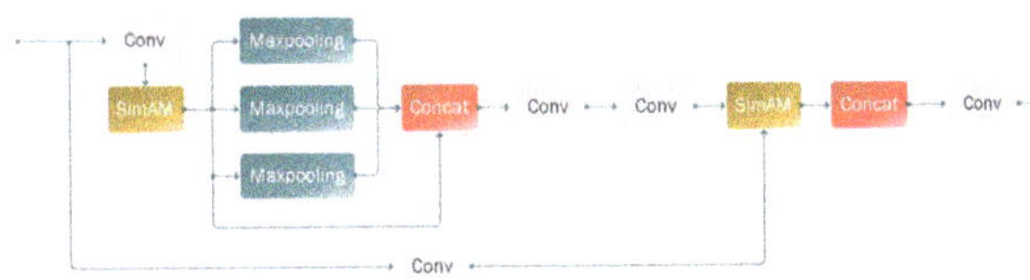

(b)

Figure 4. SPPCSPC structure diagrams: (**a**) SPPCSPC before reconstruction; (**b**) SPPCSPC after reconstruction.

Firstly, the two CBS layers located before the pooling layer are cropped out to reduce the filtering of small-target edge information by the convolutional layer. This also scales down the computational load of the neural network. Secondly, the SimAM [17] attention mechanism is introduced to address the issue of the inadequate extraction of small, defective target features. SimAM assigns a 3D attention weight to each neuron while dividing the pixels between the complex background region and the small-target region to enhance the sensitivity of the complex background and small-target regions. Adding SimAM before the pooling layer and final channel merging can help divide the defective and non-defective regions, enhancing the network's attention to the small-target region, reducing feature loss, and improving the model's robustness by reducing the influence of the tread background on the defects.

3.3. SIoU

The purpose of this function is to continuously optimise the fitting effect by defining the calculation method around the geometrical elements of the prediction and real frames. Representative methods include GIoU [18], CIoU [19], and EIoU [20]. The regression loss function of YOLOv8s adopts the CIoU loss function, which considers the overlap area, distance from the centroids, and aspect ratio. Although CIoU considers more comprehensive factors compared to the previous loss function, it does not take into account the mismatch of directions between the real and predicted frames. To improve the model's generalisation ability and accelerate convergence, the network model's border loss function is replaced by SIoU from CIoU [21]. SIoU considers the vector angles between regressions and redefines the penalty indicator. The SIoU loss function converges faster and is more accurate than the CIoU loss function. The SIoU loss function comprises angular loss, distance loss, shape loss, and IoU loss.

Figure 5 shows the angular loss map of SIOU, where σ represents the distance between the centre point of the real frame and the predicted frame, C_h represents the height difference between the centre point of the real frame and the predicted frame, α and represents the horizontal angle between the predicted frame and the real frame.

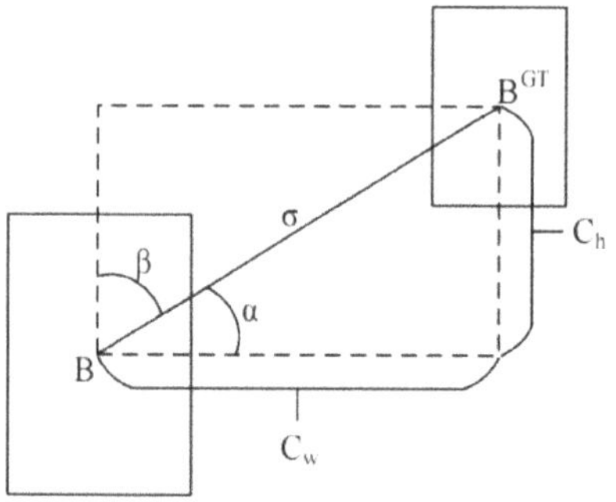

Figure 5. Calculation of angle loss.

The regression direction of the prediction frame is determined by the angular magnitude, where the angular loss is defined as follows:

$$\Lambda = 1 - 2\sin^2\left(\arcsin(x) - \frac{\pi}{4}\right) \quad (1)$$

$$\begin{cases} x = \frac{C_h}{\sigma} = \sin(\alpha) \\ \sigma = \sqrt{\left(b_{c_x}^{gt} - b_{c_x}\right)^2 + \left(b_{c_y}^{gt} - b_{c_y}\right)^2} \\ C_h = \max\left(b_{c_y}^{gt}, b_{c_y}\right) - \min\left(b_{c_y}^{gt}, b_{c_y}\right) \end{cases} \quad (2)$$

where σ is the distance between the centre point of the real frame and the predicted frame, and α is the horizontal angle between the predicted frame and the real frame

The calculation diagram for distance loss is shown in Figure 6.

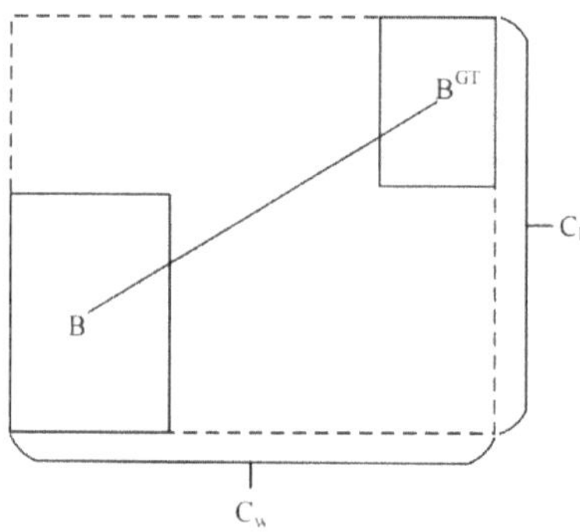

Figure 6. Calculation of distance loss.

The distance loss is defined as follows:

$$\Delta = \sum_{t=x,y} \left(1 - e^{-\gamma\rho_t}\right) \quad (3)$$

$$\begin{cases} \rho_x = \left(\frac{b_{c_x}^{gt} - b_{c_x}}{c_w}\right)^2 \\ \rho_y = \left(\frac{b_{c_y}^{gt} - b_{c_y}}{c_h}\right)^2 \\ \gamma = 2 - \Lambda \end{cases} \quad (4)$$

In the formula for distance loss, ρ is the degree of positional deviation between the true and predicted frames, and γ is the preferred distance value for time, related to angular loss. It can be seen that the contribution of the distance cost is greatly reduced when $\alpha \to 0$. On the contrary, the closer it is to $\pi/4$, the greater the contribution of the distance cost.

Shape loss is defined as follows:

$$\Omega = \sum_{t=w,h} (1 - e^{-\omega_t})^{\theta} \tag{5}$$

$$\begin{cases} \omega_w = \frac{|w - w^{gt}|}{\max(w, w^{gt})} \\ \omega_h = \frac{|h - h^{gt}|}{\max(h, h^{gt})} \end{cases} \tag{6}$$

where θ is the degree of concern for shape loss, ω_w and ω_h are the degrees of stretching of the width and height between the predicted and real frames, and w^{gt} and h^{gt} are the width and height of the real frame.

IoU cost is defined as follows:

$$\text{IoU} = \frac{|A \cap B|}{|A \cup B|} \tag{7}$$

where A denotes the prediction frame and B denotes the true frame.

SIoU is defined as follows:

$$L_{SIoU} = 1 - L_{IoU} + \frac{\Delta + \Omega}{2} \tag{8}$$

4. Experimental Section

4.1. Dataset and Metrics

The dataset comprises 6547 images of tyre treads captured by the camera, where image quality is affected by light brightness at the time of capture, and background interference outside the tread area hinders defect detection. For small defects, noise and light interference pose challenges for feature extraction by the network. Additionally, water stains on the surface can lead to misclassification as defects.

Sample images of tread defects are shown in Figure 7 with tread defects in the red box, in which there are lighting differences and water stains distributed on the tread, and the morphology of the defects is also different, which affects the recognition of defects on the tread. The tread data are partitioned into training, validation, and test sets in a ratio of 7:2:1, with the classified image datasets being organised into their respective directories to facilitate subsequent model training.

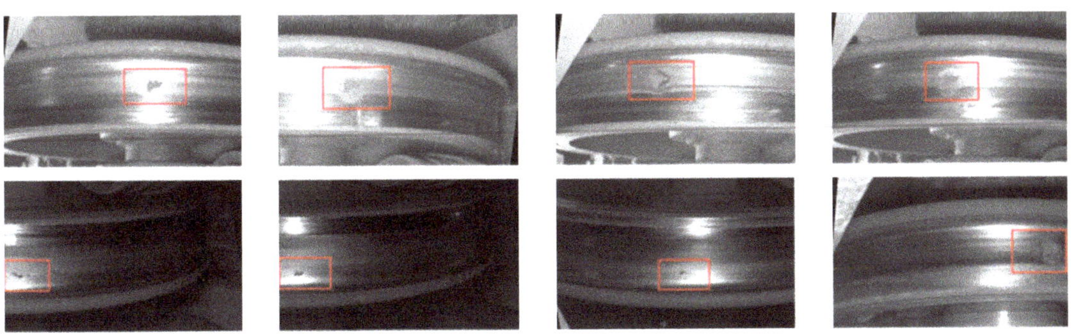

Figure 7. Sample images of tread defects.

The network's training parameters were configured according to Table 1, utilizing the Adma optimizer with a Momentum value of 0.8, Learning Rate set at 0.01, Batch Size of 16, Epochs set to 100, and Weight Decay adjusted to 0.0005. The experimental platform has a

GPU of RTX3060 (NVIDIA, Santa Clara, CA, USA) and a CPU of 12th Gen Intel® Core™ i7-12700H (Intel, Santa Clara, CA, USA).

Table 1. Experimental training parameters.

Training Parameters	Value
Optimiser	Adam
Momentum	0.8
Learning Rate	0.01
Batch Size	16
Epoch	100
Weight Decay	0.0005

Precision (P), recall (R), and mean average precision (mAP) values for all categories were selected as evaluation criteria and calculated as follows:

$$P = \frac{TP}{TP + FP} \tag{9}$$

$$R = \frac{TP}{TP + FN} \tag{10}$$

$$mAP = \frac{\sum_{i1}^{N} AP_i}{N} AP = \int_0^1 P(R)dR \tag{11}$$

$$mAP@50\% = \frac{1}{C}\sum_{i=1}^{C} AP@0.5_i \tag{12}$$

4.2. Ablation Experiment

In order to verify the effect of the improved algorithm on YOLOv8, ablation experiments were conducted, and the results are shown in Table 2. And model parameters and FPS were added as evaluation indicators.

Table 2. Results of ablation experiments.

Model	Structural Optimisations	SPPCSPC	SIoU	Precision	Recall	mAP@0.5	Parameters	FPS
YOLOv8				86.46	83.36	88.19	11.4	89
Optimisation models 1	√			93.21	90.66	94.35	12.8	86
Optimisation models 2		√		90.96	88.67	91.87	18.2	85
Optimisation models 3			√	94.05	89.06	92.92	11.5	88
Ours	√	√	√	96.30	95.31	96.95	18.9	79

The table uses a checkmark symbol (√) to indicate the corresponding improvement strategy. The same training parameters are applied to each group of experiments. Optimisation model 1 improves the detection structure by replacing the large target detection head with a shallow small-target detection head that is more sensitive to small defects. This reduces the misjudgement of water stains and improves detection accuracy. The accuracy of the detection results is improved by 6.75% and the mAP by 6.16% compared to the original model.

Optimisation model 2 replaces the original SPPC module with the SPPCSPC module, optimises its structure, trims the original CBS, and introduces SimAM. The multiscale structure of SPPCSPC, the sensory field of the neural network, and the small targets are now better matched, resulting in improved detection accuracy for the model. However, the number of parameters has slightly increased.

Optimisation model 3 uses SIoU as the loss function for border regression, resulting in more effective optimisation and improved positioning precision for the model bounding box.

Compared to the original YOLOv8, the improved version achieves a detection accuracy of 96.30%, which is 9.84% higher than the base model; a recall rate of 95.31%, which is 11.95% higher than the previous result; and an mAP of 96.95%, an improvement of 8.76%. The original YOLOv8 profile has a Missing Alarm Rate (MAR) of 16.64%. The improved YOLOv8 network reduced the false alarm rate for tread defects to 4.69%, a reduction of 11.95%.

Figure 8 shows the visualisation of the tread defect features of YOLOv8 and the improved YOLOv8, which can more clearly show the feature extraction by the network. From the figure, we can see that the improved YOLOv8 network detects defects better and with higher confidence.

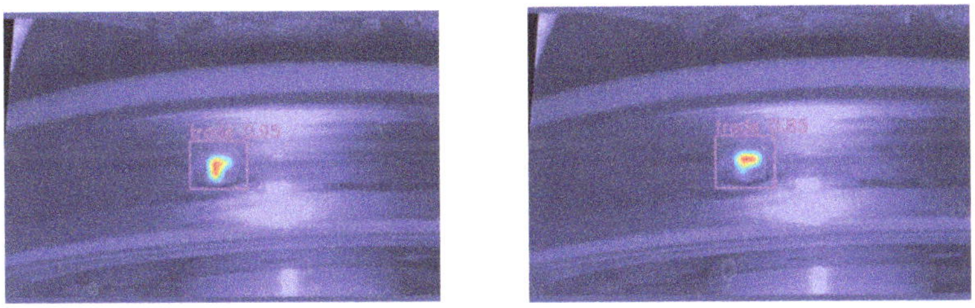

(**a**) (**b**)

Figure 8. Grad-CAM visualisation: (**a**) improved YOLOv8; (**b**) YOLOv8.

The improved YOLO8 algorithm is demonstrated in Figure 9, showcasing its proficiency in detecting minute defects on the wheel's tread surface. This improved neural network exhibits remarkable capability in identifying small defects present on the tread.

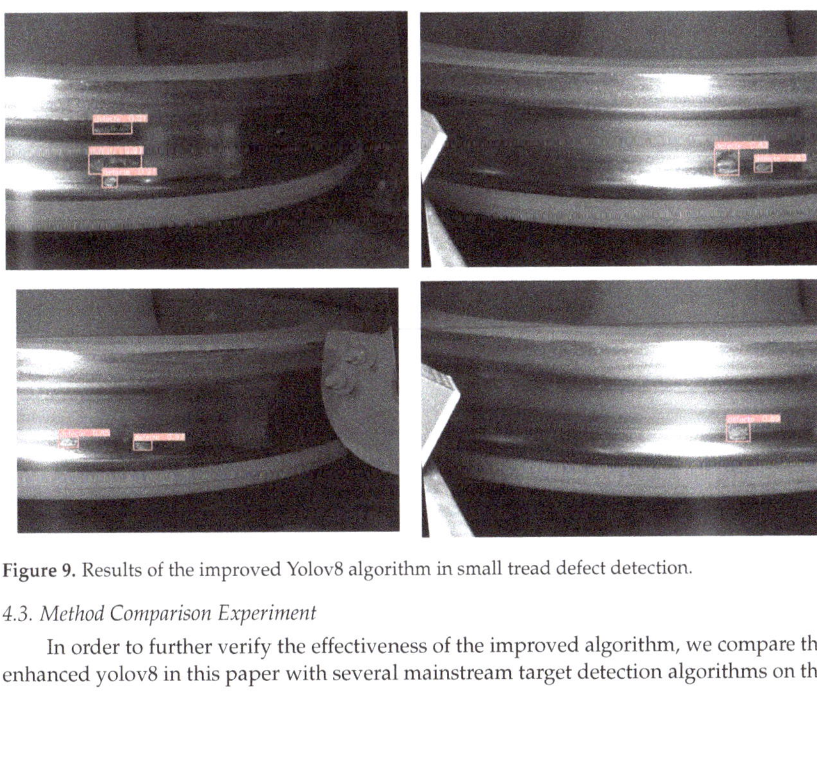

Figure 9. Results of the improved Yolov8 algorithm in small tread defect detection.

4.3. Method Comparison Experiment

In order to further verify the effectiveness of the improved algorithm, we compare the enhanced yolov8 in this paper with several mainstream target detection algorithms on the treadmill dataset, and the results are shown in Table 3. The evaluation metrics of the model include accuracy, recall, MAP and model parameters.

Table 3. Comparison results of tread defect detection.

Model	Precision	Recall	mAP@0.5	Parameters
Faster R-CNN	80.38	82.03	83.70	40.2
SSD [22]	47.02	88.76	78.12	26.3
YOLOv3 [23]	75.59	72.19	76.80	61.5
YOLOv5s	80.12	84.19	85.15	7.1
YOLOv8s	86.46	83.36	88.19	11.4
Improved YOLOv8	96.30	95.31	96.95	18.9

The metrics of this paper's model and other commonly used models for tread detection are presented in Table 3. The mAP score achieved by our proposed model surpasses that of existing algorithms, reaching an impressive 96.95%, thus demonstrating its superior performance in comparison. Notably, accurately detecting defects in the tread under challenging conditions such as varying lighting and water stains has been a persistent challenge for existing methods. However, the improved YOLOv8 algorithm proposed in this study effectively addresses these challenges.

YOLOv8 builds upon the earlier versions (YOLOv3 and YOLOv5) by incorporating advancements like the C2F module and decoupled head while abandoning the anchor-based concept in favour of an anchor-free approach. On the other hand, the SSD model suffers from a loss of target feature information due to multilayer convolution, making it less effective at detecting small targets. Faster R-CNN solves the problem of scale variation through its pyramid model, but still suffers from computational redundancy in the subsequent detection phase.

Compared with Faster R-CNN and YOLOv5s, our method showcases improvements of 10.71% and 11.5% in terms of accuracy, respectively, along with an increase in the mAP of 13.12% and 17.65%.

A comparison of our model with other models in terms of parameters and accuracy is visualised in Figure 10, where the horizontal axis represents the parameters of the model and the vertical axis represents the MAP. It can be observed that our model exhibits superior accuracy while maintaining a lower parameter count.

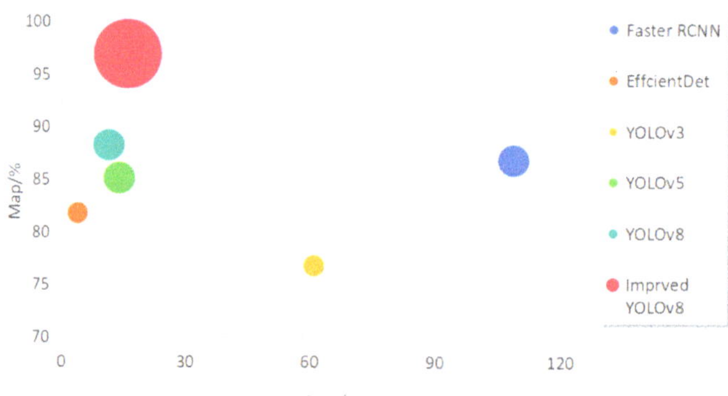

Figure 10. Comparison results with mainstream algorithms.

5. Conclusions

This paper proposes an improved YOLOv8 algorithm for detecting high-speed train wheelset treads. To avoid interference from water stains during defect detection, the detection structure in the YOLOv8 model has been improved by adding a small-target, shallow detection layer. Furthermore, the SPPC module has been replaced with the reconstructed

SPPCSPC to achieve a more efficient multiscale fusion that prioritises small-target defects. Additionally, the loss function has been substituted with SIoU to enhance the accuracy of the regression. The algorithm proposed in this study was compared to current target detection algorithms on a self-built dataset to evaluate its wheel-to-tread detection performance. The results indicate that the proposed algorithm outperforms the original YOLOv8 model, achieving a detection rate of 96.95% for tread defects, which is an improvement of 8.76%. In the future, we will pay more attention to the improvement of detection speed in real time and further improve the accuracy of detection.

Author Contributions: Conceptualization, Y.W. and X.G.; methodology, Y.W. and J.L; software, Y.W.; validation, Y.W., X.G. and L.L.; formal analysis, Y.W.; investigation, Y.W.; resources, Y.W., X.G. and J.L.; data curation, Y.W.; writing—original draft preparation, Y.W.; writing—review and editing, Y.W., J.L., X.G. and L.L.; visualization, Y.W.; supervision, X.G.; project administration, L.L. and J.L.; funding acquisition, J.L. All authors have read and agreed to the published version of the manuscript.

Funding: This research was funded by the National Natural Science Foundation of China under Grant No. 61960206010.

Institutional Review Board Statement: Not applicable.

Informed Consent Statement: Not applicable.

Data Availability Statement: Data are contained within the article.

Conflicts of Interest: The authors declare no conflict of interest.

References

1. Tang, R.; De Donato, L.; Bešinović, N.; Flammini, F.; Goverde, R.M.P.; Lin, Z.; Liu, R.; Tang, T.; Vittorini, V.; Wang, Z. A Literature Review of Artificial Intelligence Applications in Railway Systems. *Transp. Res. Part C Emerg. Technol.* **2022**, *140*, 103679. [CrossRef]
2. Girshick, R.; Donahue, J.; Darrell, T.; Malik, J. Rich Feature Hierarchies for Accurate Object Detection and Semantic Segmentation. In Proceedings of the IEEE Conference on Computer Vision and Pattern Recognition (CVPR), Columbus, OH, USA, 23–28 June 2014. [CrossRef]
3. Ren, S.; He, K.; Girshick, R.; Sun, J. Faster R-CNN: Towards Real-Time Object Detection with Region Proposal Networks. *IEEE Trans. Pattern Anal. Mach. Intell.* **2017**, *39*, 1137–1149. [CrossRef] [PubMed]
4. Chen, M.; Yu, L.; Zhi, C.; Sun, R.; Zhu, S.; Gao, Z.; Ke, Z.; Zhu, M.; Zhang, Y. Improved Faster R-CNN for Fabric Defect Detection Based on Gabor Filter with Genetic Algorithm Optimization. *Comput. Ind.* **2022**, *134*, 103551. [CrossRef]
5. Xu, X.; Zhao, M.; Shi, P.; Ren, R.; He, X.; Wei, X.; Yang, H. Crack Detection and Comparison Study Based on Faster R-CNN and Mask R-CNN. *Sensors* **2022**, *22*, 1215. [CrossRef] [PubMed]
6. Redmon, J.; Divvala, S.; Girshick, R.; Farhadi, A. You Only Look Once: Unified, Real-Time Object Detection. In Proceedings of the IEEE Conference on Computer Vision and Pattern Recognition, Las Vegas, NV, USA, 26 June–1 July 2016; pp. 779–788.
7. Bhattacharya, A.; Cloutier, S.G. End-to-End Deep Learning Framework for Printed Circuit Board Manufacturing Defect Classification. *Sci. Rep.* **2022**, *12*, 12559. [CrossRef] [PubMed]
8. Stefenon, S.F.; Seman, L.O.; Klaar, A.C.R.; Ovejero, R.G.; Leithardt, V.R.Q. Hypertuned-YOLO for Interpretable Distribution Power Grid Fault Location Based on EigenCAM. *Ain Shams Eng. J.* **2024**, *15*, 102722. [CrossRef]
9. Ding, F.; Zhuang, Z.; Liu, Y.; Jiang, D.; Yan, X.; Wang, Z. Detecting Defects on Solid Wood Panels Based on an Improved SSD Algorithm. *Sensors* **2020**, *20*, 5315. [CrossRef] [PubMed]
10. Zhou, Q.; Wang, H. CABF-YOLO: A Precise and Efficient Deep Learning Method for Defect Detection on Strip Steel Surface. *Pattern Anal. Appl.* **2024**, *27*, 36. [CrossRef]
11. Hussain, M. YOLO-v1 to YOLO-v8, the Rise of YOLO and Its Complementary Nature toward Digital Manufacturing and Industrial Defect Detection. *Machines* **2023**, *11*, 677. [CrossRef]
12. Fan, Y.; Mao, S.; Li, M.; Wu, Z.; Kang, J. CM-YOLOv8: Lightweight YOLO for Coal Mine Fully Mechanized Mining Face. *Sensors* **2024**, *24*, 1866. [CrossRef] [PubMed]
13. Sampurno, R.M.; Liu, Z.; Abeyrathna, R.M.R.D.; Ahamed, T. Intrarow Uncut Weed Detection Using You-Only-Look-Once Instance Segmentation for Orchard Plantations. *Sensors* **2024**, *24*, 893. [CrossRef] [PubMed]
14. Selvaraju, R.R.; Das, A.; Vedantam, R.; Cogswell, M.; Parikh, D.; Batra, D. Grad-CAM: Why Did You Say That? *arXiv* **2017**, arXiv:1611.07450.
15. He, K.; Zhang, X.; Ren, S.; Sun, J. Spatial Pyramid Pooling in Deep Convolutional Networks for Visual Recognition. *IEEE Trans. Pattern Anal. Mach. Intell.* **2015**, *37*, 1904–1916. [CrossRef] [PubMed]
16. Wang, C.-Y.; Mark Liao, H.-Y.; Wu, Y.-H.; Chen, P.-Y.; Hsieh, J.-W.; Yeh, I.-H. CSPNet: A New Backbone That Can Enhance Learning Capability of CNN. In Proceedings of the 2020 IEEE/CVF Conference on Computer Vision and Pattern Recognition Workshops (CVPRW), Seattle, WA, USA, 14–19 June 2020; IEEE: Piscataway, NJ, USA, 2020; pp. 1571–1580.

17. Yang, L.; Zhang, R.-Y.; Li, L.; Xie, X. SimAM: A Simple, Parameter-Free Attention Module for Convolutional Neural Networks. In Proceedings of the 38th International Conference on Machine Learning, Virtual, 18–24 July 2021.
18. Rezatofighi, H.; Tsoi, N.; Gwak, J.; Sadeghian, A.; Reid, I.; Savarese, S. Generalized Intersection Over Union: A Metric and a Loss for Bounding Box Regression. In Proceedings of the IEEE/CVF Conference on Computer Vision and Pattern Recognition (CVPR), Long Beach, CA, USA, 15–20 June 2019. [CrossRef]
19. Zheng, Z.; Wang, P.; Liu, W.; Li, J.; Ye, R.; Ren, D. Distance-IoU Loss: Faster and Better Learning for Bounding Box Regression. *Proc. AAAI Conf. Artif. Intell.* **2020**, *34*, 12993–13000. [CrossRef]
20. Zhang, Y.-F.; Ren, W.; Zhang, Z.; Jia, Z.; Wang, L.; Tan, T. Focal and Efficient IOU Loss for Accurate Bounding Box Regression. *Neurocomputing* **2022**, *506*, 146–157. [CrossRef]
21. Zhora, G. SIoU Loss: More Powerful Learning for Bounding Box Regression. *arXiv* **2022**, arXiv:2205.12740.
22. Liu, W.; Anguelov, D.; Erhan, D.; Szegedy, C.; Reed, S.; Fu, C.-Y.; Berg, A.C. SSD: Single Shot MultiBox Detector. In *Computer Vision—ECCV 2016*; Leibe, B., Matas, J., Sebe, N., Welling, M., Eds.; Lecture Notes in Computer Science; Springer International Publishing: Cham, Switzerland, 2016; Volume 9905, pp. 21–37. ISBN 978-3-319-46447-3.
23. Redmon, J.; Farhadi, A. Yolov3: An Incremental Improvement. *arXiv* **2018**, arXiv:1804.02767.

sensors

Article

A Water Environment-Based Simulated Method for Ultrasonic Testing of Slag Inclusion Weld Defects Based on Improved VMD

Jing Zhang [1,2], Guocai Zhang [3], Zijie Chen [1], Hailin Zou [1], Shuai Xue [2], Jianjie Deng [2] and Jianqing Li [1,*]

1 School of Computer Science and Engineering, Faculty of Innovation Engineering, Macau University of Science and Technology, Macau 999078, China; 2009853gii30017@student.must.edu.mo (J.Z.)
2 School of Applied Science and Civil Engineering, Beijing Institute of Technology, Zhuhai 519006, China
3 School of Mechanical and Automotive Engineering, South China University of Technology, Guangzhou 510640, China
* Correspondence: jqli@must.edu.mo

Abstract: The identification of slag inclusion defects in welds is of the utmost importance in guaranteeing the integrity, safety, and prolonged service life of welded structures. Most research focuses on different kinds of weld defects, but branch research on categories of slag inclusion material is limited and critical for safeguarding the quality of engineering and the well-being of personnel. To address this issue, we design a simulated method using ultrasonic testing to identify the inclusion of material categories in austenitic stainless steel. It is based on a simulated experiment in a water environment, and six categories of cubic specimens, including four metallic and two non-metallic materials, are selected to simulate the slag materials of the inclusion defects. Variational mode decomposition optimized by particle swarm optimization is employed for ultrasonic signals denoising. Moreover, the phase spectrum of the denoised signal is utilized to extract the phase characteristic of the echo signal from the water–slag specimen interface. The experimental results show that our method has the characteristics of appropriate decomposition and good denoising performance. Compared with famous signal denoising algorithms, the proposed method extracted the lowest number of intrinsic mode functions from the echo signal with the highest signal-to-noise ratio and lowest normalized cross-correlation among all of the comparative algorithms in signal denoising of weld slag inclusion defects. Finally, the phase spectrum can ascertain whether the slag inclusion is a thicker or thinner medium compared with the weld base material based on the half-wave loss existing or not in the echo signal phase.

Keywords: slag inclusion; ultrasonic testing; weld defects; variational mode decomposition; particle swarm optimization

1. Introduction

Austenitic stainless steel is widely preferred by industries and used in media storage, transportation, processing and production, equipment manufacturing, and other industries due to its mechanical properties (such as high tensile strength, good yield strength, and high-temperature strength), outstanding corrosion resistance, and good weldability [1]. Currently, welding is fusing metal parts to form a strong bond that can withstand harsh conditions. The quality of the weld is critical to safety and performance, as defects can lead to failures, accidents, and financial losses [2]. Weld inspection is an important task in non-destructive testing (NDT), which is necessary to detect weld defects. During the welding process, defects can occur on a weldment, such as porosity, gas pores, longitudinal cracks, lack of penetration, and slag inclusion [3]. Slag inclusion in these defects means volume defects not as visible as cracks, incomplete penetration, or incomplete fusion defects, which are called surface defects.

Defect testing of weldments has attracted much attention in recent years. Commonly used weld defect detection methods include eddy current [4], magnetic particle inspection [5],

X-ray [6], spectroscopy emission [7], and ultrasonic testing (UT) [8]. Ghubade and Kumar [9], in their review work on casting defects and methodologies for quality improvement, provided insight into casting defects and casting quality. Singh et al. [10] proposed an active pulse infrared thermal imaging technology detection method which uses a FlukeTi 32 thermal imaging camera to record the metal casting heating and temperature decay processes. Using the prior knowledge that defect-containing regions have higher temperatures than defect-free regions, defect regions with high thermal concentrations and the size and depth of defects are characterized. Li et al. [11] proposed a frequency band-selecting pulsed eddy current testing (FSPECT) method. Compared with energy equivalent square wave PEC technology, FSPECT responds better and faster to deep defects. The authors of [12] designed an active thermal imaging system with induction heating. By recording the surface temperature with an infrared thermal imaging camera, the entire infrared image sequence is evaluated using a Fourier transform, and the phase image is processed to detect surface cracks. At the same time, the signal distribution around the cracks at different depths is analyzed and calculated to determine the crack depth. Wang et al. [13] designed a digital ray detection system based on charge-coupled devices. The system can receive X-rays with a wider energy range and obtain higher imaging quality. Tkocz et al. [14] designed a low-cost four-channel phased array electromagnetic ultrasonic transducer system. The generated focused ultrasonic beam completed the defect detection experiment on a rough cast steel plate with a thickness of 32 cm. A-scan data showed that the system could detect artificial defects up to a depth of 16 cm.

The research mentioned above primarily focused on the size and depth of internal defects in materials. However, the literature on the different categories of slag inclusion defects is limited. Slag inclusion defects are typically classified as metal or non-metal inclusions. Slag defects arise from multiple factors, including suboptimal welding practices, inadequate preheating, and imprecise control over welding parameters. It is difficult to determine the accuracy of the test results for slag inclusion types since the process is irreversible and destructive to the specimen. The process of producing austenitic stainless steel is highly intricate and demands a controlled production environment. This poses a challenge when it comes to obtaining slag inclusion defect samples with known slag material. Numerical simulated methods based on finite element analysis have attracted attention from many researchers [15,16]. This simulation of the ultrasonic testing system could easily design or optimize an ultrasonic testing system by regulating the parameters of the system, which offers a theoretical foundation for detection and reduces detection blindness [17]. However, the effectiveness of this simulation approach often tends to be overly ideal. Therefore, we designed a water environment-based experiment for slag inclusion defect detection to address these issues.

Radiographic testing (RT) can be implemented utilizing either X-ray radiation or a gamma-emitting radioactive source [18]. Given the pronounced dependence of the X-ray absorption coefficient on the material density, radiographic imaging is particularly well suited for identifying volume defects characterized by an altered mass content, such as slag inclusions or porosity [19]. RT offers a more direct and precise determination of the material type of slag inclusions within welds. However, it is limited by the harmful radiation it generates, the complexity of the detection process, and the extended duration of the testing, which preclude its suitability for in situ, real-time detection in industrial environments, such as the scenario of steel structure weld inspection for a building. Compared with the RT method, UT can overcome these disadvantages and is more widely applied in weld inspections due to its safety, portability, in situ and real-time feedback, and low cost. Moreover, UT can test the depth information of a defect. UT is a commonly used method for detecting flaws in solids. It provides precise information about the position and size of defects. Therefore, we chose UT in this paper to detect the inclusion defects. However, ultrasonic signals can be easily scattered by complex factors such as the environment and the object being measured. This often leads to noise in the received echo signal, which can affect the quality and reliability of the measurement. Therefore, denoising is an essential step for removing noise and enhancing the signal. Classical signal denoising algorithms include wavelet transform (WT) [20], empirical mode decomposition (EMD) [21], and its variants such as ensemble EMD (EEMD) [22] and complementary EEMD with adaptive noise (CEEMDAN) [23]. EMD series methods may produce mode mixing or cause incomplete signal decomposition and denoising with residual noise in the denoised signal. Variational mode decomposition (VMD) is a new algorithm for decomposing signals into several intrinsic mode functions (IMFs) with different center frequencies to pick out the effective signal [24]. It overcomes the problems of endpoint effects and modal aliasing in EMD [25]. Compared with other methods, VMD has been proven to yield a better frequency resolution and noise immunity [26]. However, it requires manually setting the hyperparameters. To address this issue, researchers have investigated methods for setting parameters automatically. For instance, Hua et al. applied the grasshopper optimization algorithm (GOA) to further fine-tune the parameters in VMD for Lidar signal decomposition [27]. Similarly, Qi et al. harnessed the grey wolf optimization (GWO) method to determine the most effective VMD parameters, alongside parallel EMD for Lidar signal noise reduction [28]. Additionally, Long et al. made use of particle swarm optimization (PSO) to identify the ideal VMD parameters for filtering out disturbances from ultra-high-frequency partial discharge signals [29].

In the ultrasonic detection of slag inclusion defects in austenitic steel welds, the signal will exhibit a superior signal-to-noise ratio after effective denoising. A Fast Fourier transform (FFT) is performed on the denoised signal to obtain the phase spectrum. In accordance with the principles of wave propagation theory, it is established that mechanical and electromagnetic waves, when encountering an interface with a transition to a denser medium, experience a π phase shift upon reflection. This is in contrast to the situation where they interact with a less dense medium, where no such phase change occurs. This intrinsic relationship between the echo phase and the characteristics of the interface material is invariant across diverse experimental settings and environmental conditions. Consequently, we introduce a novel experimental methodology for the ultrasonic detection of slag inclusion defects in welds, which is based on the simulation of aquatic environments.

In a nutshell, we designed a series of experiments utilizing selected cubic specimens made of various materials immersed in a water-based environment to simulate different material types in slag inclusion defects found in weldments. Specifically, the six selected cubes were composed of different materials, including metallic materials such as iron, copper, aluminum, and magnets, as well as non-metallic materials like wood and resin. The primary objective was to analyze the phase characteristic information of ultrasonic signals and investigate the relationship between phase and slag inclusion materials by utilizing ultrasonic testing in conjunction with an improved VMD denoising method, thereby achieving the classification and detection of slag inclusion defects.

The rest of this paper is organized as follows. The methodology employed in this study is outlined in Section 2; the experiments are shown in Section 3; and the results analysis is discussed in Section 4. Finally, this work is summarized in the Section 5.

2. Theory and Methods

2.1. VMD

2.1.1. VMD Theory

VMD is a non-recursive variational mode decomposition algorithm showing excellent capacity for processing nonlinear and non-stationary signals which was proposed by Dragomiretskiy [24]. It ensures sparsity by limiting the central frequency and bandwidth of each IMF. VMD can avoid issues related to mode mixing and boundary effects, which always exists in EMD, WT, etc. The constrained model is defined as follows:

$$\begin{cases} \min\limits_{(u_k,\omega_k)} \left\{ \sum\limits_{k=1}^{K} \left\| \partial_t \left[\left(\sigma(t) + \frac{j}{\pi t} \right) \otimes u_k(t) \right] e^{-j\omega_k t} \right\|_2^2 \right\} \\ \text{s.t.} \ \sum\limits_{k=1}^{K} u_k(t) = f(t) \end{cases} \tag{1}$$

where $u_k(t)$ represents the kth IMF, ω_k represents the central frequency of $u_k(t)$, $k = 1, 2, \cdots, K$, $\sigma(t)$ is the unit pulse function, $\otimes$ represents a convolutional computation, and ∂_i represents a partial derivative operation.

Initially, the Hilbert transform was applied to each IMF to obtain a one-sided spectrum:

$$\left(\sigma(t) + \frac{j}{\pi t}\right) \otimes u_k(t). \tag{2}$$

Then, we estimated the central frequency (ω_k) of each IMF and modulated each IMF to the baseband:

$$\left[\left(\sigma(t) + \frac{j}{\pi t}\right) \otimes u_k(t)\right] e^{-j\omega_k t}. \tag{3}$$

Lastly, the bandwidth of each IMF was estimated through the H_1 Gaussian smoothness by calculating the $L^2 - norm$ of the demodulated signal gradient.

2.1.2. Solution of Variational Equations

The equation initially presented as Equation (1) was converted into a form without constraints by employing the Lagrange multiplier, denoted by λ. Furthermore, to maintain the fidelity of signal reconstruction in the presence of additive Gaussian noise, a quadratic penalty factor, represented by α, was utilized. The resulting unconstrained formulation is expressed as follows:

$$L(\{u_k\}, \{\omega_k\}, \lambda) = \alpha \sum_k \left\| \partial_t \left[\left(\sigma(t) + \frac{j}{\pi t}\right) \otimes u_k(t)\right] e^{-j\omega_k t} \right\|^2 + \left\| f(t) - \sum_k u_k(t) \right\|_2^2 + < \lambda(t), f(t) - \sum_k u_k(t) > \tag{4}$$

Next, by solving Equation (4), $\hat{f}(\omega)$ and $\hat{\omega}_k$ are updated as follows:

$$\widehat{u}_k^{n+1}(\omega) = \frac{\hat{f}(\omega) - \sum_{i \neq k} \widehat{u}_i(\omega) + \frac{\hat{\lambda}(\omega)}{2}}{1 + 2\alpha(\omega - \omega_k)^2} \tag{5}$$

$$\omega_k^{n+1} = \frac{\int_0^\infty \omega |\hat{u}_k(\omega)|^2 \, d\omega}{\int_0^\infty |\hat{u}_k(\omega)|^2 \, d\omega} \tag{6}$$

where $\hat{u}_i(\omega), \hat{f}(\omega), \hat{\lambda}(\omega)$ are the Fourier transform of $u_k(t), f(t), \lambda(t)$ respectively.

The VMD algorithm's process has four main steps:

- Step 1. Initialize u_k^1, ω_k^1, λ_k^1, and n such that all their values are zero, and set a suitable number of decomposition modes K.
- Step 2. Iteratively update u_k and ω_k using Equations (5) and (6), and stop when the preset value K is reached.
- Step 3. Update λ using the following equation:

$$\hat{\lambda}^{n+1}(\omega) \leftarrow \hat{\lambda}^n(\omega) + \tau\left(\hat{f}(\omega) - \sum_k \widehat{u}_k^{n+1}(\omega)\right) \tag{7}$$

where τ represents the noise tolerance for a given discrimination accuracy $0 < \phi < 1 \times 10^{-7}$. If this satisfies the following condition, then stop the iteration. Otherwise, return to Step 2:

$$\sum_{k=1}^{K} \frac{||\hat{u}_k^{n+1} - \widehat{u}_k^n||_2^2}{||\hat{u}_k^n||_2^2} = \phi \tag{8}$$

where $\hat{u}_k$ is the updated expression of u_k.

- Step 4. Arrange the obtained K IMF components from low to high frequencies, calculate the correlation coefficient between each component and the original signal, eliminate the IMF components with smaller correlation coefficients, and reconstruct the remaining components.

2.2. Improved VMD through Particle Swarm Optimization

Efficient and precise parameter selection for VMD hinges on the choice of (K, α). This study employs particle swarm optimization to ascertain the optimal (K, α) values for VMD, thereby eliminating subjective judgment and automating the search for the best parameters. The fundamental concept behind employing the PSO algorithm to discover the best solution lies in the collaborative effort and exchange of information among the various members of the population [30]. The algorithm can be described as follows.

In an N-dimensional space, a swarm consists of m particles, where $x_i = (x_{i1}, x_{i2}, \cdots, x_{in})$ represents the position of the ith particle in the N-dimensional space, $v_i = (v_{i1}, v_{i2}, \cdots, v_{in})$ represents the motion of a particle, $p_i = (p_{i1}, p_{i2}, \cdots, p_{in})$ represents the local optimal solution that the particle has experienced, and $p_g = (p_{g1}, p_{g2}, \cdots, p_{gn})$ is the best position among the entire swarm as the global optimum.

In PSO, the update of velocity and position of a particle can be calculated as follows:

$$v_{id}(t+1) = \omega v_{id}(t) + c_1 r_1 [p_{id} - x_{id}(t)] + c_2 r_2 \left[p_{gd} - p_{gd}(t)\right] \tag{9}$$

$$x_{id}(t+1) = v_{id}(t) + v_{id}(t+1) \tag{10}$$

where ω is the inertia weight, c_1, c_2 are the learning factors, and r_1, r_2 are random numbers distributed in the range [0, 1].

Specifically, the optimal parameter (K, α) is selected in parallel by introducing a two-dimensional particle swarm. According to Equations (9) and (10), there will be the following equations:

$$\begin{bmatrix} v_k(t+1) \\ v_\alpha(t+1) \end{bmatrix} = \omega \begin{bmatrix} v_k(t) \\ v_\alpha(t) \end{bmatrix} v_{id}(t) + c_1 r_1 \left(\begin{bmatrix} p_k \\ p_a \end{bmatrix} - \begin{bmatrix} x_k(t) \\ x_\alpha(t) \end{bmatrix} \right) + c_2 r_2 \left(\begin{bmatrix} g_k \\ g_\alpha \end{bmatrix} - \begin{bmatrix} x_k(t) \\ x_\alpha(t) \end{bmatrix} \right) \tag{11}$$

$$\begin{bmatrix} x_K(t+1) \\ x_\alpha(t+1) \end{bmatrix} = \begin{bmatrix} x_K(t) \\ x_\alpha(t) \end{bmatrix} + \begin{bmatrix} v_K(t+1) \\ v_\alpha(t+1) \end{bmatrix}. \tag{12}$$

The intricacy of components indicates their regularity, with reduced complexity signifying a higher degree of regularity. The sample entropy serves as a statistical instrument for assessing the complexity and unpredictability of time series data [31]. Essentially, the sample entropy measures the probability of recurrence of comparable data patterns in a time series. Consequently, a greater sample entropy value suggests a more intricate, less predictable, and more irregular time series.

The equation for the sample entropy is given below:

$$SampEn(m, r, N) = -ln\left[\frac{B^{m+1}(r)}{B^m(r)}\right] \tag{13}$$

where r represents the signal tolerance, m is the embedding dimension, N is the signal length, and $B^m(r)$ is the mean pattern count.

The fitness function described in this paper, represented by Equation (14), is calculated as the average of the squared error (SE) for all intrinsic mode functions (IMFs) obtained from a single variational mode decomposition (VMD). A $fitness$ value of less than 0.1 indicates that the optimal solution has been achieved:

$$fitness(K, \{u_k\}) = \frac{1}{K}\sum_{k=1}^{K} SampEn(u_i) \tag{14}$$

where K is the total number of IMFs decomposed via VMD and $\{u_k\} = \{u_1, \ldots, u_K\}$ are IMFs.

PSO can be optimized by minimizing the fitness function with K and alpha as the optimization-seeking parameters. The following steps should be taken:

- Step 1. Firstly, input the original series (ultrasonic signal). Then, set the range of the VMD parameter pairs (K, α), where K takes integer values in the range of [2, 12] and α's range is within [100, 30,000]. After that, initialize the positions and velocities of the swarm's particles. Finally, calculate the local and global optimal solutions using Equation (14).
- Step 2. Apply Equations (11) and (12) to update the particle positions. Refresh the local and global solutions accordingly.
- Step 3. Output the global optimal solution of the IMF's number K and the penalty factor α when the fitness value is less than 0.1 or the maximum number of iterations is reached.
- Step 4. Using the optimized results, apply the VMD method to break down the vibration signal into different IMFs. Filter out the IMFs that fall outside the frequency range of the ultrasonic processing equipment, and combine the remaining effective IMFs to obtain a precise ultrasonic vibration signal.

2.3. Ultrasonic Wave Propagation Theory

Classical wave propagation theory posits that a wave transitioning from a less dense to a denser medium will undergo reflection at the point of intersection between the two media. The cause of this reflection is the abrupt phase shift of the reflected wave at the midpoint of its cycle, a concept referred to as half-wave loss [32]. Half-wave loss is a phenomenon that occurs when a mechanical wave reflects at the interface between two media with different properties. This typically happens when the wave transitions from a less dense medium to a denser one, such as from air to a solid. In the case of mechanical waves like ultrasound, this occurs when the wave transitions from a medium with a lower refractive index to one with a higher refractive index. The reflected wave undergoes a phase shift of *pi*, which is equivalent to a half-wavelength change in the spatial period. This phase shift leads to a so-called phase inversion, where the wave phase becomes negative after reflection if it is positive before reflection at the interface between the media.

Half-wave loss in the phase of the ultrasonic echo will occur at the interface when propagation is constituted in a wave-dense medium compared to steel, such as the tungsten inclusion defect. However, the phase of ultrasonic echo will change continuously when ultrasonic waves propagate through the steel matrix and encounter porosity or non-metallic inclusions, and a reflection will occur at the interface between the steel and non-metallic inclusion without the half-wave loss. Depending on the phase of the ultrasonic echo, we can determine whether the inclusions of the slag are a thin medium or thick medium relative to the steel matrix.

Welding slag inclusions are embedded within the austenitic stainless steel matrix, serving as a model for an environment with a foreign object included. Foreign objects can be categorized as either wave-thin or wave-dense media relative to the matrix material [33]. According to wave propagation theory, ultrasonic waves propagate through the environment and reflect off foreign objects. The presence or absence of a half-wave loss in the echo signal's phase can indicate whether the foreign material is a wave-thin or wave-dense medium. In our experimental set-up, water and diverse material samples were employed to simulate the environment and slag inclusions, respectively.

3. Experiment

3.1. Experimental Set-Up

As it is not easy to observe the material of slag inclusion in weld defects, this experiment utilized a simulated experiment with a water environment to simulate defects inside the weldment, as shown in Figure 1a. The ultrasound experimental system consisted of an SIUI CTS-1002® digital ultrasound machine (Shantou, China), a 5Z14N transmitting probe (SIUI longitudinal wave straight probe, 5 MHz, 14 mm, shown in Figure 1b), and a signal acquisition computer (Tektronix MDO4024C oscilloscope, Beaverton, OR, USA). The ultrasonic probe was responsible for transmitting and receiving the ultrasonic signals, producing pulsed ultrasonic signals at a nominal center frequency of 5 MHz. Theoretically, its operating frequency would be equal to the nominal frequency. However, a downshift in frequency could be potentially caused by the manufacturing technology of the piezoelectric crystal probe and aging of the detector with a served time duration. Figure 2 illustrates the loaded excitation waveform of the probe when the SIUI CTS-1002 was operating at a voltage of 100 V. The oscilloscope operated at a sampling frequency of 2.5 GHz, which was employed to visualize the temporal disparity of ultrasonic waves as they traversed various positions within austenitic stainless steel. It filled the tank with water, fixed a cylindrical pillar on the bottom, and placed the cubic sample on top of the pillar in this experiment. We maintained a certain distance between the sample and the bottom of the water tank to distinguish the echo signals reflected from the sample surface and the bottom of the tank. Figure 1c shows that the vertical distances were 9.00 cm and 10.20 cm between the upper surface of the slag specimen and the transmitting probe surface and the bottom surface of the specimen and the bottom of the water tank, respectively. The specimen was a cubic sample with a side length of 2.00 cm. The ultrasonic probe was placed on the top of the water tank to transmit ultrasound vertically and directly immerse it in the water, which eliminated the need for a coupling agent and resulted in reduced attenuation of the ultrasound wave. The reflected ultrasonic echo signals of the 6 cubic specimens were collected by our ultrasound system.

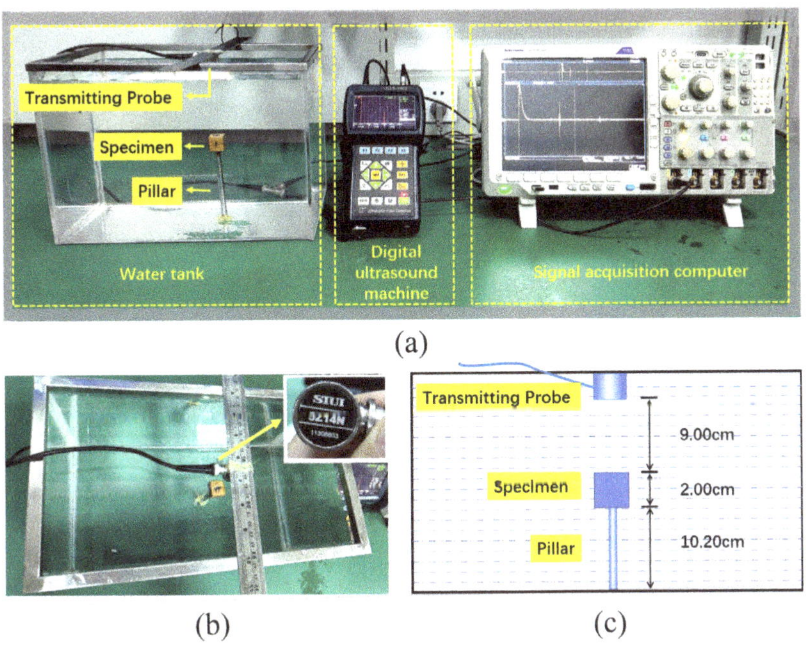

Figure 1. (**a**) Experimental system of ultrasonic testing based on water environment simulation. (**b**) Top view of the water tank and ultrasonic probe. (**c**) Schematic diagram of vertical distances between objects placed inside the water tank.

The original echo signals were collected by the oscilloscope shown in Figure 3, where (a) is the reference original echo signal obtained when no slag cube block was placed in the water environment and (b) is the original echo signal obtained when a wood specimen was immersed in the water tank. The signal encircled by a red box on the left in Figure 3a is the echo from the bottom surface of the water tank, and the following weaker one encircled by the right red box is the secondary echo. It can be seen that the first echo's occurrence time was approximately 2.8×10^{-4} s, given an ultrasonic wave propagation velocity of approximately 1500 m/s. This corresponded to a traveled distance of approximately 42 cm, which was consistent with twice the distance between the ultrasonic probe and the bottom surface of the water tank (21.20 cm), as shown in Figure 1c. Upon placement of a slag specimen in the water, the ultrasonic waves underwent multiple traversals between the ultrasonic detector and specimen. As depicted in Figure 3b, the red rectangles highlight the initial and secondary echoes reflected from the water–wood cubic block interface, occurring at approximately 1.2×10^{-4} s and 2.4×10^{-4} s, respectively. These echo arrival times corresponded to the distances traversed by the ultrasonic waves, in accordance with the experimental distance set-up of 9.00 cm between the transmitting probe and slag inclusion specimen.

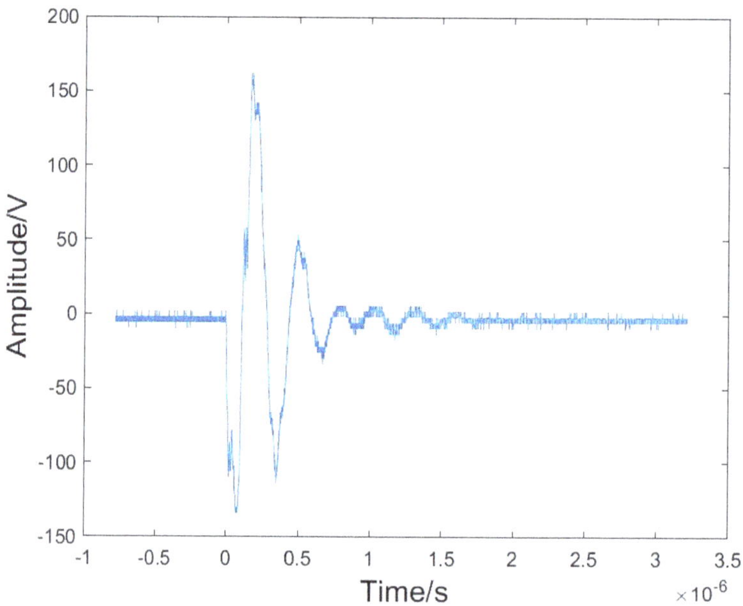

Figure 2. The loaded excitation waveform of the probe with the SIUI CTS–1002 operating at a working voltage of 100 V.

The original echo signal reflected from the sample surface was intercepted and used for analysis in this experiment. We chose the experiment of the wood specimen immersed in water as an example to illustrate the experimental procedures and subsequent analysis of the results. The chosen white birch wood specimen had a density of approximately 0.599 g/cm^3 in a dry state and approximately 0.676 g/cm^3 after being submerged in water for half an hour. An original ultrasound echo signal from the sample of wood is shown in Figure 4, VMD was performed on the ultrasonic echo signal. As can be seen in the figure, there was significant noise in the signal, especially at the beginning and end of the signal.

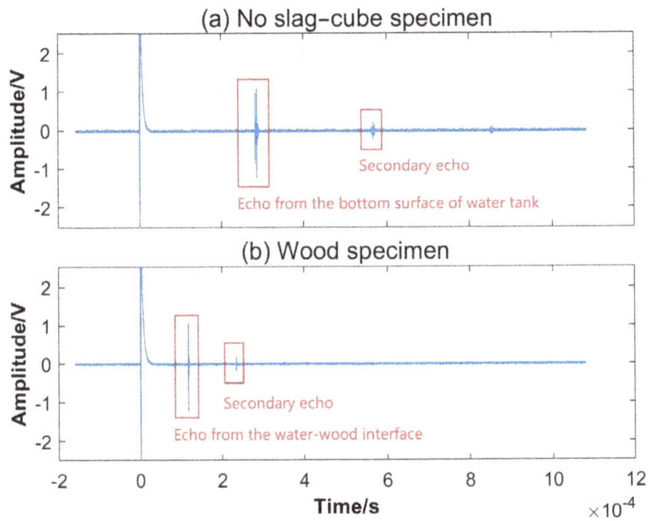

Figure 3. Original echo signals measured by the oscilloscope: (**a**) no slag cube specimen and (**b**) wood cube specimen placed in the water environment.

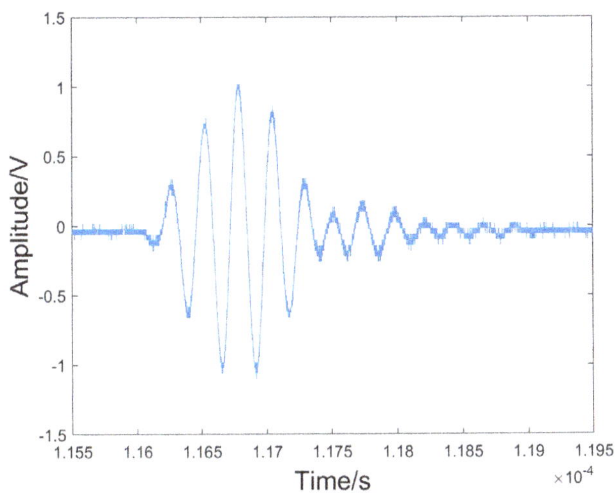

Figure 4. The original echo signal reflected from the water–wood interface.

3.2. Evaluating Indicator

Signal denoising is the extraction of effective information from the original signal. VMD is an adaptive decomposition method. After decomposition by these algorithms, a series of IMFs with a certain center frequency can be obtained. One or more layers of the decomposed IMF contain noise for noisy signals. Moreover, it is particularly important to judge whether the IMF is noise or a useful signal. Usually, in the signal processing process, some researchers will directly discard IMFs with smaller orders. At the same time, some useful signals contained in IMFs with smaller orders will also be discarded. However, the signal-to-noise ratio (SNR) is reduced, and the denoising effect is poor. This paper first obtains the normalized cross-correlation (NCC) function of each modal function IMF and the original signal and then uses the obtained NCC to determine the useful signal mode and noise mode.

In this case, we needed to select effective IMFs for reconstruction to obtain denoising signals. The SNR and RMSE were used as evaluation indices of the denoising effect. The SNR shows an energy relationship between the signal and noise. The higher the SNR, the more useful the information, and the less noise there is in the signal. Therefore, the SNR can be a quite intuitive method for evaluating the effect of denoised signals by analyzing whether the SNR improves. The definition of the SNR is as follows:

$$SNR = 10 \cdot \log_{10}\left(\frac{\sum_{i=1}^{N} \|x(i)\|^2}{\sum_{i=1}^{N} \|\hat{x}(i) - x(i)\|^2}\right) \tag{15}$$

where x is the original signal, $\hat{x}$ is the denoised signal, and $\| * \|$ indicates the norm.

The root mean square error shows the difference between the denoised signal and original signal in numerical terms, and the smaller the RMSE, the better the noise reduction effect. The RMSE is defined as follows:

$$RMSE = \sqrt{\frac{\sum_{i=1}^{N} \|\hat{x}(i) - x(i)\|^2}{\text{length}(x)}} \tag{16}$$

where $\text{length}(*)$ represents the length of the signal.

In order to effectively identify and separate the noise component in the IMFs decomposed by VMD, a normal correlation coefficient was introduced based on time–frequency analysis of the IMF component to further determine the degree of correlation between the IMF and the original signal. The larger the NCC, the better the correlation between the corresponding IMFs and the original signal. The NCC is defined by

$$NCC = \frac{\sum_{i=1}^{N} (u_k(i) \cdot \hat{x}(i))}{\sqrt{\left(\sum_{i=1}^{N} u_k^2(i)\right) \times \left(\sum_{i=1}^{N} \hat{x}^2(i)\right)}} \tag{17}$$

where u_k and $\hat{x}$ are the kth IMF and denoised signal, respectively.

4. Results

4.1. PSO-VMD Denoising

This section presents the results of the echo signal decomposition received from the water–slag specimen interface through VMD optimized by PSO. To validate the superiority of our proposed vibration signal processing method, we compared it with other methods. The comparative methods included WOA [34], GA [35] and ALO [36] algorithms used to optimize the two core parameters (K, α) of VMD. The defect signal processing of ultrasonic testing data from the water–wood interface served to illustrate the application of the IMF's section approach in the proposed PSO-VMD method. We applied these methods to process the original ultrasonic echo signals individually. The fundamental concept underlying these methods is to decompose the original defect signal and eliminate non-ultrasonic components. By employing the PSO technique for parameter optimization adaptively, we deployed VMD on the ultrasonic defect testing data.

To compare the accuracy and efficiency of the different signal processing methods, Table 1 lists the main parameters K and α for the improved VMD algorithms, SNR, RMSE, and processing time. The SNR and RMSE indicators were employed to assess the approaches' accuracy, while the processing time refers to quantitatively analyzing the efficiency of denoising. It can be seen that all of the improved VMD methods could decompose the original signal by setting the parameters K and α automatically. The proposed PSO-VMD method provided the least IMFs (seven) with the highest SNR and the lowest RMSE. The SNR reflects the ratio of the overall signal to the noise component, where the higher the SNR, the better the noise reduction effect. The RMSE quantitatively represents the deviation between the denoised signal and the original signal, where a lower RMSE indicates more effective noise reduction. PSO-VMD could effectively denoise and extract features from ultrasonic testing signals to highlight the slag inclusion defect features. Compared with PSO-VMD, the WOA-VMD, GA-VMD and ALO-VMD methods demonstrated varying extents of over-decomposition phenomena, which decomposed the signal into 10, 8, and 10 IMFs, respectively.

Table 1. Comparative results for parameters, evaluation index, and processing time for different VMD optimization algorithms.

Methods	WOA-VMD	GA-VMD	ALO-VMD	PSO-VMD
K	10	8	10	7
α	5000	4589	27,595	2962
SNR/dB	22.854	22.8964	22.908	23.0279
RMSE	0.02139	0.02129	0.02126	0.02097
Processing time/s	79.7	1155.9	78.9	536.7

Additionally, a comparison experiment on the denoising efficiency of the improved VMD algorithms was conducted on a PC equipped with an Intel i7-3700K CPU (Santa Clara, CA, USA), Nvidia Geforce RTX 3090 GPU, and 128 GB of memory (Santa Clara, CA, USA). Each algorithm was configured with the maximum number of iterations set to 10, and each was executed 10 times on the echo signals of wood specimen. The average processing times for each algorithm are presented in Table 1. The experimental results indicate that both the WOA and ALO optimization algorithms exhibited superior efficiency, with processing times of 79.7 s and 78.9 s, respectively, in comparison with the GA method's 1155.9 s, demonstrating the lowest efficiency. Our proposed PSO algorithm held a middle ground at 536.7 s in terms of efficiency. Consequently, there exists a considerable scope for enhancement in the noise reduction efficacy of PSO-VMD when compared with the WOA-VMD and ALO-VMD algorithms. It is anticipated that advancements in computational power will mitigate this discrepancy, rendering it a non-obstacle in the foreseeable future.

The components were sorted by frequency from low to high in Figure 5 for IMFs in the time domain and in Figure 6 for IMFs in the frequency domain. The low-frequency component IMF1 demonstrated strong periodic fluctuations, and the high-frequency component appeared to be random. IMF1 symbolized effective signals from slag inclusion defects, while other IMF components corresponded to interference or noise. The echo signals from the slag specimens could be effectively distinguished from the interference components.

Table 2 shows the normalized cross-correlation between the IMF components of the denoised signal and the original signal. The NCC of IMF1 was 0.9975, which was significantly higher than those of the other modes. It can be seen from the spectrogram presented in Figure 6 that the center frequencies among the IMFs were distinct and separate without any overlap, reflecting the efficacy of PSO in the context of VMD parameter tuning. Figure 7 shows the comparison between the denoised signal and the original signal, which exhibited characteristics similar to the original data but with less noise and a higher signal-to-noise ratio.

Table 2. NCC between the IMF component of the denoised signal and the original signal.

IMFs	IMF1	IMF2	IMF3	IMF4	IMF5	IMF6	IMF7
NCC	0.9975	0.0330	0.0291	0.0288	0.0280	0.0275	0.0271

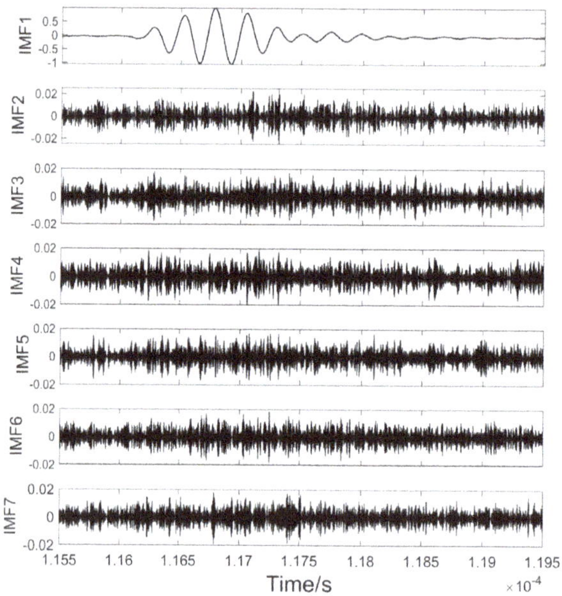

Figure 5. PSO–VMD decomposition results in time domain.

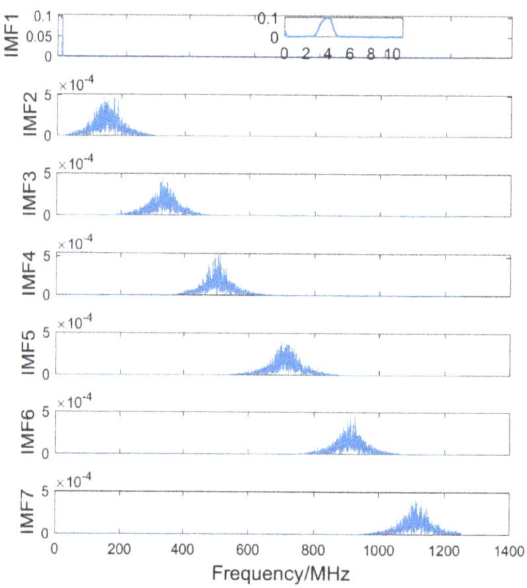

Figure 6. The decomposed IMF spectrum under PSO–VMD in the frequency domain. The inset in IMF1 is a local enlargement of the IMF component.

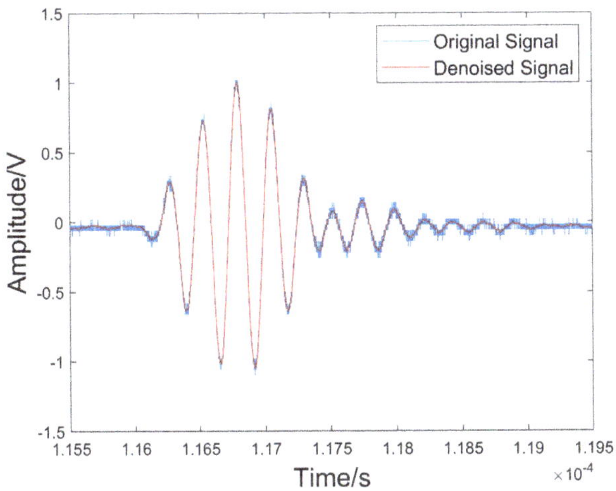

Figure 7. Denoised signal and original signal from the water–wood interface.

4.2. Phase of the Echo Signal

In this experiment, water was the original medium for ultrasonic wave propagation, and the surfaces, those of cubic blocks of iron, aluminum, copper, wood, resin, and other materials placed in the water environment to simulate slag inclusion defects, were the interfaces for sound wave reflection. Among the six materials selected, metallic blocks such as iron, aluminum, copper, and magnets are thick media compared with water, while non-metallic wood and resin are thin media. By analyzing the phase spectrum of the echo signal, we could determine whether the simulated inclusion was a thin medium or a thick one based on the half-wave loss theory.

The phase spectrum describes the phase information of the frequency components within a signal. In practical applications, the phase spectrum helps with understanding the phase relationships among different frequency components of a signal, which is quite important for signal analysis, processing, and interpretation. According to the theory of half-wave loss, in the propagation of sound waves, the phase spectrum serves as a critical means for analyzing reflected signals. We could determine whether the material at the reflective interface was a denser medium by assessing whether the reflected signal had undergone a π phase shift.

The phase spectrum can be either wrapped or unwrapped. The wrapped phase spectrum limits the phase values to the range from $-\pi$ to π, which helps with visualizing and analyzing phase changes, especially when there is a phase shift (such as a half-wave loss). The unwrapped phase spectrum, on the other hand, shows continuous phase changes from $-\infty$ to ∞, which may be more intuitive in some cases but may not be as clear as the wrapped phase spectrum when analyzing phase jumps. Therefore, the wrapped phase spectrum was employed to analyze the phase of the echo signals reflected from the slag specimen surfaces. Figure 8a,b depicts the wrapped full-phase spectrum of the echo signal from the wood specimen interface before and after the denoising process. Figure 8c presents an expanded view of the denoised phase spectrum from 1 to 5 MHz, highlighting the specific frequency range of interest.

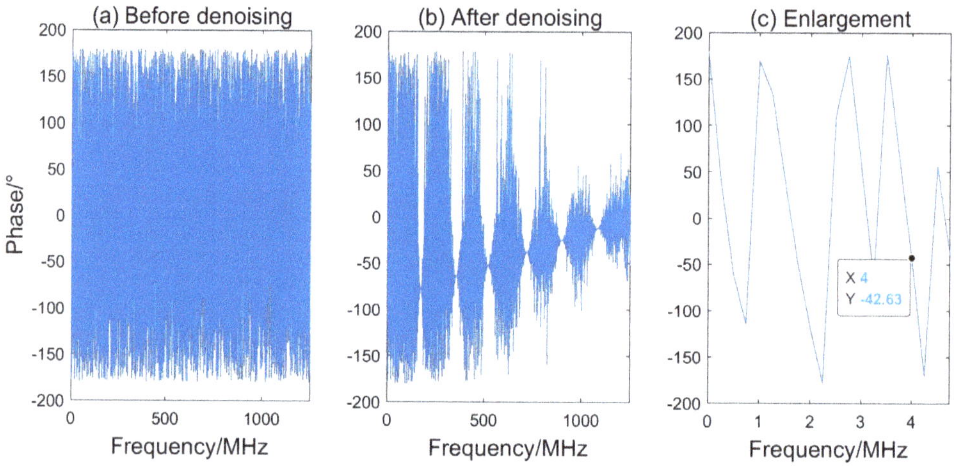

Figure 8. Phase spectrum (**a**) before denoising and (**b**) after denoising of echo signal from the wood specimen's surface. (**c**) Expanded view of (**b**) within the 1–5 MHz frequency range.

The central frequency of the denoised signal was 4 MHz, which was obtained from the IMF1 component shown in Figure 6. Then, the corresponding phase of −42.63° could be picked out from the enlargement of the denoised signal's phase spectrum in Figure 8c. This experimental result shows that the detector's operational frequency deviated from its nominal value of 5 MHz, which is potentially attributable to variations in crystal attachment, manufacturing tolerances, and probe aging.

The decomposition and phase analysis of PSO-VMD were performed for six different material samples, and the results are presented in Table 3. The table shows that the ultrasonic signal phase of the non-metallic materials was negative, while that of the metallic materials was positive, indicating opposite signs. Therefore, it can be concluded that the phase of the ultrasonic echo signal can reflect whether a slag inclusion is metallic or non-metallic.

Table 3. Frequencies and phases of 6 samples' echo signals.

Material	Frequency (MHz)	Phase (°)
Wood	4	−43.36
Resin	4	−24.81
Iron	4.25	97.11
Copper	4.25	64.87
Magnet	4.25	161.7
Aluminum Alloy	4.25	155.6

In austenitic stainless steel, if the ultrasonic testing signals of slag inclusion defects in a weld show indications of half-wave loss, then this is typically indicative of tungsten inclusions. This type is a denser medium relative to the austenitic stainless steel matrix. The tungsten inclusion defect in austenitic steel mainly occurs during the process of tungsten inert gas (TIG) welding. If there is no half-wave loss in the phase change of an echo reflected from the slag interface, then this indicates that the slag inclusion is a thin medium, which is always non-metallic compounds or other metallic inclusions like aluminum particles. Identifying the category of a slag inclusion defect is critical for finding the reasons for defects developing, improving production efficiency, and reducing the rate of defective products, ultimately reducing production costs.

5. Conclusions

This paper designed a water environment-based simulation method for weld inclusion material classification using ultrasonic testing and introduced the PSO-VMD denoising and phase processing method with an ultrasonic signal. The classification of slag inclusion material is effectively carried out using the phase spectrum of an echo signal. To summarize this paper, we can draw the following conclusions:

- A water environment experiment with six categories of cubic samples, including four metallic and two non-metallic materials, was used to simulate the slag inclusion defects with different materials in the weldments of austenitic stainless steel.
- The proposed method with PSO-VMD denoising the ultrasonic signals utilizes PSO to optimize the VMD hyperparameters, including the suitable number of decomposition modes K and penalty factor α. This was proven to effectively denoise the noisy signal and to be superior to other comparative algorithms, namely WOA-VMD, GA-VMD and ALO-VMD. PSO-VMD's results show that it decomposed the minimum number of modes (only seven IMFs) but with the highest SNR and lowest RSME among all methods, which demonstrates that PSO-VMD has superior performance in the noise reduction of slag inclusion defect detection signals.
- The phase spectrum was proposed as a valuable tool for analysis of the phase characteristics of ultrasound echo signals. By analyzing the phase spectrum, we can ascertain whether a slag inclusion has a thicker or thinner medium compared with the austenitic stainless steel base material, based on the presence or absence of a half-wave loss in the echo signal phase.

Our work provides a referential identification method for identifying material types with slag inclusion defects for practical applications.

Author Contributions: Conceptualization, J.Z. and G.Z.; methodology, J.Z. and G.Z.; software, Z.C.; validation, J.Z., J.D., H.Z., and Z.C.; formal analysis, J.Z. and S.X.; investigation, J.D.; resources, J.L.; data curation, G.Z. and J.L.; writing—original draft preparation, J.Z. and H.Z.; writing—review and editing, J.Z., H.Z., and J.L.; visualization, J.Z., G.Z., and S.X.; supervision, J.L.; project administration, J.L.; funding acquisition, S.X. and J.L. All authors have read and agreed to the published version of the manuscript.

Funding: This work was supported by the Fund of Guangdong-Hong Kong-Macao Joint Laboratory for Intelligent Micro-Nano Optoelectronic Technology (Grant No. 2020B1212030010) and the Youth Talent Innovation Project of the Guangdong Education Department in China (Grant No. 2021KQNCX149).

Institutional Review Board Statement: Not applicable.

Informed Consent Statement: Not applicable.

Data Availability Statement: Data are available upon request from the authors.

Conflicts of Interest: No potential conflict of interests were reported by the authors.

References

1. Kaçar, R.; Baylan, O. An investigation of microstructure/property relationships in dissimilar welds between martensitic and austenitic stainless steels. *Mater. Des.* **2004**, *25*, 317–329. [CrossRef]
2. Haghshenas, M.; Gerlich, A.P. Joining of automotive sheet materials by friction-based welding methods: A review. *Eng. Sci. Technol. Int. J.* **2018**, *21*, 130–148. [CrossRef]
3. Malarvel, M.; Singh, H. An autonomous technique for weld defects detection and classification using multi-class support vector machine in X-radiography image. *Optik* **2021**, *231*, 166342. [CrossRef]
4. Dmitriev, S.F.; Malikov, V.N.; Sagalakov, A.M.; Shevtsova, L.I. Flaw inspection of welded joints in titanium alloys by the eddy current method. *Weld. Int.* **2017**, *31*, 608–611. [CrossRef]
5. Abolfazl Zolfaghari, A.Z.; Kolahan, F. Reliability and sensitivity of magnetic particle nondestructive testing in detecting the surface cracks of welded components. *Nondestruct. Test. Eval.* **2018**, *33*, 290–300. [CrossRef]

6. Azeez, S.T.; Mashinini, P.M. Radiography examination of friction stir welds of dissimilar aluminum alloys. *Mater. Today Proc.* **2022**, *62*, 3070–3075. [CrossRef]
7. Dorafshan, S.; Maguire, M.; Collins, W. Infrared Thermography for Weld Inspection: Feasibility and Application. *Infrastructures* **2018**, *3*, 45. [CrossRef]
8. Hwang, Y.I.; Park, J.; Kim, H.J.; Song, S.J.; Cho, Y.S.; Kang, S.S. Performance Comparison of Ultrasonic Focusing Techniques for Phased Array Ultrasonic Inspection of Dissimilar Metal Welds. *Int. J. Precis. Eng. Manuf.* **2019**, *20*, 525–534. [CrossRef]
9. Ghubade, A.B.; Kumar, A. Review on casting defects and methodologies for quality improvement. *J. Emerg. Technol. Innovat. Res.* **2019**, *6*, 1008–1019.
10. Singh, R.R.B.; Sasikumar, T.; Suresh, S.; Ramanan, G. A Novel Detection of Defects in Al–SiC Composite by Active Pulsed Infrared Thermography Using Data and Image Processing. *Trans. Indian Inst. Met.* **2020**, *73*, 2767–2783. [CrossRef]
11. Li, P.; Xie, S.; Wang, K.; Zhao, Y.; Zhang, L.; Chen, Z.; Uchimoto, T.; Takagi, T. A novel frequency-band-selecting pulsed eddy current testing method for the detection of a certain depth range of defects. *NDT Int.* **2019**, *107*, 102154. [CrossRef]
12. Oswald-Tranta, B.; Schmidt, R. Crack depth determination with inductive thermography. In Proceedings of the Thermosense: Thermal Infrared Applications XXXVII, Baltimore, MD, USA, 18–21 April 2015; Volume 9485, pp. 98–106. [CrossRef]
13. Wang, K.Y.; Cheng, Y.Y. Design of X-ray digital imaging and data acquisition system. *Russ. J. Nondestruct. Test.* **2009**, *45*, 362–366. [CrossRef]
14. Tkocz, J.; Greenshields, D.; Dixon, S. High power phased EMAT arrays for nondestructive testing of as-cast steel. *NDT Int.* **2019**, *102*, 47–55. [CrossRef]
15. Li, X.; Li, B.; Liu, Z.; Niu, R.; Liu, Q.; Huang, X.; Xu, G.; Ruan, X. Detection and Numerical Simulation of Non-Metallic Inclusions in Continuous Casting Slab. *Steel Res. Int.* **2019**, *90*, 1800423. [CrossRef]
16. Mukhtar, M.F.H.; Mahmod, M.F. Simulation Analysis of Ultrasonic Testing in Steel-based Butt Weld Joint. *Res. Prog. Mech. Manuf. Eng.* **2021**, *2*, 136–144. [CrossRef]
17. Song, Y.; Hua, L.; Wang, X.; Wang, B.; Liu, Y. Research on the Detection Model and Method for Evaluating Spot Welding Quality Based on Ultrasonic A-Scan Analysis. *J. Nondestruct. Eval.* **2015**, *35*, 4. [CrossRef]
18. Kasban, H.; Zahran, O.; Arafa, H.; El-Kordy, M.; Elaraby, S.M.; Abd El-Samie, F. Welding defect detection from radiography images with a cepstral approach. *NDT Int.* **2011**, *44*, 226–231. [CrossRef]
19. Ditchburn, R.; Burke, S.; Scala, C. NDT of welds: State of the art. *NDT Int.* **1996**, *29*, 111–117. [CrossRef]
20. Daubechies, I. The wavelet transform, time-frequency localization and signal analysis. *IEEE Trans. Inf. Theory* **1990**, *36*, 961–1005. [CrossRef]
21. Huang, N.E.; Shen, Z.; Long, S.R.; Wu, M.C.; Shih, H.H.; Zheng, Q.; Yen, N.C.; Tung, C.C.; Liu, H.H. The empirical mode decomposition and the Hilbert spectrum for nonlinear and non-stationary time series analysis. *Proc. R. Soc. Lond. Ser. Math. Phys. Eng. Sci.* **1998**, *454*, 903–995. [CrossRef]
22. Wu, Z.; Huang, N.E. Ensemble empirical mode decomposition: A noise-assisted data analysis method. *Adv. Adapt. Data Anal.* **2009**, *1*, 1–41. [CrossRef]
23. Torres, M.E.; Colominas, M.A.; Schlotthauer, G.; Flandrin, P. A complete ensemble empirical mode decomposition with adaptive noise. In Proceedings of the 2011 IEEE International Conference on Acoustics, Speech and Signal Processing (ICASSP), Prague, Czech Republic, 22–27 May 2011; pp. 4144–4147. [CrossRef]
24. Dragomiretskiy, K.; Zosso, D. Variational Mode Decomposition. *IEEE Trans. Signal Process.* **2014**, *62*, 531–544. [CrossRef]
25. Li, H.; Liu, T.; Wu, X.; Chen, Q. An optimized VMD method and its applications in bearing fault diagnosis. *Measurement* **2020**, *166*, 108185. [CrossRef]
26. Wang, D.; Yue, C.; Wei, S.; Lv, J. Performance Analysis of Four Decomposition-Ensemble Models for One-Day-Ahead Agricultural Commodity Futures Price Forecasting. *Algorithms* **2017**, *10*, 108. [CrossRef]
27. Hua, T.; Dai, K.; Zhang, X.; Yao, Z.; Wang, H.; Xie, K.; Feng, T.; Zhang, H. Optimal VMD-Based Signal Denoising for Laser Radar via Hausdorff Distance and Wavelet Transform. *IEEE Access* **2019**, *7*, 167997–168010. [CrossRef]
28. Qi, B.; Yang, G.; Guo, D.; Wang, C. EMD and VMD-GWO parallel optimization algorithm to overcome Lidar ranging limitations. *Opt. Express* **2021**, *29*, 2855–2873. [CrossRef] [PubMed]
29. Long, J.; Wang, X.; Dai, D.; Tian, M.; Zhu, G.; Zhang, J. Denoising of UHF PD signals based on optimised VMD and wavelet transform. *IET Sci. Meas. Technol.* **2017**, *11*, 753–760. [CrossRef]
30. Eberhart, R.; Kennedy, J. A new optimizer using particle swarm theory. In Proceedings of the MHS'95 Sixth International Symposium on Micro Machine and Human Science, Nagoya, Japan, 4–6 October 1995; pp. 39–43. [CrossRef]
31. Richman, J.S.; Lake, D.E.; Moorman, J. Sample Entropy. In *Numerical Computer Methods, Part E*; Methods in Enzymology; Academic Press: Cambridge, MA, USA, 2004; Volume 384, pp. 172–184. [CrossRef]
32. Cao, M.; Yuan, J.; Liu, H.; Fang, X.; Zhu, J. A simulation of the quasi-standing wave and generalized half-wave loss of electromagnetic wave in non-ideal media. *Mater. Des.* **2003**, *24*, 31–35. [CrossRef]
33. Wojciech, J.; Jacek, G. Detection of slag inclusions using infrared thermal imagining system. *MATEC Web Conf.* **2021**, *338*, 01012. [CrossRef]
34. Liu, Z.; Peng, Y. Study on Denoising Method of Vibration Signal Induced by Tunnel Portal Blasting Based on WOA-VMD Algorithm. *Appl. Sci.* **2023**, *13*, 3322. [CrossRef]

35. Liu, Z.; Liu, H. A novel hybrid model based on GA-VMD, sample entropy reconstruction and BiLSTM for wind speed prediction. *Measurement* **2023**, *222*, 113643. [CrossRef]
36. Shi, L.; Wen, J.; Pan, B.; Xiang, Y.; Zhang, Q.; Lin, C. Dynamic Characteristics of a Gear System with Double-Teeth Spalling Fault and Its Fault Feature Analysis. *Appl. Sci.* **2020**, *10*, 7058. [CrossRef]

sensors

MDPI

Article

Structural Plastic Damage Warning and Real-Time Sensing System Based on Cointegration Theory

Qiang Gao, Junzhou Huo *, Youfu Wang, Xiaotian Wang and Chongru Wang

School of Mechanical Engineering, Dalian University of Technology, Dalian 116024, China; gao13qiang88@163.com (Q.G.); wangyoufu@mail.dlut.edu.cn (Y.W.); wangxiaotian1@126.com (X.W.); wangchongru1@163.com (C.W.)

* Correspondence: huojunzhou@dlut.edu.cn

Abstract: Structural damage can affect the long-term operation of equipment. Real-time damage warning for structures can effectively avoid accidents caused by structural damage. In this paper, a real-time warning method of structural plastic damage based on the cointegration theory is proposed. This method calculates the cointegration relationship between the strain signals at different measuring points, and the stability of the strain signal relationships is also evaluated. The problem of inaccurate detection caused by the error of strain measurement and environmental influence can be eliminated by the comprehensive judgment of strain between asymmetrical measuring points. A real-time damage sensing system is developed in this paper. In order to improve the real-time and practicability of the system, this paper proposes and determines the residual warning coefficient by analyzing the proportion of the strain residuals exceeding the residual threshold. The research on this sensing system has certain value for the engineering application of damage monitoring methods.

Keywords: cointegration theory; real-time warning; sensing system; warning coefficient; damage alarming

1. Introduction

The health of structural parts directly affects the service life of equipment [1–3]. Under the long-term action of loads, structural parts will suffer a certain degree of damage, which will affect their service life and even cause engineering accidents. Therefore, the accurate monitoring and warning of structural damage can provide a timely reference for equipment maintenance [4,5], which can effectively avoid losses caused by structural damage.

Structural damage monitoring has always been a hot topic in the field of engineering research [6,7]. Common structural damage mainly includes cracks, holes, plastic deformation, structural fractures and other forms. Researchers have conducted relevant research on different types of structural damage using different types of sensors and signal processing methods [8]. Detection methods based on eddy current [9], acoustic emission [10], piezoelectric ultrasonic guided wave [11], electrical impedance [12], visual recognition [13], fiber bragg grating [14] and other technologies each have their respective advantages in the field of structural health monitoring. In recent years, ultrasonic detection technology and machine learning methods have been widely introduced into structural health monitoring [15]. Among them, ultrasonic detection is widely used to monitor cracks in structures [16,17]. Liu et al. studied a wireless sensor based on a nonlinear ultrasonic modulation method to achieve the real-time online monitoring of fatigue cracks [18]. Kim et al. proposed a reference-free ultrasonic guided-wave crack detection method using the polarization characteristics of piezoelectric wafers [19]. With the continuous optimization and development of machine learning methods, they have been applied to monitor crack damages in structures [20,21]. Wang et al. predicted the growth law of fatigue cracks based on the particle filter model [22]. Chen et al. identified cracks in bridge structures based on a deep learning algorithm and realized the integrated application of the algorithm [23]. Zhou et al. used laser ultrasound and the YOLOv5 framework to monitor and identify structural cracks in interference fits [24]. The continuous development of sensor technology has also provided new methods for monitoring crack damage. Xu et al. studied a new layer sensor that monitors the crack extension of steel bridges by different signals at different fatigue stages [25]. Pang et al. studied a sensor based on microstrip antennas that can detect cracks under layers [26]. Strain can accurately reflect changes in structural stress, and the method of identifying damage through strain in a structure has certain advantages. Yu et al. studied full-optical strain FBG with tunable DFB laser demodulation for monitoring tensile and fatigue damage [27]. Pauw et al. used embedded optical fiber sensors to monitor the fatigue damage of structures [28]. Different detection technologies have their own advantages, and the monitoring of structural damage through the fusion of multiple sensors is also being studied by scholars [29]. Qi et al. monitored crack growth during fatigue by integrating strain sensors and piezoelectric guided waves [30]. Wang et al. used a weighted adaptive Kalman filtering-based method to integrate optical fiber and guided-wave information to identify structural fatigue cracks [31]. In summary, most studies on structural damage monitoring focus on monitoring macroscopic damage such as cracks. In actual engineering applications, the plastic deformation of materials often occurs before cracks initiate, and even before obvious cracks appear, the structure has already been destroyed. Therefore, the monitoring and warning of structural damage before crack initiation are of great significance for early maintenance of structures.

Based on the cointegration theory, an efficient and real-time plastic damage warning method is proposed in this paper, which can accurately identify the plastic damage of structures. The method identifies structural plastic damage through the change in the cointegration residual of the strain series between the measuring points. By analyzing the proportion of residuals exceeding the residual threshold, this paper proposes a warning coefficient. And the stability of the real-time alarming of the damage sensing system is improved by comparing the residual and warning coefficient. This method further realizes the application of damage monitoring theory in practical engineering.

2. Cointegration Theory

Time series can be divided into two types: unstable and stable time series. Unstable time series refers to a series in which the mean, variance and other statistical data change with time, while stable time series refers to a series in which the statistical data do not change with time. Most of the strain data for structural damage monitoring are unstable series that change with time, and their linear combination may become a series that does not change with time.

For an unstable time series y_t, after d times of difference, it becomes a stationary time series, which is integrated of order d, denoted by $y_t \sim I(d)$. For an unstable variable $Y = [y_1, y_2, \ldots, y_n]^{\mathrm{T}}$, its components are all integrated of order 1. If there is a set of linear combination coefficients $\boldsymbol{\beta} = [\beta_1, \beta_2, \cdots \beta_n]$, which makes the previous $\boldsymbol{Y}$ become a stationary series, then $y_1, y_2, \ldots, y_n$ satisfies the cointegration relationship, and the equation is shown below.

$$\boldsymbol{\beta}^{\mathbf{T}}\boldsymbol{Y} = \beta_1 y_1 + \beta_2 y_2 + \ldots + \beta_n y_n \sim I \tag{1}$$

where $\boldsymbol{\beta} = [\beta_1, \beta_2, \cdots \beta_n]$, $\boldsymbol{\beta}^{\mathrm{T}}\boldsymbol{Y}$ is residual covariance, and it is a Gaussian noise series with a mean equal to zero.

From the above formula, we can see that the main purpose of cointegration is to seek a long-term stable equilibrium relationship between two sets of unstable time series. If two data series become stable after the first-order difference calculation and are also stable after linear combination, it means that there is a cointegration relationship between them.

An unstable series is more likely to cause pseudo-regression. The significance of the cointegration test is to determine whether the relationship described by its regression equation is pseudo-regression. Therefore, before solving the cointegration residual, it is necessary to test the stationarity of the time series. Commonly used methods include the ADF (Augmented Dickey–Fuller) test and E-G (Engel–Granger) test.

2.1. Augmented Dickey–Fuller (ADF) Method

The ADF method is used to test the stability of time series. If there is a time series y_t, its ADF model is as follows:

$$\Delta y_t = \rho y_{t-1} + a + \sum_{i=1}^{p-1} b_i \Delta y_{i-1} + \varepsilon_t \quad (2)$$

where Δy_{i-1} is the difference component of y_t and p and b_i are coefficients. The a represents the constant term. The ρ is the lag order, and ε_t is the residual series. The coefficient ρ is used to determine the stationarity of the time series y_t. When each series or its difference series is integrated of order 1, the requirements of the cointegration equation are met, and the next step is to continue the E-G test to complete the subsequent solution.

2.2. Engel–Granger (E-G) Method

The theoretical basis of the E-G test is to perform a unit root test on the residuals of the regression equation. If the dependent variable can be described by various linear combinations of the independent variables, it means that there is a relatively stable equilibrium relationship between the two. The part that cannot be described forms a residual series. There is no need for serial correlation between the residual series, so the residuals must be stationary. Therefore, the purpose of the E-G method is to verify whether its residual series is stable. The steps of this method are as follows.

Assuming that x_t and y_t are both integrated of order 1, the residual difference between them is calculated as follows.

$$\varepsilon_t = y_t - \beta x_t - \alpha \quad (3)$$

After the ADF test, if the residual series ε_t is stable, it means that there is a cointegration relationship between the original series x_t and y_t, where α and β are the corresponding cointegration coefficients.

The existence of a cointegration relationship means that the two sets of time series have a stable relationship in the long term, that is, their long-term equilibrium states are interdependent. The cointegration residual is a stable series obtained after eliminating unstable factors. If the cointegration residual changes suddenly, it means that the cointegration relationship has changed, which can be used as a key indicator for damage warning.

3. Process of Structural Damage Warning Method

By loading the sample, the strain signals at different measuring points are tested. The residual threshold of damage warning is calculated by the strain signals of the measuring points in the undamaged state. The residual threshold is calculated according to the following method:

$$\begin{cases} \mathrm{RTU} = \mu + 3\sigma \\ \mathrm{RTL} = \mu - 3\sigma \end{cases} \quad (4)$$

where the RTU is the upper limit of the residual threshold, the RTL is the lower limit of the residual threshold, μ is the residual mean and σ is the residual variance. The flowchart of the damage warning method is shown in Figure 1. First, the stability of the strain data is tested by the ADF method. If the two sets of test strains are unstable, this method is used to evaluate the stability of the corresponding difference series. If the first-order difference of the strain series is stable, the series is integrated of order 1. It meets the requirements of the cointegration equation, and then the E-G test is performed on it. If the E-G result shows that the residual is stable, the residual equation can be constructed. After calculating the residuals of the two series, the ADF stability test is performed on the residual series. If the residuals are unstable, it means that the stability of the two series is destroyed, which also means that structural damage has occurred. If the residuals are stable, the strain series will continue to be calculated to realize the warning of structural damage.

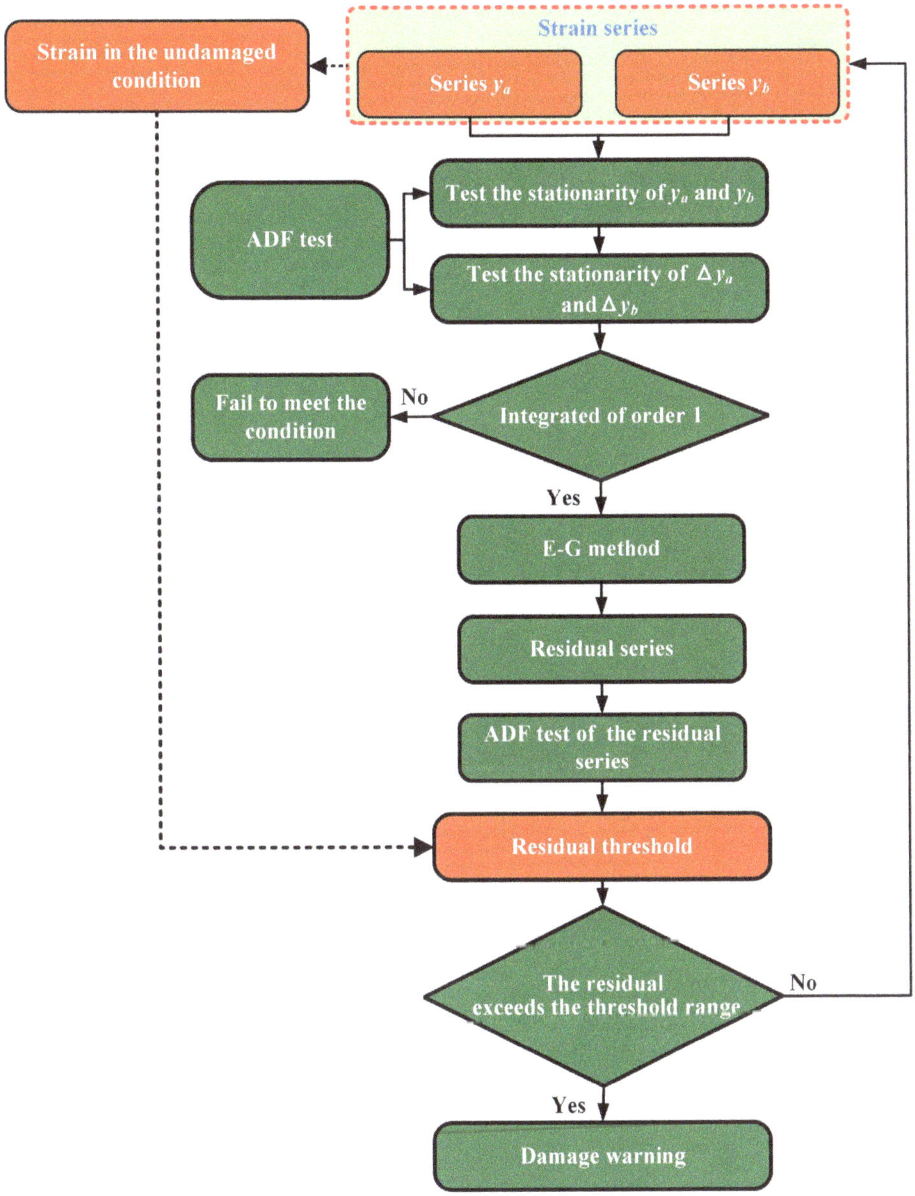

Figure 1. Flowchart of plastic damage warning method based on cointegration theory.

4. Test System and Experiment Analysis

4.1. Test System and Samples

As shown in Figure 2, the test system consists of a sample, testing machine, load control terminal, strain collecting instrument and sensing system. The load control terminal controls the testing machine to apply the load to the sample. The strain collecting instrument transmits the collected strain data to the sensing system. The sensor system analyzes the residual of the strain series of multiple channels through the co-arrangement theory and finally realizes the alarm of sample damage.

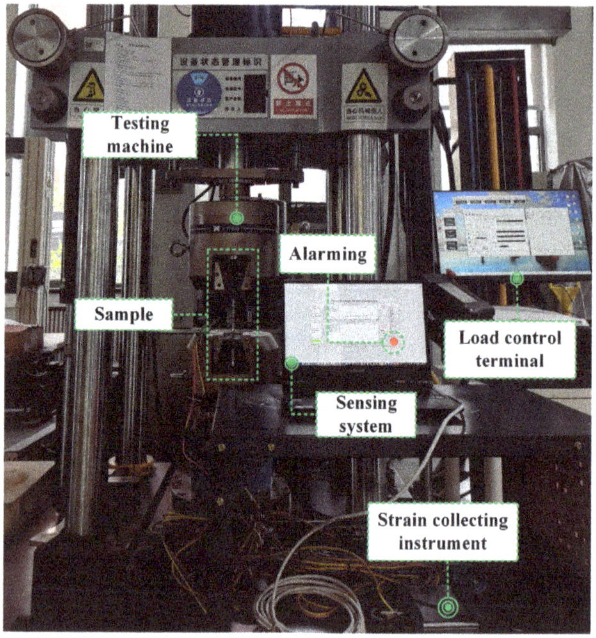

Figure 2. Test system diagram.

Figure 3 shows the sample, which is made of Q345D structural steel. Sample 1 is the stress concentration sample, and sample 2 is the standard sample. Strain gauges are pasted in the center and on both sides of samples 1 and 2. The detailed dimensions of the structure and strain gauge pasting are shown in Figure 3a. The thickness of the samples is 5 mm. The actual pictures of the two samples are shown in Figure 3b.

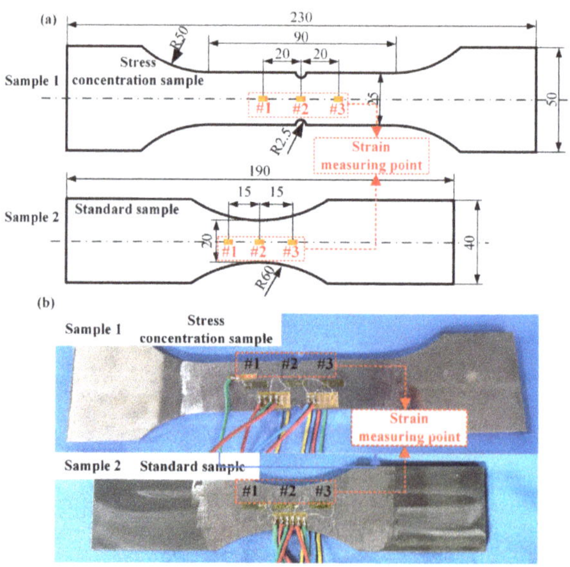

Figure 3. Strain test structure and strain gauge installation position: (**a**) sample structure dimensions and strain gauge installation position; (**b**) sample actual picture (unit: mm).

4.2. Analysis of Experimental Results

By applying tensile force to the sample, the strain signal of the sample measuring point is tested. Each strain channel is synchronously sampled with a sampling rate of 200 Hz. The strain test results of different measuring points of the two samples are shown in Figures 4 and 5, respectively. The signal of the entire test process is divided into three stages: the no-load stage, the elastic deformation stage and the plastic deformation stage. As shown in Figure 4a, 0–45 s is the no-load stage, and the strain signal tested at this time is the noise of the system. Next, 45–82 s is the elastic deformation stage. At 45 s, the test machine starts to load the sample with a loading rate of 0.5 kN/s. The enlarged curve of the no-load stage and the elastic deformation stage is shown in Figure 4b. It can be seen from the figure that the strain signals of each measuring point from 45 to 82 s show a linear growth trend. It can be judged by Hooke's law that this stage belongs to the elastic deformation stage. After 82 s of continuous loading, the sample slowly enters the plastic deformation stage. It can be seen from Figure 4a that the strain of each measuring point increases rapidly after entering the plastic zone. The enlarged view of Figure 4b shows that the strain growth rate of measuring point 2 located at the stress concentration position is greater than the strain growth rate of the symmetrically arranged measuring points 1 and 3.

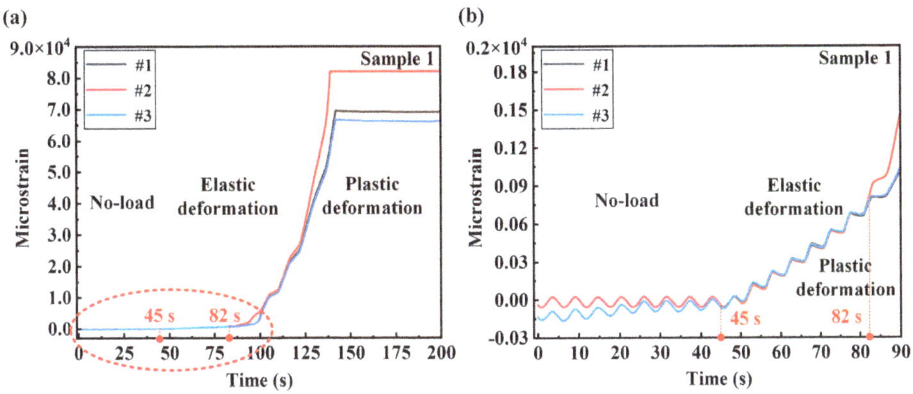

Figure 4. Strain test results of measuring points 1, 2 and 3 of sample 1: (**a**) strain at different measuring points during the entire loading process; (**b**) strain enlarged diagram of the no-loading stage and the plastic deformation stage.

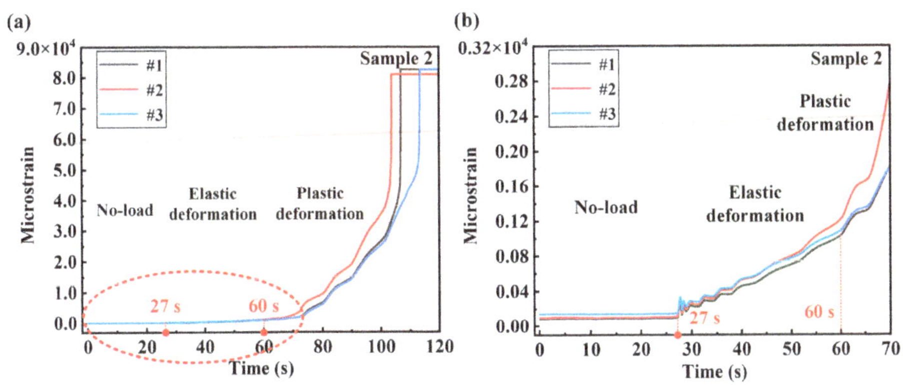

Figure 5. Strain test results of measuring points 1, 2 and 3 of sample 2: (**a**) strain at different measuring points during the entire loading process; (**b**) strain enlarged diagram of the no-loading stage and the plastic deformation stage.

The strain changes of sample 2 during the entire loading process are also analyzed based on the strain test data in Figure 5, and the loading parameters are consistent with those of sample 1. In Figure 5a, the test machine starts loading at 27 s, the strain and load of the sample are linearly related before 60 s and the sample is in the elastic deformation area. It can be clearly seen from Figure 5a,b that after 60 s, with the continuous loading of the load, the strains of the three measuring points of sample 2 gradually change and increase. After 60 s, the sample begins to slowly undergo plastic deformation, and the strain also increases rapidly until the strain gauge fails and the structure is damaged. Similarly, the strain growth rate of measuring point 2 at the center position is greater than the strain growth rate of symmetrical points 1 and 3 when entering the plastic deformation stage. For engineering structural parts, they can work normally in the elastic deformation stage and can still return to the initial state after unloading. Once the structure undergoes plastic deformation, it will cause irreversible damage to the structure and cannot return to the initial state after unloading. Irreversible structural damage will occur after the structural part enters the plastic stage, so accurate monitoring of it can provide timely warning of structural damage.

According to the cointegration theory, the strain series needs to be integrated of order 1 in order to conduct a stability assessment. The strain and force changes of the structural parts are linear in elastic deformation. Figure 6 is a first-order difference series of the strain signal at each measuring point. Sample 1 is in the no-load and elastic deformation stages from 0–82 s, and the first-order difference series of the two stages shows the same law. The strain first-order difference series of sample 2 also maintains the same law of the first-order difference in the no-load and elastic deformation stages from 0 to 60 s. After calculating the difference series, the stability of the difference series needs to be checked.

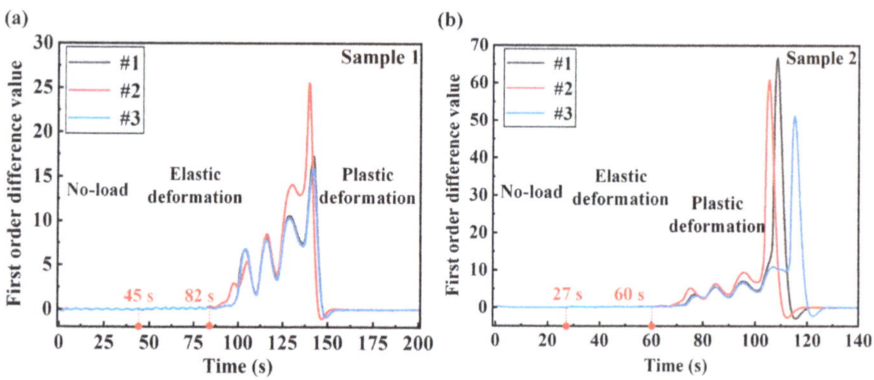

Figure 6. First-order difference series of strain at measuring points 1, 2 and 3: (**a**) strain difference series of sample 1; (**b**) strain difference series of sample 2.

The strain signals of the first 25 s of sample 1 and sample 2 are selected as the stability test data. The ADF test is performed on each strain series and the first-order difference series of the strain, respectively, and the verification results are shown in Tables 1 and 2. Among them, y_1, y_2 and y_3 are the strain series corresponding to different measuring points of the sample, and Δy_1, Δy_2 and Δy_3 are the first-order difference series of the corresponding strains. By comparing the ADF check value with the ADF critical value at the 1% and 5% significance levels, if the ADF check value of the series is less than the critical value, it means that the series is stable. If the ADF check value of the series is greater than the critical value, the series is unstable. It can be seen from the ADF test results in Tables 1 and 2 that the ADF test values of series y_1, y_2 and y_3 are all greater than the critical values of 1% and 5%, so the strain series are all unstable. The first-order difference series Δy_1, Δy_2 and Δy_3 are all stable after verification. The strain series of sample 1 and sample 2 are both unstable after verification, and their first-order series are stable. Therefore, the strain series of the two samples are all integrated of order 1, which meets the cointegration requirements.

Table 1. ADF check results of strain series and first-order difference series of sample 1.

Series	ADF Test Value	1% Critical Value	5% Critical Value	Stability
y_1	−1.0398	−2.5690	−1.9416	Unstable
Δy_1	−99.1729	−2.5690	−1.9416	Stable
y_2	−1.0791	−2.5690	−1.9416	Unstable
Δy_2	−101.1411	−2.5690	−1.9416	Stable
y_3	0.8419	−2.5690	−1.9416	Unstable
Δy_3	−95.9264	−2.5690	−1.9416	Stable

Table 2. ADF check results of strain series and first-order difference series of sample 2.

Series	ADF Test Value	1% Critical Value	5% Critical Value	Stability
y_1	−0.1069	−2.5690	−1.9416	Unstable
Δy_1	−121.2684	−2.5690	−1.9416	Stable
y_2	−0.2294	−2.5690	−1.9416	Unstable
Δy_2	−123.6009	−2.5690	−1.9416	Stable
y_3	−1.1644	−2.5690	−1.9416	Unstable
Δy_3	−111.5629	−2.5690	−1.9416	Stable

The threshold of the residual is calculated using the selected data. The cointegration equation of the strain series between different measuring points is constructed by the E-G method, and then the residual is calculated by the cointegration equation. The parameters of the cointegration equation of the strain series of each measuring point are shown in Tables 3 and 4. Strain series #1–#2 represents the combination of data from measuring points 1 and 2. The thresholds of the strain residual of sample 1 and sample 2 are calculated by Formula (4), and its parameters are shown in Tables 5 and 6. In addition, the threshold mean value calculated by the three groups of data is calculated in the table, and it is used to judge whether the limit is exceeded.

Table 3. Cointegration equation parameters for different strain series of sample 1.

Strain Series	β	α
#1–#2	1.0217	1.7590
#1–#3	0.6183	45.1140
#2–#3	0.6055	42.4706

Table 4. Cointegration equation parameters for different strain series of sample 2.

Strain Series	β	α
#1–#2	1.5264	22.3674
#1–#3	−0.1026	−126.8029
#2–#3	−0.0400	−95.5583

Table 5. Residual mean, variance and upper and lower limits of different strain series of sample 1.

Strain Series	μ	σ	RTU	RTL
#1–#2	-1.3747×10^{-4}	0.4373	1.3118	−1.3121
#1–#3	−0.0048	0.5002	1.4958	−1.5054
#2–#3	−0.0046	0.4943	1.4785	−1.4876
Mean value	N/A	N/A	1.4287	−1.4350

Table 6. Residual mean, variance and upper and lower limits of different strain series of sample 2.

Strain Series	μ	σ	RTU	RTL
#1–#2	-7.7404×10^{-4}	0.7447	2.2332	−2.2348
#1–#3	5.5210×10^{-4}	0.4254	1.2769	−1.2658
#2–#3	6.5017×10^{-4}	0.4179	1.2543	−1.2530
Mean value	N/A	N/A	1.5881	−1.5845

The ADF method is used to test the stability of the first-order different residual series. From the test results of Tables 7 and 8, by comparing with the 1% and 5% critical values, the ADF test results of the residuals between each series are all less than the critical value, satisfying the stable condition. And it can be seen that the cointegration residuals between different strain series are all stable. Therefore, there is a cointegration relationship between the strains of the three measurement points of sample 1 and sample 2.

Table 7. ADF test results of strain residual series of sample 1.

Strain Series	ADF Test Value	1% Critical Value	5% Critical Value	Stability
#1–#2	−71.4768	−2.5821	−1.9416	Stable
#1–#3	−119.5111	−2.5821	−1.9416	Stable
#2–#3	−120.6987	−2.5821	−1.9416	Stable

Table 8. ADF test results of strain residual series of sample 2.

Strain Series	ADF Test Value	1% Critical Value	5% Critical Value	Stability
#1–#2	−106.9975	−2.5821	−1.9416	Stable
#1–#3	−119.1516	−2.5821	−1.9416	Stable
#2–#3	−116.0702	−2.5821	−1.9416	Stable

The corresponding residual series is calculated for the strains of the three stages. By comparing the residual calculation results with the residual threshold, the warning of plastic deformation damage is realized. Figures 7 and 8 are the residual calculation results of the strain series between the measuring points of sample 1 and sample 2, respectively. For sample 1, the stage of 0–82 s corresponds to the no-load and elastic deformation stages, and its curve is enlarged, as shown in Figure 7b. It can be seen from the enlarged residual diagram that the residual calculation results of the two groups #1–#2 and #2–#3 do not exceed the RTU and RTL before the plastic deformation stage, showing a stable state. With the continuous loading of force, the structural parts began to slowly enter the plastic deformation stage after 82 s. At this time, the relationship between strain and force at each measuring point becomes a nonlinear relationship. Since measuring points 1 and 3 are arranged on the symmetrical side of the stress concentration part of sample 1, the cointegration residual of #1–#3 changes little after plastic deformation. The residual verification results show that the strain cointegration relationship between the stress concentration measuring point and the symmetrical measuring point is very sensitive to plastic deformation. After plastic deformation occurs, the residual value will deviate from the original stable state. As the degree of plastic deformation increases, the residual value gradually increases. The change in the structural damage state can be clearly seen through the change in the residual value.

The residual calculation results of sample 2 are analyzed. As shown in Figure 8, before 60 s of loading, the residual results calculated for the sample in the no-load and elastic deformation stages do not exceed the RTU and RTL. As shown in Figure 8b, after entering the plastic deformation stage, the residual calculation results increased significantly, and the residual stability is destroyed at this time. After the material enters the plastic deformation, the strain relationship changes at the center measuring point and the symmetrical measuring point are very obvious. However, for the symmetrically arranged #1–#3 combination, the residual changes are small just after entering the plastic deformation. In the later stage of plastic deformation, the residual changes significantly. The structural damage warning through residual values needs to be calculated under the action of dynamic loading. After 160 s and 130 s in Figures 7 and 8, the loading stops, the strains are all stable values of the mean and the residual cannot be judged at this time.

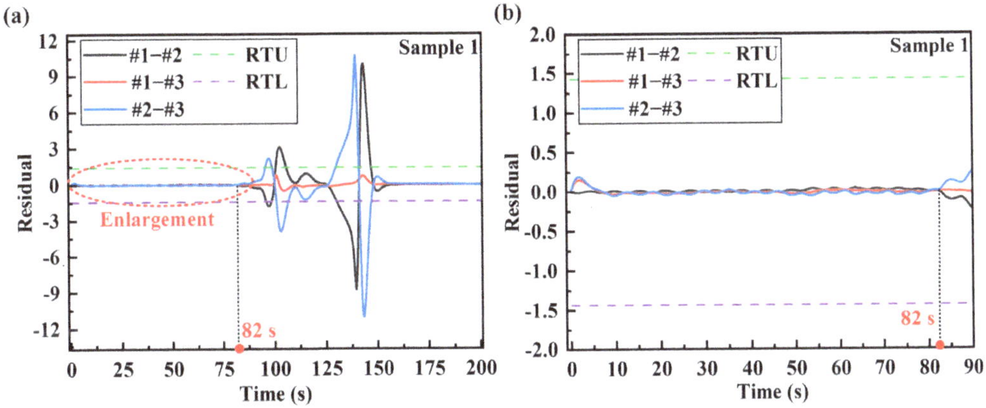

Figure 7. Calculation results of residual at the measuring point of sample 1: (**a**) calculation results of residual during the whole process; (**b**) enlarged residual during the no-load stage and the elastic deformation stage.

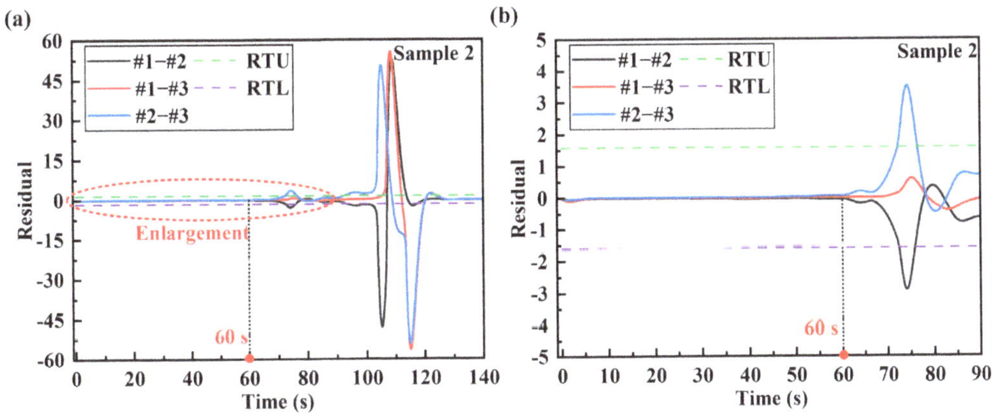

Figure 8. Calculation results of residual at the measuring point of sample 2: (**a**) calculation results of residual during the whole process; (**b**) amplified residual during the unloaded stage and the elastic deformation stage.

Accurate warning of structural damage through the cointegration method requires calculation and analysis based on the relatively stable relationship between the strains of the two measuring points. For the measuring points of the structural parts, the strains of the asymmetric points of the structure should be selected to evaluate the stability of the cointegration relationship. Within the elastic deformation range of the structure, the strains of the two points are linearly related, and their residuals show a certain stability. After the structure enters the plastic deformation, the strains of the asymmetric measuring points will obviously destroy their original stable relationship, and then the warning of structural damage is performed by comparison with the residual threshold.

5. Real-Time Damage Sensing System

In order to ensure the stability of the system alarm, the warning coefficient q is proposed in this paper. The warning coefficient is obtained by analyzing the data. In addition, the warning coefficient should be determined according to the sampling data and sampling frequency. The warning coefficient is the upper warning limit of the proportion of data exceeding the residual threshold in the sampled data. A too-large warning coefficient will lead to no warning after damage occurs in the structure. A too-small warning coefficient may be disturbed by unstable factors such as noise in the actual load, resulting in a false alarm. In Figures 9 and 10, the regions in which the residuals exceed the residual thresholds are marked. In 3000 data samples, the proportion of residuals that exceed the threshold is calculated, as shown in Table 9. According to the data analysis in the table, the smallest proportion is 0.19. In practical applications, the residual threshold setting of the damage sensing system needs to be less than 0.19. Since the sampling data of the real-time sensing system is phased sampling, in order to improve the stability of the system alarm, the warning coefficient of the two samples is set to 0.05.

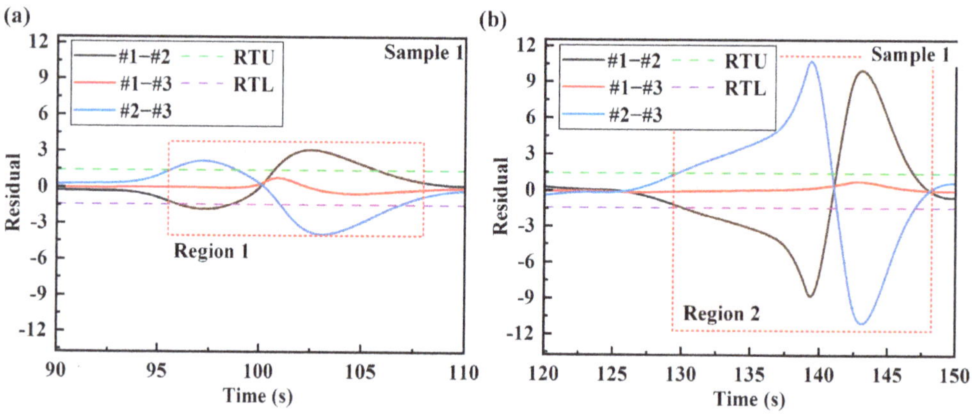

Figure 9. The regions of sample 1 in which the residual exceeds the residual thresholds: (**a**) the labeled figure of region 1; (**b**) the labeled figure of region 2.

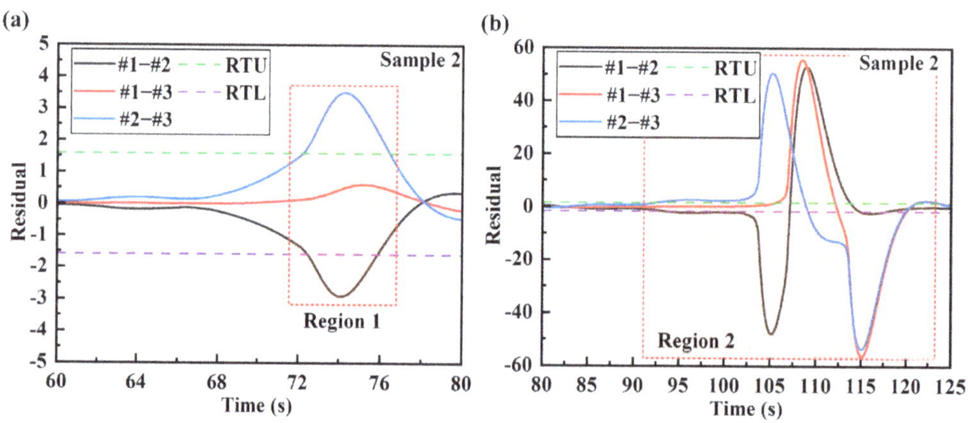

Figure 10. The regions of sample 2 in which the residual exceeds the residual thresholds: (**a**) the labeled figure of region 1; (**b**) the labeled figure of region 2.

Table 9. The number and proportion of residuals exceeding the threshold in different regions.

Samples	Regions	#1–#2	Proportion	#1–#3	Proportion	#2–#3	Proportion
Sample 1	Region 1	570	0.19	0	0	652	0.22
		977	0.33	0	0	1141	0.38
	Region 2	1086	0.36	0	0	2253	0.75
		2172	0.72	0	0	1113	0.37
Sample 2	Region 1	0	0	0	0	814	0.27
		709	0.24	0	0	0	0
	Region 2	2857	0.95	1304	0.43	3276	1.09
		1518	0.51	1649	0.55	2357	0.79

The warning of structural damage is inseparable from the real-time calculation and interactive display of a sensing system. In this paper, an intelligent damage sensing system is developed by combining the hybrid programming of LabVIEW 2015 and MATLAB R2023b. The workflow of the sensing system is shown in Figure 11a. The strain-collecting instrument collects the strain data of the sample in real time and transmits the data to the sensing system. And the sensing system packages the strain data of three measuring points into n sampling data and transmits them to the MATLAB module. According to the analysis conclusion of this paper, the strain series of measuring points 1 and 2 (#1–#2) and measuring points 2 and 3 (#2–#3) are combined. In order to improve the stability of the alarming function of the sensing system, the system counts the number of m_1 (#1–#2) and m_2 (#2–#3) of the calculated residuals exceeding the threshold in n sampling data in real time. The accuracy of the residual calculated for small sampling data will decrease, and too much sampling data will affect the real-time alarming of the sensing system. Through testing and analysis, the n of the sensing system is 3000. The MATLAB module calculates the corresponding residuals based on the two sets of data sampled from #1–#2 and #2–#3. Then, the proportion of the two sets of residuals in n data is counted in LabVIEW. If the proportion of any set exceeding the threshold is greater than q, the result will be reported to the sensor system for damage alarming. Through the continuous sampling and analysis of the sensing system, the warning of structural damage can be achieved in real time. Figure 11b,c are the interface displays of the system damage warning for sample 1 and sample 2, respectively.

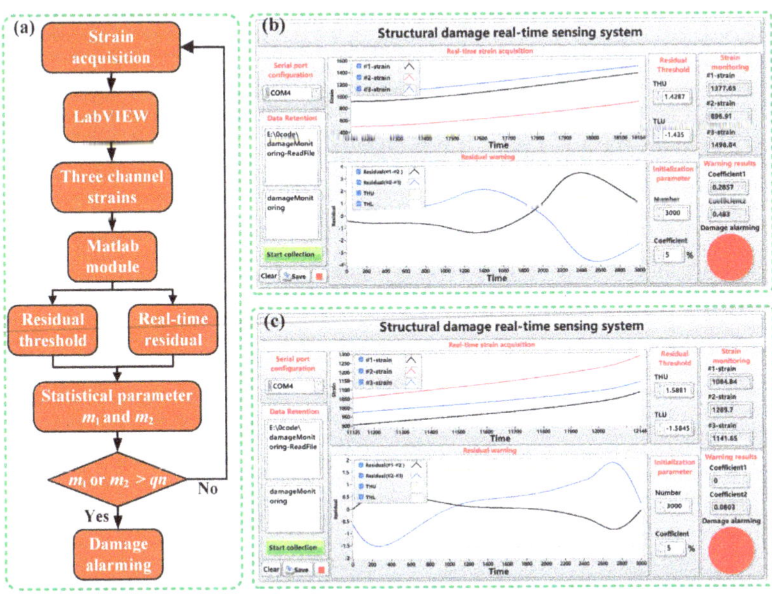

Figure 11. Structural damage sensing system: (**a**) system workflow diagram; (**b**) damage warning result of sample 1; (**c**) damage warning result of sample 2.

Figure 12 shows the monitoring value of the warning coefficient of the structural damage sensing system. In the no-load and elastic deformation stages, the damage warning coefficient is 0, and the system does not issue an alarm. In Figure 12, all 3000 data samples correspond to a sampling stage. From the beginning of the damage to structural destruction, the damage warning coefficient basically increases gradually. Once the structure is destroyed, the stress change in the structure is a constant value, and the damage warning coefficient becomes 0. It is worth noting that in the third warning stage of sample 1, both damage coefficients become 0, indicating that after the damage occurs in the first stage, the strain in the structure produces a stable fluctuation, causing the residual to return to a stable state. After continuous loading, the warning coefficient of the structure is greater than the warning coefficient of the second stage, and the warning of damage can still be achieved. It can be seen from the curve of the warning coefficient that under the condition of a sampling number of 3000, a sampling frequency of 200 Hz and a warning coefficient of 0.05, the damage warning accuracy of sample 1 is 75%, and the damage warning accuracy of sample 2 is 100%. Therefore, different structural forms will have a certain impact on the accuracy of the warning system. A comparison of existing damage monitoring studies is presented in Table 10.

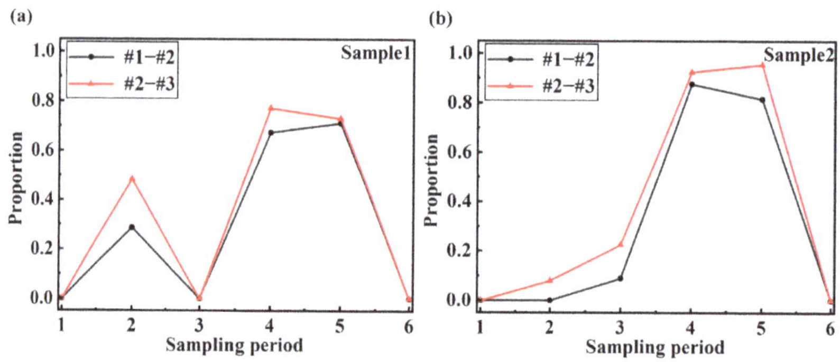

Figure 12. Warning coefficient of structural damage sensing system: (**a**) warning coefficient of sample 1; (**b**) warning coefficient of sample 2.

Table 10. Comparison of studies on damage monitoring.

Studies	Sensor	Monitoring Damage Type	Monitoring System	Real Time
Qian et al. [32]	Strain sensor	Crack	No	No
Kontsos et al. [33]	Piezoelectric sensor	Crack initiation	No	No
Xiang et al. [34]	Piezoelectric sensor	Fatigue crack	No	No
Chen et al. [35]	Eddy current sensor	Fatigue crack	No	Yes
Yuan et al. [36]	Piezoelectric sensor	Impact	Yes	Yes
This paper	Strain sensor	Early plastic damage	Yes	Yes

6. Conclusions

Based on the cointegration theory, this paper proposes a structural damage warning method. This method realizes the warning of structural damage by calculating the cointegration residual between two measuring points in the structure, avoiding the uncertainty of the warning of the single strain threshold method. The strain cointegration residuals of stress-concentrated samples and standard samples are analyzed. The residual analysis results show that the residual of the strain series between two asymmetric measuring points is more sensitive to the plastic deformation damage of the structure. The calculated residual is compared with the threshold value in the residual steady state to realize the identification of structural damage. In order to improve the engineering practicality of the method, this paper combines LabVIEW and MATLAB to develop a real-time structural damage sensing system. The system calculates the residual warning coefficients of the two groups of strain series #1–#2 and #2–#3 in real time and comprehensively considers the two prediction results to warn of structural damage. The test results show that under the conditions of a sampling number of 3000, a sampling frequency of 200 Hz and a warning coefficient of 0.05, the system alarm accuracy for the stress concentration sample is 75%, and for the standard sample it is 100%. Due to different structural characteristics, in actual applications, the system prediction accuracy for different types of structures will vary. The proposed method further realizes the engineering application of structural damage monitoring.

Author Contributions: Conceptualization, Q.G. and J.H.; methodology, Q.G. and Y.W.; software, X.W. and C.W.; writing—original draft, Q.G.; writing—review and editing, J.H. All authors have read and agreed to the published version of the manuscript.

Funding: This research was supported by the National Natural Science Foundation of China under grant number 52275236, Liaoning Province major science and technology grant number 2022HJ1/10400031, and Liaoning Province science and technology plan joint plan grant number 2023JH2/101700286.

Institutional Review Board Statement: Not applicable.

Informed Consent Statement: Not applicable.

Data Availability Statement: Data are contained within the article.

Acknowledgments: Thanks to the authors of this paper for their work on the research content. The authors are thankful for the experimental platform and test conditions provided by the project.

Conflicts of Interest: The authors declare no conflicts of interest.

References

1. Achenbach, J.D. Structural Health Monitoring—What is the prescription? *Mech. Res. Commun.* **2009**, *36*, 137–142. [CrossRef]
2. Li, S.; Liu, F.; Peng, G.; Cheng, F.; Zhao, B.; Ji, M. A lightweight SHM framework based on adaptive multisensor fusion network and multigeneration knowledge distillation. *IEEE Trans. Instrum. Meas.* **2022**, *71*, 3526919. [CrossRef]
3. Kralovec, C.; Schagerl, M. Review of structural health monitoring methods regarding a multi-sensor approach for damage assessment of metal and composite structures. *Sensors* **2020**, *20*, 826. [CrossRef] [PubMed]
4. Moradi, M. MEMS multisensor intelligent damage detection for wind turbines. *IEEE Sens. J.* **2015**, *15*, 1437–1444. [CrossRef]
5. Etxaniz, J. Ultrasound-based structural health monitoring methodology employing active and passive techniques. *Eng. Fail. Anal.* **2023**, *146*, 107077. [CrossRef]
6. Arcadius Tokognon, C.; Gao, B.; Tian, G.Y.; Yan, Y. Structural health monitoring framework based on internet of things: A survey. *IEEE Internet Things J.* **2017**, *4*, 619–635. [CrossRef]
7. Janeliukstis, R.; Mironovs, D. Smart composite structures with embedded sensors for load and damage monitoring—A review. *Mech. Compos. Mater.* **2021**, *57*, 131–152. [CrossRef]
8. Broer, A.A.R.; Benedictus, R.; Zarouchas, D. The need for multi-sensor data fusion in structural health monitoring of composite aircraft structures. *Aerospace* **2022**, *9*, 183. [CrossRef]
9. Ha, N.; Lee, H.-S.; Lee, S. Development of a wireless corrosion detection system for steel-framed structures using pulsed eddy currents. *Sensors* **2021**, *21*, 8199. [CrossRef]
10. Joseph, R.; Giurgiutiu, V. Analytical and experimental study of fatigue-crack-growth ae signals in thin sheet metals. *Sensors* **2020**, *20*, 5835. [CrossRef]
11. Na, W.S.; Baek, J. A review of the piezoelectric electromechanical impedance based structural health monitoring technique for engineering structures. *Sensors* **2018**, *18*, 1307. [CrossRef] [PubMed]
12. Migot, A.; Saaudi, A.; Giurgiutiu, V. Delamination depth detection in composite plates using the lamb wave technique based on convolutional neural networks. *Sensors* **2024**, *24*, 3118. [CrossRef] [PubMed]
13. Castro-Toscano, M.J.; Rodriguez-Quinonez, J.C.; Sergiyenko, O.; Flores-Fuentes, W.; Ramirez-Hernandez, L.R.; Hernandez-Balbuena, D.; Lindner, L.; Rascon, R. Novel sensing approaches for structural deformation monitoring and 3D measurements. *IEEE Sens. J.* **2021**, *21*, 11318–11328. [CrossRef]
14. Soman, R. Sequential niching particle swarm optimization algorithm for localization of multiple damage locations using fiber bragg grating sensors. *NDT E Int.* **2024**, *143*, 103069. [CrossRef]
15. Khani, M.M.; Vahidnia, S.; Ghasemzadeh, L.; Ozturk, Y.E.; Yuvalaklioglu, M.; Akin, S.; Ure, N.K. Deep-learning-based crack detection with applications for the structural health monitoring of gas turbines. *Struct. Health Monit.* **2019**, *19*, 1440–1452. [CrossRef]
16. Jiao, P.; Egbe, K.-J.I.; Xie, Y.; Matin Nazar, A.; Alavi, A.H. Piezoelectric sensing techniques in structural health monitoring: A state-of-the-art review. *Sensors* **2020**, *20*, 3730. [CrossRef]

17. Qing, X.; Li, W.; Wang, Y.; Sun, H. Piezoelectric transducer-based structural health monitoring for aircraft applications. *Sensors* **2019**, *19*, 545. [CrossRef]
18. Liu, P.; Lim, H.J.; Yang, S.; Sohn, H.; Lee, C.H.; Yi, Y.; Kim, D.; Jung, J.; Bae, I. Development of a "stick-and-detect" wireless sensor node for fatigue crack detection. *Struct. Health Monit.* **2017**, *16*, 153–163. [CrossRef]
19. Kim, S.B.; Sohn, H. Instantaneous reference-free crack detection based on polarization characteristics of piezoelectric materials. *Smart Mater. Struct.* **2007**, *16*, 2375–2387. [CrossRef]
20. Yoon, J. Deep neural network-based structural health monitoring technique for real-time crack detection and localization using strain gauge sensors. *Sci. Rep.* **2022**, *12*, 20204. [CrossRef]
21. Azimi, M.; Eslamlou, A.; Pekcan, G. Data-driven structural health monitoring and damage detection through deep learning: State-of-the-art review. *Sensors* **2020**, *20*, 2778. [CrossRef] [PubMed]
22. Wang, L.; Zhang, C.; Tao, C.; Ji, H.; Yang, Y.; Qiu, J. Prediction of multiple fatigue crack growth based on modified paris model with particle filtering framework. *Mech. Syst. Signal Process.* **2023**, *190*, 110124. [CrossRef]
23. Chen, W.; He, Z.; Zhang, J. Online monitoring of crack dynamic development using attention-based deep networks. *Autom. Constr.* **2023**, *154*, 105022. [CrossRef]
24. Zhou, Y.; Chen, X.; Li, H.; Hao, S.; Yan, X.; Zhao, Y. Automatic identification and location method for shaft-hole interference fit crack based on deep learning. *Mech. Syst. Signal Process.* **2023**, *192*, 110194. [CrossRef]
25. Xu, W.; Cui, C.; Luo, C.; Zhang, Q. Fatigue crack monitoring of steel bridge with coating sensor based on potential difference method. *Constr. Build. Mater.* **2022**, *350*, 128868. [CrossRef]
26. Pang, Q.; Dong, G.; Yang, X. Metal crack detection sensor based on microstrip antenna. *IEEE Sens. J.* **2023**, *23*, 8375–8384. [CrossRef]
27. Yu, J.; Li, C.; Qiu, X.; Chen, H. A full-optical strain fbg sensor for in-situ monitoring of fatigue stages via tunable DFB laser demodulation. *Opt. Quantum Electron.* **2021**, *53*, 156. [CrossRef]
28. De Pauw, B.; Hinderdael, M.; Moonens, M.; De Baere, D.; Geernaert, T.; Berghmans, F.; Guillaume, P. Fatigue failure monitoring of 316L stainless steel coupons using optical fibre based distributed strain sensing. *Smart Mater. Struct.* **2019**, *28*, 105054. [CrossRef]
29. Yu, Y.; Liu, X.; Yan, J.; Wang, Y.; Qing, X. Real-time life-cycle monitoring of composite structures using piezoelectric-fiber hybrid sensor network. *Sensors* **2021**, *21*, 8213. [CrossRef]
30. Qi, C.; Weixi, Y.; Jun, L.; Heming, G.; Yao, M. A research on fatigue crack growth monitoring based on multi-sensor and data fusion. *Struct. Health Monit.* **2021**, *20*, 848–860. [CrossRef]
31. Wang, Y.; He, M.; Sun, L.; Wu, D.; Wang, Y.; Qing, X. Weighted adaptive kalman filtering-based diverse information fusion for hole edge crack monitoring. *Mech. Syst. Signal Process.* **2022**, *167*, 108534. [CrossRef]
32. Feng, L.; Qian, X. Enhanced real-time crack monitoring and updating in welded structural components with limited measurement data. *Mech. Syst. Signal Process.* **2023**, *198*, 110403. [CrossRef]
33. Malik, S.; Kontsos, A. A novel information entropy approach for crack monitoring leveraging nondestructive evaluation sensing. *Mech. Syst. Signal Process.* **2024**, *214*, 111207. [CrossRef]
34. Zhu, W.; Xu, Z.; Xiang, Y.; Liu, C.; Deng, M.; Qiu, X.; Sun, D.; Xuan, F. Nonlinear ultrasonic detection of partially closed cracks in metal plates using static component of lamb waves. *NDT E Int.* **2021**, *124*, 102538. [CrossRef]
35. Fan, X.; He, Y.; Chen, T. Research on fatigue crack quantitative monitoring based on eddy current sensor with an interactive induction coil layout. *Smart Mater. Struct.* **2023**, *32*, 035013. [CrossRef]
36. Qiu, L.; Lin, X.; Wang, Y.; Yuan, S.; Shi, W. A mechatronic smart skin of flight vehicle structures for impact monitoring of light weight and low-power consumption. *Mech. Syst. Signal Process.* **2020**, *144*, 106829. [CrossRef]

sensors

MDPI

Article

Automatic Detection and Classification of Natural Weld Defects Using Alternating Magneto-Optical Imaging and ResNet50

Yanfeng Li [1,*], Pengyu Gao [2], Yongbiao Luo [1], Xianghan Luo [1], Chunmei Xu [3], Jiecheng Chen [1], Yanxi Zhang [2], Genxiang Lin [1] and Wei Xu [1]

1 School of Automobile and Transportation Engineering, Guangdong Polytechnic Normal University, Guangzhou 510632, China; 15815380750@163.com (Y.L.); x2234140821@163.com (X.L.); 25996@random.com (J.C.); lingengxiang@gpnu.edu.cn (G.L.); xuy_wei@126.com (W.X.)

2 Guangdong Provincial Welding Engineering Technology Research Center, Guangdong University of Technology, Guangzhou 510006, China; perrygao99@gmail.com (P.G.); yanxizhang@126.com (Y.Z.)

3 School of Computer Science and Technology, Southwest University of Science and Technology, Mianyang 621010, China; xchm@swust.edu.cn

* Correspondence: lyflwl88@163.com

Abstract: It is difficult to detect and identify natural defects in welded components. To solve this problem, according to the Faraday magneto-optical (MO) effect, a nondestructive testing system for MO imaging, excited by an alternating magnetic field, is established. For the acquired MO images of crack, pit, lack of penetration, gas pore, and no defect, Gaussian filtering, bilateral filtering, and median filtering are applied for image preprocessing. The effectiveness of these filtering methods is evaluated using metrics such as peak signal–noise ratio (PSNR) and mean squared error. Principal component analysis (PCA) is employed to extract column vector features from the downsampled defect MO images, which then serve as the input layer for the error backpropagation (BP) neural network model and the support vector machine (SVM) model. These two models can be used for the classification of partial defect MO images, but the recognition accuracy for cracks and gas pores is comparatively low. To further enhance the classification accuracy of natural weld defects, a convolutional neural network (CNN) classification model and a ResNet50 classification model for MO images of natural weld defects are established, and the model parameters are evaluated and optimized. The experimental results show that the overall classification accuracy of the ResNet50 model is 99%. Compared with the PCA-SVM model and CNN model, the overall classification accuracy was increased by 7.4% and 1.8%, and the classification accuracy of gas pore increased by 10% and 4%, respectively, indicating that the ResNet50 model can effectively and accurately classify natural weld defects.

Keywords: magneto-optical imaging; natural weld defect; alternating magnetic field; convolutional neural network; ResNet50

1. Introduction

Welding technologies play a crucial role in modern manufacturing, especially in high-precision assembly and electronic component fabrication [1–4]. With advancements in welding processes, the demand for high-quality welds has increased, making defect detection particularly critical. Traditional methods such as ultrasonic testing and radiographic testing are already mature but have limitations, including an insufficient ability to identify complex defects and radiation risks during the detection process [5,6]. Magneto-optical (MO) imaging technology, with its advantages of no radiation and high sensitivity to minor defects, is increasingly becoming a key technique in weld defect detection [7,8]. This technology has significant advantages in detecting subsurface defects, with a detection depth of up to 6 mm, and its experimental costs are low [9]. The conventional MO detection method uses a direct current (DC) magnetic field to excite the weldments; because the magnetic field strength and direction of the direct current magnetic field are constant, some useful weld information may be lost, and it is easy to saturate [10]. The acquisition method of MO images of weld defects under alternating magnetic field excitation can collect dynamic MO images, thereby improving the problem of magnetic field information loss. Independent component analysis is used to achieve denoising of MO images and compare the denoising performance with dynamic filtering [11]. A dynamic MO imaging system is used to detect weld defects, and an MO image detection model for weld defects is established through support vector machine (SVM) models. The defect classification accuracy is only 91.5%, with which, it is difficult to meet the detection requirements [12]. The processing and analysis of MO imaging data remains challenging, especially in efficiently and accurately extracting and classifying defect features from complex images [13]. To address these issues, researchers are exploring solutions that combine advanced deep learning techniques to improve the effectiveness of MO imaging in defect detection.

The core of deep learning lies in neural network models, which require the construction of various neural network models to solve different problems [14,15]. Through self-learning, it can obtain highly abstract feature information that cannot be extracted manually. All neural network models will encounter the problem of generalization. The problem of generalization mainly focuses on enriching the sample space, limiting the hypothesis space, and changing the optimization objective. To enrich the sample space is to expand the sample space and increase the inductive bias of the model. Limiting the hypothesis space refers to reducing the risk of model overfitting by restricting the parameters and structure of the model. Changing the optimization objective means changing the traditional method to minimize the empirical risk on the training set as the optimization objective. By enhancing data through methods such as rotation, cropping, flipping, and brightness adjustment, the small sample problem is solved through data augmentation, so that the generalization ability of the weld image analysis model is improved [16,17]. An artificial neural network (ANN) with modified performance function is studied for automatic recognition and detection of weld defects in radiographic images [18,19]. A defect classification method based on direct multi-class support vector machine is proposed, which aims to minimize structural risk as the optimization objective and improve the generalization ability of the weld seam radiographic image analysis model in the case of a small training set [20,21]. Transfer learning in deep learning models can avoid overfitting, solve the problem of using small datasets, and extract the features of the images to be classified [22]. Several pre-trained networks, such as VGG16, ResNet50, DenseNet [23], AlexNet [24], and others, have served as feature extractors in classification problems [25]. The introduction of VGG-Net [26] and ResNet [27] further promotes the performance improvement of classification networks. As the depth of the network increases, ResNet improves the problem of gradient vanishing during backpropagation. A new CNN model based on ResNet50 is proposed to classify the defects in the radiographic images. Techniques such as stratified cross-validation, data augmentation, and regularization will be chosen to improve the model generalization ability and avoid overfitting [28]. A TOFD image weld defect detection method based on multi-image fusion and feature hybrid enhancement is proposed by combining deep learning techniques with domain knowledge in the field of TOFD detection [29]. A defect classification model for SMT welding images based on an improved ResNet model, namely the ResNet-34-ECA model, was proposed, with an overall classification accuracy of 98.2% [30]. In the field of aircraft engines, deep learning algorithms are also widely used. A hierarchical health monitoring model called the adaptive thresholding and coordinate-attention-based tree-inspired network (ATCATN) has been developed for the health monitoring of aero-engine bearings under strong background noise. The experimental results show that this model can accurately identify the fault locations and sizes of aero-engine bearings even under strong noise interference [31]. A data-driven time–frequency analysis (TFA) technology for CTNet was developed, which combines a fully convolutional auto-encoder network with the convolutional block attention module (CBAM). The experimental results show that the CTNet has good ability to detect wind turbine faults [32]. Deep learning algorithms can automatically learn and extract the features in images to effectively complete the tasks such as image classification and object detection [33].

At present, there is increasing international attention on the application of MO imaging technology. One study proposed using an improved remanence/magneto-optical imaging combined with a cost-sensitive convolutional neural network (CNN) for automatic defect detection and classification of low-carbon steel WAAM components [34]. The study demonstrated that this method significantly improved detection accuracy for surface defects in low-carbon steel products. Another study developed an MO imaging nondestructive testing system based on alternating magnetic field excitation, which effectively addressed the issue of detecting hidden weld defects. By establishing a three-dimensional finite element model and extracting texture features using gray-level co-occurrence matrix (GLCM) and Tamura methods, the researchers successfully applied a backpropagation (BP) neural network for defect classification, achieving an overall classification accuracy of 91.1% [35]. Additionally, research has focused on image fusion technology, using pixel standard-deviation-based image fusion methods to enhance the visual effect and detection accuracy of MO imaging. This method significantly improved image quality and showed promising results in weld defect detection [36].

There are still relatively few MO imaging weld defect detection methods based on deep learning, and they do not take into account the fact that images need to be "dynamically" observed during the evaluation process, resulting in low accuracy and credibility in the defect detection results. This paper aims to explore automatic detection and classification methods for natural weld defects by combining alternating MO imaging technology with neural network algorithms. Classification models including BP neural networks, SVM model, CNN, and ResNet50 are utilized for the recognition and classification of natural weld defects. An alternating magnetic field excitation imaging system based on Faraday MO effect is constructed to capture MO images of various weld defects. The collected MO images were preprocessed using a Gaussian filter, a bilateral filter, and a median filter, and the effectiveness of the different filtering techniques was evaluated. Image features were extracted using principal component analysis (PCA), and these features were input into the BP neural network and the SVM model for classification. These two models can be used for the classification of partial defect MO images, but the recognition accuracy of cracks and pores is relatively low. To improve the classification accuracy, the CNN-based model and the ResNet50 model were established, and the model parameters were optimized. The experimental results show that the overall classification accuracy of the CNN and ResNet50 models is 97.2% and 99%, respectively. Compared with different detection methods, the deep learning methods used for natural welded defects detection and recognition demonstrates stronger noise resilience, and can maintain high detection and recognition accuracy even under the significant deformation of welded defects.

The paper is organized as follows. Section 2 introduces the weld defect test system based on MO imaging and analyzes the mechanism of MO imaging. Section 3 presents the PCA method to extract the principal components of the column vector, establishes the PCA-BP classification model and the PCA-SVM classification model, and compares the classification performance of the models. Section 4 studies the CNN classification model and the ResNet50 classification model, explores the parameter settings of the two models, and compares and analyzes the classification performance of the models. Section 5 draws conclusions.

2. Experimental Methods

2.1. Experimental Setup

The MO imaging detection device based on the leakage magnetic characteristics of natural weld defects primarily consists of the tested weldment, an excitation mechanism, an MO imaging sensor, a testing platform, and computer storage equipment. The schematic diagram of the MO imaging detection system for welded defects excited by alternating magnetic field is shown in Figure 1. The welding materials were carbon steel plate (No. 45) and high-strength steel (HSS). The excitation mechanism is an alternating current (AC) excitation system that generates an alternating magnetic field by connecting AC power to the excitation coil. The magnitude and direction of the alternating magnetic field will change periodically with the change in time. The MO testing platform uses a three-axis linkage motion platform, which can achieve motion and detection of the three directions of the X, Y, and Z axes, as shown in Figure 2. The MO sensor is mainly composed of a light-emitting diode (LED) light, an optical conduction system, a CMOS, and an MO film, with a size of 20×15 mm^2, and its surface is coated with a mirror layer. The primary parameters of the MO sensor are detailed in Table 1.

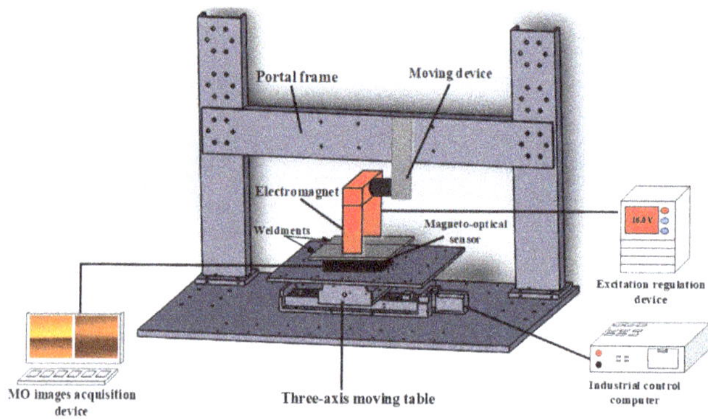

Figure 1. Schematic diagram of MO imaging detection system for welded defects excited by alternating magnetic field.

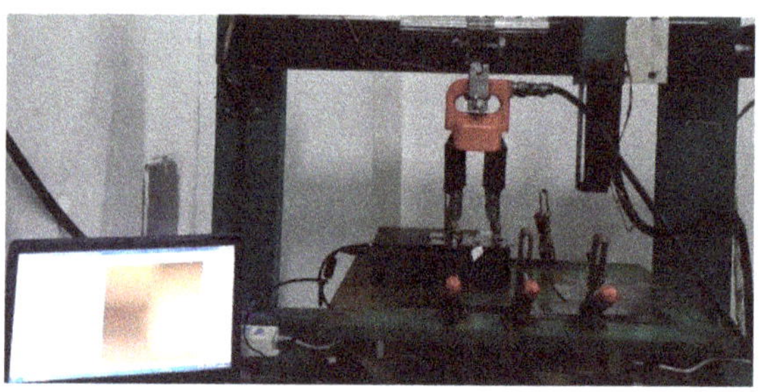

Figure 2. Physical diagram of MO imaging nondestructive testing platform for weld defects excited by alternating magnetic field.

Table 1. Main parameters of MO imaging sensor.

Light Source Wavelength	Sampling Frequency	Maximum Resolution	Pixel Equivalent	Magnetic Field Range
590 nm	[8, 100] fps	2592×1944 pixel^2	102 pixel/mm	[−2, 2] kA/m

The weldment material was selected from the HSS plate with a length of 150 mm, a width of 50 mm, and a thickness of 12 mm. Lack of penetration and pit were formed by the YAG laser welding machine on the abutting steel plate. The laser power is 10 kW. The defocus amount is −1 mm and the welding speed is 3 m/s. The shielding gas is argon, and the gas flow rate is 30 L/min. The angle between the gas nozzle and the torch is 45°. Under high-power laser welding conditions, there is a deviation of 2 mm and −2 mm between the starting and ending positions of the laser beam and the weld seam during welding, resulting in pits and incomplete penetration of the test weldment. The size and shape of surface pits on welded parts can be measured, but the depth and location of incomplete penetration cannot be predicted. Natural welding defects were formed on a carbon steel plate (No. 45) by TIG welding. The welding current of the TIG machine was adjusted within the range of 80 A–130 A, and the welding speeds for the initial welding tests were set to 1, 1.5, 2, and 2.5 mm/s. At the same time, different types of natural weld defects were obtained by rapidly cooling the high-temperature-welded specimens that had completed the welding operation, including crack, pit, and incomplete penetration.

2.2. Principle of MO Imaging for Weld Defects

The principle of MO imaging for natural weld defects mainly involves the concepts of leakage magnetic fields [37], the Faraday MO effect [38], and the application of external excitation magnetic fields, as illustrated in Figures 3 and 4. When an external magnetic field is applied to a ferromagnetic weldment with high magnetic permeability, it becomes magnetized. If there are defects on the surface or inside of the weldment, the local magnetic permeability of the weldment material will decrease, and the magnetic resistance will increase, resulting in a non-uniform magnetic flux distribution. The magnetic field lines that pass through the weld defect are distorted, with a portion of the magnetic field lines passing through the weldment and another portion overflowing into the air and returning to the weldment, forming a leakage magnetic field in the local area of the weld defect's surface, as shown in Figure 3. As the shape and size of the weld defect change, the magnitude of the leakage magnetic field varies accordingly. MO imaging equipment is used to detect changes in the leakage magnetic field and convert these signals into corresponding MO images. By applying image processing algorithms to extract defect information from MO images, the category of defects can be determined.

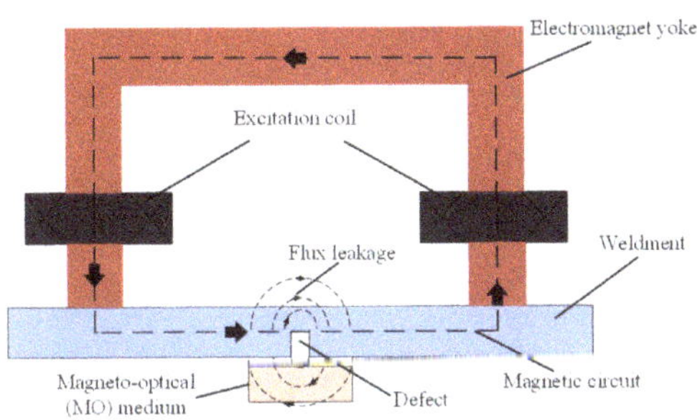

Figure 3. Schematic diagram of leakage magnetic field at welded defect.

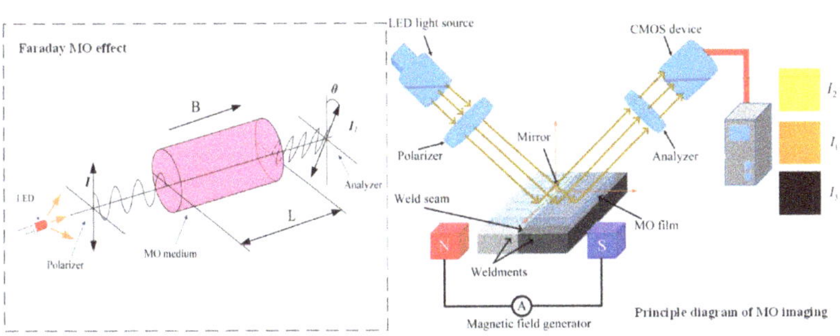

Figure 4. Principle diagram of MO imaging detection for weld defects.

Faraday MO rotation effect means that when a beam of linearly polarized light travels through an MO medium. If an external magnetic field is applied along the direction of light propagation, the vibration surface of the polarized light rotates by a Faraday rotation angle $\boldsymbol{\theta}$. Although the trajectory of linearly polarized light does not change in a strong magnetic field, it will deflect in the presence of a leakage magnetic field generated by defects. A polarizer detects this deflection angle, and a CMOS camera captures the light intensity. The left side of Figure 4 is a schematic diagram of Faraday MO rotation effect, where the Faraday rotation angle $\boldsymbol{\theta}$ is proportional to the magnitude of the external magnetic field, the effective path of the light through the MO medium and the material properties of the MO medium. The rotation direction of the incident polarized light is solely dependent on the direction of the external magnetic field. The magnitude of $\boldsymbol{\theta}$ can be expressed as follows:

$$\boldsymbol{\theta} = V\boldsymbol{B}L_{\mathrm{m}} \tag{1}$$

where $\boldsymbol{B}$ is the magnetic flux density and L_{m} is the effective path length of the polarized light through the MO medium. V is the Verdet constant of the MO medium.

The vertical component of the leakage magnetic field $\boldsymbol{B}_{z}$ perpendicular to the weldment is the primary cause of deflection. The weldment is located in the external magnetic field generated by the magnetic field generator. In the defect-free area, the weldment will form a complete magnetic circuit, and there will be no vertical magnetic field along the propagation direction of polarized light. The vibration surface of polarized light will not deflect. When there are defects, the weldment cannot form a complete magnetic circuit, and the vertical magnetic field component at the defect below the reflective mirror surface changes. If there is a vertical magnetic field along the propagation direction of the polarized light, the vibration surface of the polarized light will be deflected. The linearly polarized light containing welding quality information is received by the CMOS sensor after passing through a polarizer, forming an MO image of the weld defect, as shown on the right side of Figure 4. When the direction of the external magnetic field is toward the north pole, the plane of polarization rotates clockwise by a Faraday rotation angle along the direction of light propagation. When the direction of the magnetic field is toward the south pole, the plane of polarization rotates counterclockwise by a Faraday rotation angle $\boldsymbol{\theta}$.

By applying AC power to the magnetic field generator can generate an alternating magnetic field at both ends of the magnetic yoke that changes in direction and magnitude at a certain frequency, and the Faraday rotation angle also changes accordingly. The light intensity detected by the analyzer is represented as follows:

$$I_1 = I_0 cos^2\alpha \tag{2}$$

$$I_2 = I_0 cos^2(\alpha - \boldsymbol{\theta}) \tag{3}$$

$$I_3 = I_0 cos^2(\alpha + \boldsymbol{\theta}) \tag{4}$$

where I_0 is the intensity of the incident linearly polarized light, and α is the inherent rotation degree of the linearly polarized light without the application of an external magnetic field. These equations show how the intensity of light detected by the polarizer varies with the direction and intensity of the magnetic field, reflecting the corresponding variations in the Faraday rotation angle [39].

2.3. Preprocessing of MO Images for Weld Defect Analysis

Each image filtering algorithm has its applicability. It is extremely difficult to judge the filtering effect of MO images by observing the visual effect of the filtered image through human visual perception, and this method is time-consuming and labor-intensive [40]. Therefore, by comparing the mean square error (MSE), the peak signal–noise ratio (PSNR), the structural similarity (SSIM), and the filtering time between the filtered image and the original image, the quality of the image filtering effect can be determined. This section studied Gaussian filtering, bilateral filtering, and median filtering algorithms, and used these three algorithms to filter MO images. The filtering effect was evaluated by MSE, PSNR, and SSIM.

MSE is an indicator used to evaluate the difference between the original image and the filtered image. MSE calculates the average of the squared differences in pixel values between the original image and the filtered image, as follows:

$$\text{MSE} = \frac{1}{mn}\Sigma_{i=0}^{m-1}\Sigma_{j=0}^{n-1}\left(I_{(i,j)} - K_{(i,j)}\right)^2 \tag{5}$$

where m and n denote the number of rows and columns of the image, respectively, $I_{(i,j)}$ is the pixel value of the original image at (i,j), and $K_{(i,j)}$ is the pixel value of the filtered image at (i,j). Therefore, a smaller MSE indicates that the filtered image approximates the original image more closely, reflecting a higher image quality.

PSNR is the ratio between the maximum possible value (peak signal) of an image and the distortion of the image, and the formula is:

$$\text{PSNR} = 10\log_{10}\left(\frac{MAX^2}{MSE}\right) \tag{6}$$

where MAX represents the maximum value of the image pixels, typically 255. The unit of PSNR is decibels (dB), and the higher the value, the smaller the distortion and the higher the image quality.

SSIM can be used to compare the structural similarity between two images, and its formula is:

$$\text{SSIM}(x,y) = \frac{(2u_x u_y + C_1)(2\sigma_{xy} + C_2)}{\left(\mu_x^2 + \mu_y^2 + C_1\right)\left(\sigma_x^2 + \sigma_y^2 + C_2\right)} \tag{7}$$

where x and y represent the two images to be compared; u_x and u_y represent the mean of images x and y, respectively. μ_x^2 and μ_y^2 are the variance of images x and y and σ_{xy} is the covariance of images x and y. C_1 and C_2 are constants. The range of SSIM values is [−1,1], and the closer the SSIM is to 1, the more similar the two images are. Taking the example of incomplete penetration MO image, Table 2 shows the evaluation results of three different filtering algorithms.

Table 2. Filtering results and indicators of different filtering algorithms for incomplete penetration.

Type	Filtering Result	MSE	PSNR (dB)	SSIM	Filtering Time (s)
Gaussian filtering		0.39	52.18	0.9956	0.1009
Bilateral filtering		2.74	43.75	0.9656	0.8810
Median filtering		3.10	43.22	0.9684	0.0154

It can be seen from the table that the MSE value of Gaussian filtering is the smallest, the PSNR value is the largest, and the SSIM value is the closest to 1. Although its filtering time is slightly longer than that of median filtering, its noise-reduction performance is superior. Additionally, after multiple comparisons of different MO images using various filtering algorithms, Gaussian filtering consistently produced the best results. Therefore, for the purpose of filtering MO images, Gaussian filtering stands out as the optimal choice.

3. Detection and Classification by BP Neural Network and SVM

3.1. Feature Extraction Based on PCA

Multi-angle weld defects were formed on the weld seam using the experimental method described in Section 2.1, including pit, crack, lack of penetration, gas pore, and no defects. The excitation voltage of the alternating magnetic field was set to 120 V, and the lift height of the magnetic field generator was consistently maintained at 20 mm. The sampling frequency of the MO sensor was set to 75 Hz. MO images of natural weld defects under alternating magnetic field excitation, including three consecutive frames of dynamic MO images, are shown in Table 3. After image denoising, the collected MO images of weld defects have a size of 400 pixel × 400 pixel. Its pixels are too large, which may affect the efficiency of classification as input to the classification model. To reduce the amount of information and still retain key features, a nearest neighbor method can be used to downsample the Gaussian filtered MO images.

Table 3. MO images of five natural weld defects under alternating magnetic field excitation.

Defects	Frame 1	Frame 2	Frame 3
Pit			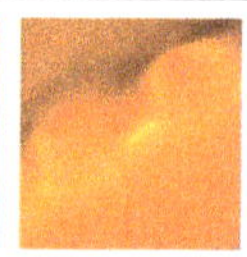
Crack			
Lack of penetration			
Gas pore			
No defects			

The downsampled images still contain the main original information, serving as input for the subsequent principal component analysis to enhance feature extraction efficiency. Figure 5 shows the downsampling results of the MO image; it can be seen from Figure 5d that the image distortion is quite obvious. Therefore, a downsampling image with a size of 50 pixel × 50 pixel was ultimately selected.

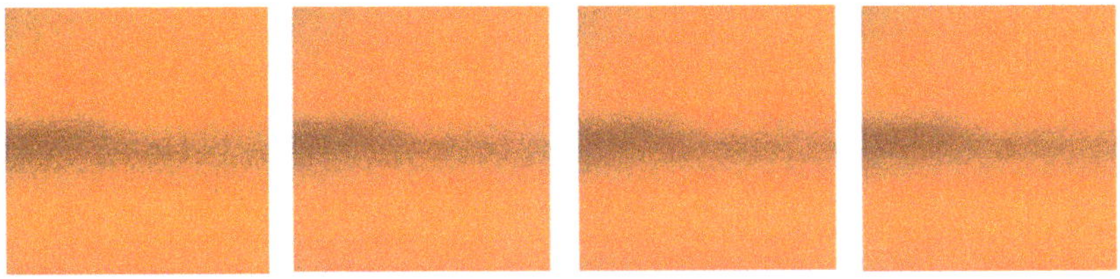

Figure 5. Downsampling of MO image. (a) 200 × 200; (b) 100 × 100; (c) 50 × 50; (d) 25 × 25.

PCA is a commonly used data dimensionality reduction technique that aims to transform original data into fewer features to extract the main information of the data [41,42]. The downsampled image is represented as a matrix $\boldsymbol{X}$ with dimensions 50 × 50. The data need to be centralized by subtracting the mean of each feature from its value, resulting in a standard normal distribution with a mean of 0 and a variance of 1 for each feature, which can be expressed as:

$$\mathbf{Z}_{ij} = \frac{\mathbf{X}_{ij} - \mu_j}{\sigma_j} \tag{8}$$

where $\mathbf{Z}$ denotes the standardized matrix, μ_j represents the mean of the j column pixels, and σ_j indicates the standard deviation of the j column pixels. The standardized MO image is represented as $Z = (z_1, z_2, z_3, \cdots z_{50})$, with $E(z_j) = 0$, $D(z_j) = 1$, $j = \{1, 2, 3, \cdots 50\}$.

The principal components X_i and the variance contribution rates P_i of the column vector pixels in the MO image are given by:

$$X_i = \sum\nolimits_{k=1}^{50} \alpha_{ik} Z_k \tag{9}$$

$$P_i = \frac{\lambda_i}{\sum_{k=1}^{50} \lambda_k} \tag{10}$$

Let P represent the sum of the variance contribution rates of the first m principal components, which can be expressed as:

$$P = \frac{\sum_{i=1}^{m} \lambda_i}{\sum_{k=1}^{50} \lambda_k} \tag{11}$$

where $\alpha_{ik} Z_k$ denotes the product of the principal component's feature vector and the standardized matrix, and λ refers to the eigenvalue of the correlation coefficient matrix of the downsampled MO images. P is set to 95%.

After processing a downsampled MO image through image processing methods to extract the variance contribution of PCA column vectors as shown in Figure 6, it can be observed that the sum of the variance contribution rates of the preceding three principal components in the MO image of weld defects accounts for more than 95% of the information in the original image. Considering the varying contributions of principal components across different downsampled MO images, to adequately extract the original information of the image, $\boldsymbol{m}$ is set to 4.

After applying PCA, the image feature matrix is transformed to 50 × 4, converting the original feature matrix with a size of 200 × 1. The first 50 rows correspond to the first principal component of the MO image, followed by the subsequent 50 rows that correspond to the second principal component; then, another 50 rows correspond to the third principal component; finally, the last 50 rows correspond to the fourth principal component. The information of the image can be compressed into the four principal components, thus realizing the dimension reduction processing. A total of 1250 MO images of five types of natural weld defects were collected under the excitation of an alternating magnetic field. After Gaussian filtering and downsampling, the principal components of the column vectors of each defect MO image were reduced. The four-dimensional principal components of the column vectors were selected to be retained, forming an image input matrix of 200 × 1250, which was used as the input set for the BP neural network classification model and SVM classification model.

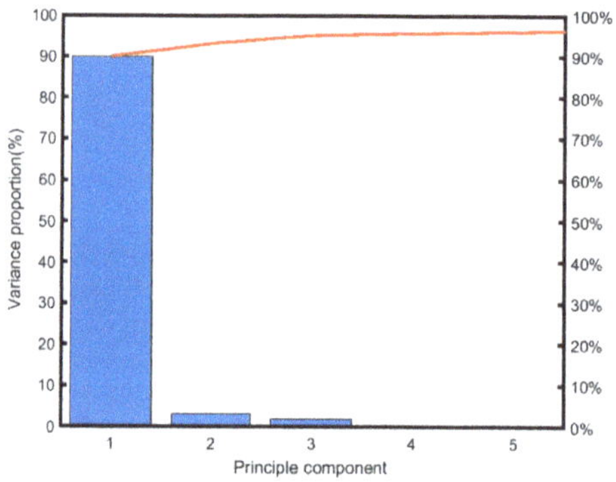

Figure 6. Principal components proportion of a downsampling MO image of weld defects.

To expand the sample, the MO image is improved as follows:

(1) Resize: Resize the original image either horizontally or vertically.
(2) Rotation: Make the original image rotate clockwise or counterclockwise according to a certain angle.
(3) Flip: Flip the original images along the horizontal or vertical axes at their center coordinates.
(4) Brightness change: Change the brightness of the original image.
(5) Contrast adjustment: Change the intensity of the brightness difference in the original image.

The original number of defect's MO images was expanded using the above method, and the expanded dataset is shown in Figure 7.

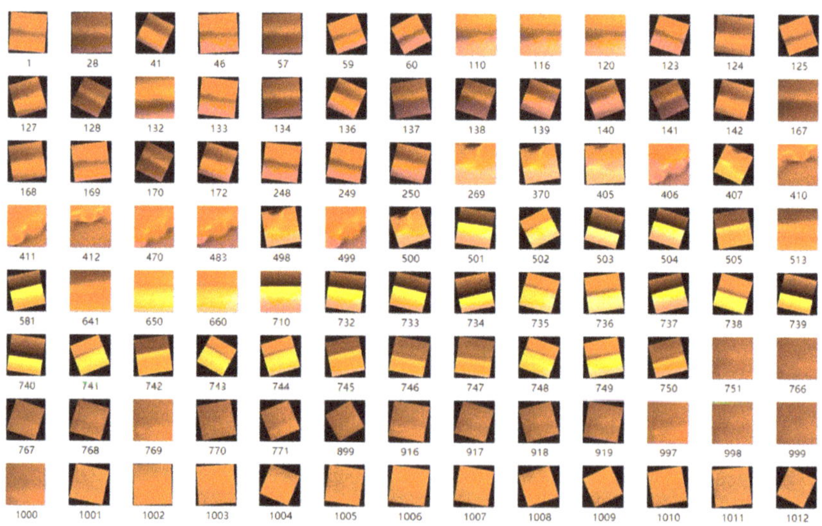

Figure 7. Part of weld defect's MO images dataset.

3.2. Classification by the PCA + BP Model

The principle of the BP neural network is to train the network through the back-propagation algorithm, enabling it to learn the mapping relationship between inputs and outputs [43,44]. PCA is a commonly used dimensionality reduction technique that helps improve the training efficiency and classification performance of subsequent BP by removing redundant features from the data. The use of PCA effectively reduces the number of input features, thereby reducing the complexity of the model, preventing overfitting, and improving the performance of the model on small sample datasets. As shown in Figure 8, the training set D_T is defined as a dataset containing m samples, which can be expressed as:

$$D_T = \{(x_1, y_1), (x_2, y_2), \ldots (x_m, y_m)\} \tag{12}$$

where $x_i \in R^d, y_i \in R^l$, meaning there are d input attributes and an output l- dimensional vector.

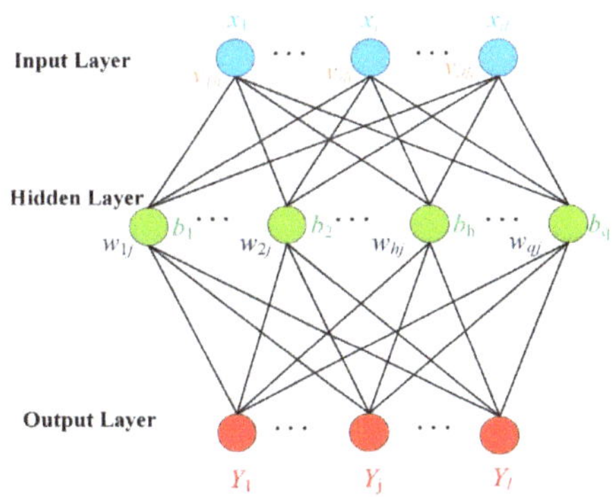

Figure 8. Topological structure of the BP neural network.

A multilayer forward feedback neural network defining an input neuron as d, hidden layer neurons as q, and output neurons as l. v and w are the weights of input layer and hidden layer and implied layer and output layer, respectively. For the hidden layer, the input and output of the h-th neuron can be represented by Equations (13) and (14), respectively:

$$z_h = \sum_{i=1}^{d} v_{ih} x_i + b_h \tag{13}$$

$$a_h = f_1(z_h) \tag{14}$$

where x_i is the output value of the i neuron in the input layer, b_h represents the deviation, and $f_1(z)$ is the activation function for the hidden layer. The hyperbolic tangent function is used as the activation function, which can be expressed as:

$$f(x) = \frac{e^x - e^{-x}}{e^x + e^{-x}} \tag{15}$$

For the output layer, the input and output of the j neuron can be represented by Equation (16) and Equation (17), respectively.

$$z_j = \sum_{h=1}^{q} w_{hj} a_h + c_j \tag{16}$$

$$a_j = f_2(z_j) \tag{17}$$

where c_j represents the deviation, and $f_2(z)$ denotes the activation function of the output layer, with the SoftMax function being utilized as the activation function in this study. During the training process, the weights and biases are updated through the BP algorithm to minimize the error. This includes the weights w_{hj} and deviations c_j from the hidden layer to the output layer, as well as the weights v_{ih} and deviations b_h from the input layer to the hidden layer.

The experiment used 1250 MO images of natural weld defects as classification samples, of which 1000 were used as the training set. In the training set, there were 200 samples each of crack, pit, lack of penetration, gas pore, and no defects. PCA is employed to extract the preceding four principal components, P_1, P_2, P_3, and P_4, from the vector pixel representations of the images. These components represent the primary features of the weld defects and form the input set for the classification model, denoted as $= \{X_{1j}, X_{2j}, \cdots X_{200j}\}$, where $(j = 1,2,3,\cdots 1250)$. The number of neurons in the hidden layer is set to $q = 20$, and the output set is defined as $Y_{1j} = 1$, $Y_{2j} = 2$, $Y_{3j} = 3$, $Y_{4j} = 4$, $Y_{5j} = 5$, with $(j = 1,2,3,\cdots 250)$ representing the output labels of the BP network model corresponding to crack, pit, lack of penetration, gas pore, and no defects, respectively. The learning rate is configured to 0.001, and the training employs the optimized Levenberg–Marquardt backpropagation algorithm.

To prevent issues with the neural network during training, such as falling into local optima, the random initialization of weights and deviations are set. Figure 9 shows the classification model of the BP neural network for MO imaging of weld defects.

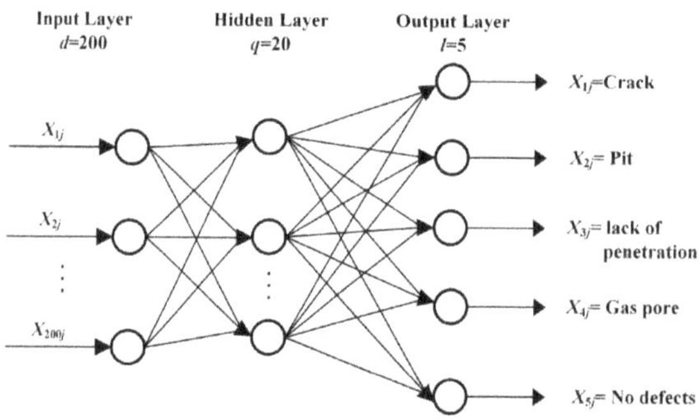

Figure 9. Classification model of the BP neural network for MO imaging of weld defects.

Table 4 shows the classification results of the BP neural network model for natural weld defects based on MO imaging. According to the table, the overall classification accuracy of the BP neural network model is 90.8%. Among the 250 images in the test set, 23 images were misclassified. Specifically, out of 50 MO images of cracks, 2 were classified as pit. Out of 50 MO images of pits, 2 were classified as cracks and 2 were classified as incomplete fusion. Out of 50 MO images of incomplete fusion, 5 were classified as gas pore. Out of 50 MO images of gas pore, 2 were classified as pit, 5 were classified as incomplete fusion, and 4 were classified as no defects. Out of 50 MO images of no defects, 1 was misclassified as gas pore. Thus, it is evident that the BP neural network classification model has poor performance in classifying pits and gas pores.

Table 4. Classification results of the PCA-BP model for natural weld defects under alternating magnetic field excitation.

Defect Types	Number of Images	Train Samples	Test Samples	Recognition Result	Classification Accuracy/%
Crack	250	200	50	48	96
Pit	250	200	50	46	92
Lack of penetration	250	200	50	45	90
Gas pore	250	200	50	39	78
No defects	250	200	50	49	98
Total	1250	1000	250	227	90.8

3.3. Classification by the PCA + SVM Model

SVM is a commonly used machine learning algorithm, mainly to find an optimal hyperplane given a dataset, so as to classify or regress the data, especially suitable for high-dimensional feature spaces [45,46]. The use of PCA provides SVM with a more effective feature set, thereby improving classification accuracy and helping SVM find the best decision boundary in high-dimensional space. The hyperplane in SVM can be represented using the linear Equation (18):

$$\omega^T + b = 0 \tag{18}$$

where ω denotes the normal vector of the hyperplane, ω^T is the transpose of the vector ω, and b represents the bias term.

Let the sample set be $\{(x_i, y_i), i = 1, \ldots, n$, where n represents the sample size, which in this study is $n = 1250$, and y_i is the label for the i sample x_i.

To solve the optimal hyperplane of the SVM, an optimization function is constructed to find the maximum value of the Lagrangian multiplier, which can be expressed as:

$$max(\alpha) = \Sigma_{i=1}^{n}\alpha_i - \frac{1}{2}\sum_{\substack{i=1\\ j=1}}^{n} y_i y_j \alpha_i \alpha_j K(x_i \cdot x_j) \tag{19}$$

where α_i corresponds to the Lagrange multiplier for the i sample, and $K(x_i, x_j)$ is the kernel function. The term $\frac{1}{2}\sum_{\substack{i=1\\ j=1}}^{n} y_i y_j \alpha_i \alpha_j K(x_i \cdot x_j)$ represents the similarity between samples, as computed by the kernel function.

The commonly used types of kernel functions currently include linear kernel function, polynomial kernel function, and radial basis function (RBF). The linear kernel function is applicable only to linearly separable datasets and is incapable of addressing nonlinear issues. The polynomial kernel function has multiple undetermined coefficients and requires parameter tuning through methods such as cross-validation, which is quite complex. Conversely, The RBF does not require parameter adjustment and is suitable for processing nonlinear data. Therefore, the RBF is used, which can be expressed as:

$$K(x_i, x_j) = e^{-\gamma||x_i - x_j||^2} \tag{20}$$

where x_i and x_j are the feature vectors of the data points, and γ is a parameter of the kernel function that governs the distribution of sample points after mapping to the feature space.

The discriminant function of the SVM classifier is obtained by solving it, which can be expressed as:

$$f(x) = sign\left(\sum_{i=1}^{n} \alpha_i^* y_i K(x_i \cdot x_j) + b^*\right) \tag{21}$$

where α_i^* and b^* are the optimal solutions for the Lagrange multipliers and bias terms obtained through grid search, respectively. The sign function is a squashing function that maps a real number to a binary output value of either 1 or −1, used for the final classification decision.

The experiment utilized 1250 MO images of weld defects as classification samples, of which 1000 were used as the training set. There were 200 samples each of crack, pit, lack of penetration, gas pore, and no defects in the training set. These images were subjected to noise reduction and downsampling to pixels. PCA was conducted to extract the preceding four principal components, P_1, P_2, P_3, and P_4, which represent the main characteristics of weld defects. This resulted in a total of 200 × 1250 inputs being classified using an SVM multi-classification model. The classification results of the model are shown in Table 5, with an overall classification accuracy of 91.6%.

Table 5. Classification results of the PCA-SVM model for natural weld defects.

Defect types	Number of Images	Train Samples	Test Samples	Recognition Result	Classification Accuracy/%
Crack	250	200	50	43	86
Pit	250	200	50	46	92
Lack of penetration	250	200	50	47	94
Gas pore	250	200	50	44	88
No defects	250	200	50	49	98
Total	1250	1000	250	229	91.6

For the crack images, five were misclassified as pores and two were misclassified as having no defects. For the pit image, one image was misclassified as a crack and three images were misclassified as lack of penetration. For the lack of penetration image, one image was misclassified as a crack, one image was misclassified as a pit, and one image was misclassified as having no defects. For gas pores, six were misclassified as cracks. For no defects, one image was identified as having a lack of penetration. This section will utilize the reduced dimensional feature vectors obtained through PCA as classification samples, establishing a classification model for weld defects based on the BP neural network and SVM. The experimental results showed that the overall recognition rate of the BP classification model was 90.8%, and the overall recognition rate of the SVM classification model was 91.6%. During the experiment, MO images were continuously acquired along the weld seam, sometimes only local characteristics of the weld defects were captured, which also led to some pores to be misclassified as cracks, there is an obvious misclassification between these two types of samples. Therefore, the classification accuracy of pores is relatively low in the BP and SVM models. Due to the significant differences between no defects and pit and other samples, the recognition rate in both BP and SVM models can reach over 92%.

4. Detection and Classification by CNN and ResNet50

4.1. The Architecture of CNN

CNN is a type of deep learning neural network that applies convolution operations on different parts of the input image to extract features and generate output, and then downsamples through pooling operations to reduce computational complexity and overfitting [47–51]. Then, the extracted features are transported to a fully connected layer, which classifies or regresses these features. The structural flowchart of a CNN model is shown in Figure 10. The primary components of a CNN consist of the input layers, convolutional layers, pooling layers, fully connected layers, and the output layers. CNN is a mainstream method in modern image processing and computer vision fields. Compared with traditional machine learning methods, CNN can automatically extract hierarchical features to improve efficiency.

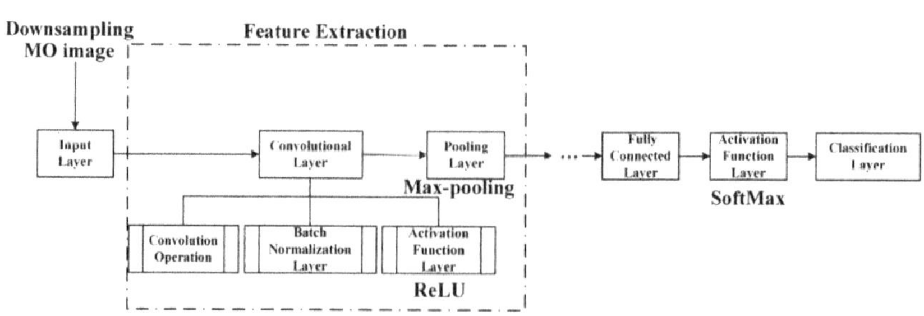

Figure 10. The structure flowchart of a CNN model.

Let an image of size $[W_1 \times H_1 \times C_1]$ be input, where H_1, W_1, and C_1 represent the height, width, and number of channels of the image, respectively, defining the image matrix as $[W_1 \times H_1 \times C_1]$. Consider a pixel located at (i, j), that feeds into the convolutional layer; this layer consists of multiple convolutional kernels. Each kernel performs convolution operations on the input image to extract various features. Let the size of each convolutional kernel be $K \times K \times C_1$, where $C_1 = 3$ means that the input has three channels. The convolution kernel has C_2, and the size of the output feature map is $W_2 \times H_2 \times C_2$. The convolution operation is expressed by Equation (22):

$$z_{i,j,k} = \sum_{c=1}^{C_1} \sum_{p=1}^{K} \sum_{q=1}^{K} w_{(p,q,c,k)} x_{(i+p,j+q,c)} + b_k \tag{22}$$

where $z_{i,j,k}$ denotes the value at position (i, j) in the k-th channel of the output feature map, and k corresponds to the number of kernels and $k \leq C_2$. The term $w_{(p,q,c,k)}$ represents the weight for the k-th channel of the convolutional kernel at row p and column q of the c-th input channel, and $x_{(i+p,j+q,c)}$ indicates the pixel value at position $(i + p, j + q)$ in the c-th channel of the input image. Consequently, for an input image with c channels and k convolutional kernels, there will be a total of $c \times k$ weight tensors utilized in the convolution operation. The term b_k signifies the bias for the k channel.

The image after convolution operation enters a batch normalization (BN) layer, which standardizes the output of the convolution layer to improve training efficiency and model stability. The function of the activation layer is to introduce nonlinear factors and increase the expressive power of the neural network, which can be expressed as follows:

$$a_{i,j,k} = f\left(y_{i,j,k}\right) \tag{23}$$

where $f(\cdot)$ is the activation function, and the commonly used activation functions include sigmoid, tanh, ReLU, etc. The ReLU is used as the activation function of the convolution layer, which is expressed as follows:

$$f(x) = \begin{cases} 0, x < 0 \\ x, x \geq 0 \end{cases} \tag{24}$$

After the convolution layer operation, it is followed directly by a pooling layer, which performs downsampling on the feature map, thereby reducing both the number of parameters and computational complexity while enhancing the model's robustness. The pooling layer is typically categorized into max pooling and average pooling. This article adopts max pooling, and the pooling layer is represented as follows:

$$P_{i,j,k} = \max_{p,q}\left(a_{i \times s+p, j \times s+q, k}\right) \tag{25}$$

where $P_{i,j,k}$ represents the feature value at the i-th row and j-th column of the k-th channel output after pooling, with k as the size of the pooling window. The max function extracts the maximum value from a specified region. Variables p and q represent the row and column offsets of the pooling kernel in the input tensor, respectively. s is the step size, usually taken as 2.

After being processed by multiple convolutional and pooling layers, the output feature map will be flattened into a one-dimensional vector and entered into the fully connected layer for classification or regression tasks. The output formula of the fully connected layer is expressed as:

$$F = WP + b \tag{26}$$

where W represents the weight matrix with dimensions $(N,\ M)$, and N is the number of neurons in the previous layer, and M denotes the number of neurons in the current layer. P corresponds to the vector obtained from the pooling layer after flattening into $(N,\ 1)$, and b signifies the bias vector with a size of $(M,\ 1)$, while F is the output vector.

Following the fully connected layer, an activation function layer is applied, and this article uses the SoftMax activation function. Finally, the process culminates in the output layer for classification tasks. In convolutional neural networks, the cross-entropy loss function is used to measure the difference between the model's predicted values and the true values. By minimizing the cross-entropy loss function, the prediction results of the model can be made closer to the true labels, thereby improving the accuracy of the model.

During training, a batch of data is used to calculate the loss function, and the gradient of the loss function for each trainable parameter is calculated for each image in this batch. This article uses the stochastic gradient descent (SGD) algorithm, which updates parameters using only one batch of data at a time [52]. For each trainable parameter, w_i, update its current value, $w_{i,t}$, as follows:

$$w_{i,t+1} = w_{i,t} - \eta \frac{1}{m} \sum_{i=1}^{m} \frac{\partial L_i}{\partial w_i} \tag{27}$$

where η represents the learning rate, m is the number of images in the batch, and L_i is the loss function for the i-th image, while $\frac{\partial L_i}{\partial w_i}$ is the gradient of the loss function of the i-th image with respect to w_i.

This paper adopts a learning rate scheduling method, that is, the learning rate is dynamically adjusted during the training to improve the performance of the model, which is expressed as follows:

$$\eta = \eta_0 d_r^{floor(\frac{g_s}{d_s})} \tag{28}$$

where η_0 represents the initial learning rate, d_r denotes the decay rate, which is constrained between 0 and 1. g_s signifies the number of steps in the current iteration, and d_s refers to the count of decay steps. The floor( ) function represents the process of rounding down. By managing the decay rate and the number of decay steps, it is feasible to reduce the learning rate, allowing for a higher initial learning rate to be established.

4.2. Parameter Evaluation of CNN Models

To achieve optimal image classification performance, it is necessary to conduct experiments on the various structural parameters of the CNN prediction model in the classification of weld defect's MO image, as well as to evaluate its accuracy and adaptability in classification tasks. The size of the convolution kernel directly influences the receptive field and the quality of the output features in deep convolutional neural networks. A larger convolution kernel size can obtain a smaller receptive field, and the output single feature can contain more information. However, a large size of the convolution kernel may lead to a surge in model computation, and the network is more difficult to train and optimize. Therefore, a network model with five convolutional layers, five pooling layers, and one fully connected layer was used in the experiment, as shown in Figure 11.

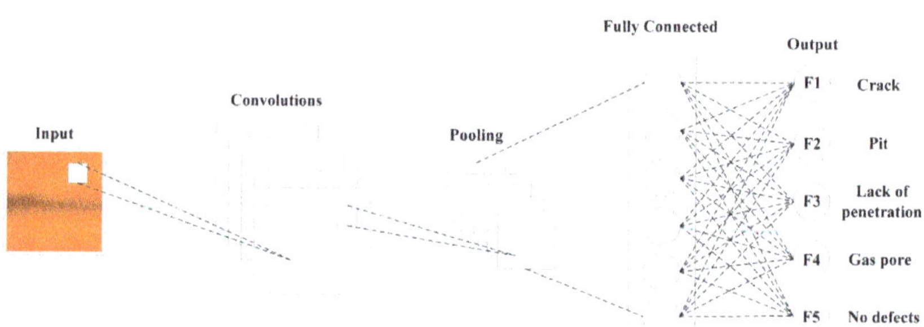

Figure 11. The CNN model for defect MO images.

The effect of different structural parameters on the model is analyzed by comparing the effect of the convolution kernel size $K \times K$ in the first layer on the classification results. The odd-size convolution kernel has the advantages of facilitating the aggregation of image features, improving the robustness of the model, and easily expanding the network, so $K = [3, 5, 7, 9, 11]$ is selected. In deep learning, accuracy and loss value are two commonly used evaluation indicators, which are used to measure the quality of training and model performance. The higher the accuracy and the smaller the loss value, the better the fit of the model. Figures 12 and 13 show the training accuracy and loss values obtained under different convolution kernel sizes, respectively.

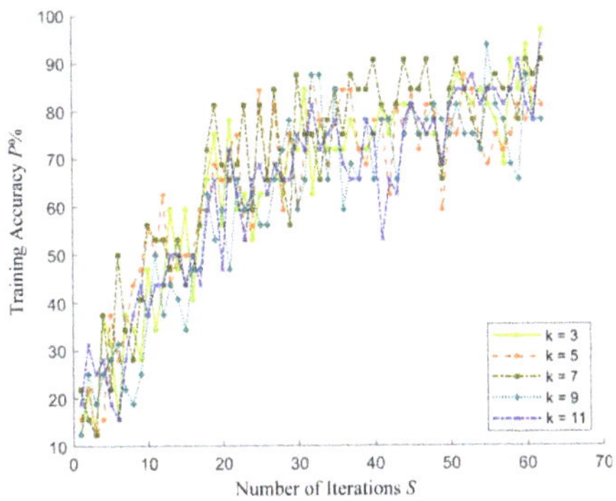

Figure 12. Training accuracy for different convolution kernel sizes.

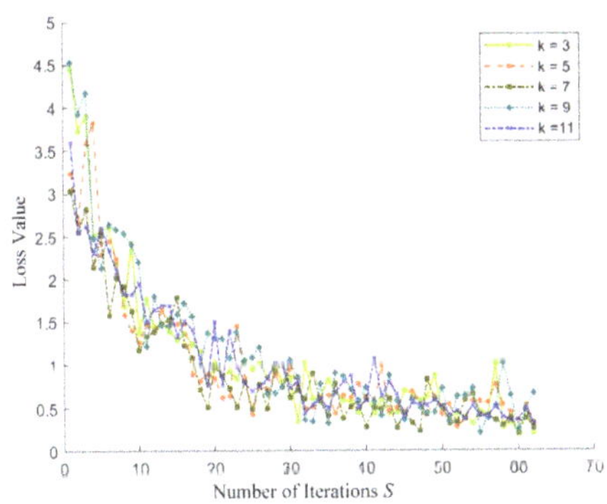

Figure 13. The training loss values associated with different convolution kernel sizes.

When the number of iterations, S, reaches 30, the accuracy of training reaches more than 70%, and the loss value also drops to around 1. Therefore, the intercept interval 30–62 is taken, and the standard deviation of the accuracy and loss value of training with different convolution kernel sizes is calculated, respectively, as shown in Table 6.

Table 6. Training results of different convolution kernels.

Convolution Kernel Size	Accuracy Standard Deviation	Loss Value Standard Deviation	Training Set Accuracy %
3×3	0.073	0.207	93.2
5×5	0.073	0.160	96.4
7×7	0.067	0.175	97.2
9×9	0.082	0.216	92.8
11×11	0.084	0.170	97.6

The standard deviation of accuracy for the 7 × 7 kernel size is the lowest, while the 5 × 5 kernel size shows the lowest standard deviation for loss values. A smaller standard deviation in accuracy and loss indicates a more stable optimization process, reflecting the stability and reliability of the model. Although the 11 × 11 convolution kernel has the highest accuracy, its standard deviation of accuracy is high, and the fitting effect between the recognition model and the training set is poor. Therefore, considering all factors, the optimal size for the convolution kernel in the first layer should be 7 × 7.

In a neural network architecture, the excitation layer plays a key role in performing nonlinear transformation on the output of the convolutional layer. Without the excitation layer, no matter how complex the network architecture is, its final output will only be a linear superposition of the input, and it is difficult to capture the inherent law contained in the complex dataset. The essence of the excitation layer lies in the application of activation functions, such as Sigmoid function, Tanh function, and ReLU function, which are widely used activation functions. The slope of the activation function directly affects the speed of parameter iteration update, and the higher the slope, the faster the parameter update. The Sigmoid function and Tanh function have saturation zones, which can prevent parameter explosion during the training process. The Relu function has no saturation zone, and its advantage is that it can ensure the update iteration speed within the positive interval. Different activation functions have their own characteristics in various classification models, and the differences are significant. Therefore, based on the above three typical activation functions, the performance of them in the MO image classification model of weld defects is analyzed, and the corresponding training accuracy and loss values are shown in Figures 14 and 15. As can be seen from the figures, the loss value and accuracy of Sigmoid are poor. Although Tanh can achieve faster convergence, it is less stable and will produce large fluctuations. Therefore, ReLU activation function is more suitable for this classification model.

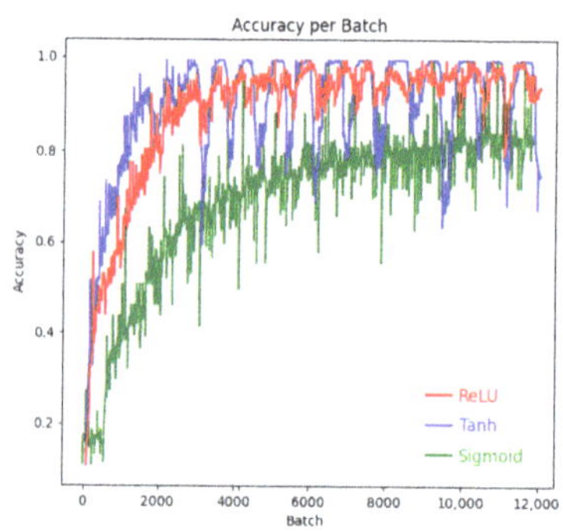

Figure 14. Training accuracies for three activation functions.

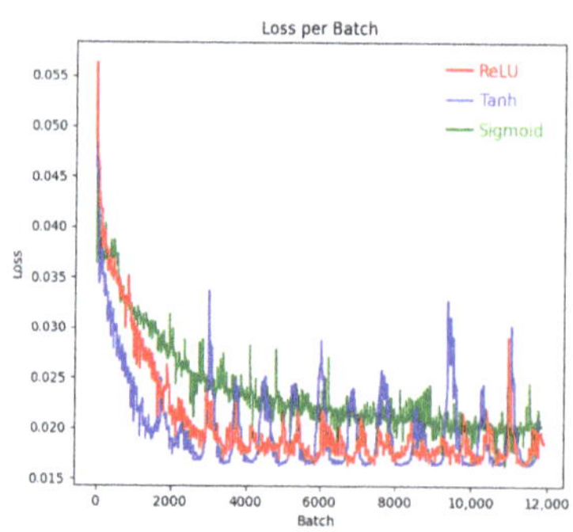

Figure 15. The loss values for the three activation functions.

4.3. Classification by the CNN Model

According to the analysis in Section 4.2, the size of the first convolution kernel is 7 × 7, and the weights and biases of the convolution and fully connected layers are initialized using a normal distribution with a mean of 0 and a variance of 1. The scaling factor of the batch normalization layer is initialized to 1, and the bias is initialized to 0. The retention probability of the dropout layer is initialized to 0.5. Due to the significant impact of training parameters on model prediction performance, the optimal training parameters for the model were determined through experiments. This article uses the Adam optimizer for model training. The Adam optimizer combines momentum and adaptive learning rates, allowing for rapid convergence and effectively reducing the oscillation of parameter updates. Its adaptive mechanism provides each parameter with a unique learning rate, which is especially advantageous when handling high-dimensional data and complex models. Additionally, the bias correction function of the Adam optimizer enhances the stability during the initial training phase. Therefore, the choice of the Adam optimizer ensures efficient and stable optimization of the model throughout the entire training process. The primary initial parameters of the model are shown in Table 7.

Table 7. Initial parameters of the convolution neural network model.

Batch Size	Initial Learning Rate	Learning Rate Decay Factor	Number of Learning Rate Decay Steps	Number of Iterations
32	0.0002	0.1	1	62

Table 8 shows the operational configuration parameters of the computer. The structure of the CNN model for classifying these MO images of weld defects is illustrated in Table 9.

Table 8. Parameters of computer operation configuration.

Deep Learning Framework	CPU	RAM	GPU	Operating Environment	Programming Language
Pytorch2.1.1	Intel(R)Core (TM)i7-10875 H	16 GB	NVIDIA GeForceRTX 2060	Anaconda 3	Python3.11.5

Table 9. Structural parameters of CNN model.

Layers	Types	Input Size	Filter Size	Number of Filters	Stride	Weights	Biases
I0	Input Layer	50 × 50 × 3	—	—	—		—
C1	Convolution Layer 1	50 × 50 × 3	7 × 7 × 3	32	1	7 × 7 × 3 × 32	1 × 1 × 32
P2	Pooling Layer 1	50 × 50 × 32	2 × 2	—	2	—	—
C3	Convolution Layer 2	25 × 25 × 32	3 × 3 × 32	64	1	3 × 3 × 32 × 64	1 × 1 × 64
P4	Pooling Layer 2	25 × 25 × 64	2 × 2	—	2	—	—
C5	Convolution Layer 3	12 × 12 × 64	3 × 3 × 64	128	1	3 × 3 × 64 × 128	1 × 1 × 128
P6	Pooling Layer 3	12 × 12 × 128	2 × 2	—	2	—	—
C7	Convolution Layer 4	6 × 6 × 128	3 × 3 × 128	256	1	3 × 3 × 128 × 256	1 × 1 × 256
P8	Pooling Layer 4	6 × 6 × 256	2 × 2	—	2	—	—
C9	Convolution Layer 5	3 × 3 × 256	3 × 3 × 256	512	1	3 × 3 × 256 × 512	1 × 1 × 512
P10	Pooling Layer 5	3 × 3 × 512	2 × 2	—	2	—	—
F11	Fully Connected Layer	1 × 1 × 512	—	—	—	5 × 512	5 × 1
S12	Classification Layer	1 × 1 × 5	—	—	—	—	—

The experiment used 5000 MO images of weld defects as classification samples, of which 4000 were used as training set. In the training set, there were 800 samples each of crack, pit, lack of penetration, gas pore, and no defects. After undergoing noise reduction and downsampling to 50 × 50 pixels, these images were used as the input for training a CNN model. The training accuracy and loss value of the CNN model are shown in Figure 16. In order to improve the generalization ability of the model, K-fold cross-validation method was adopted to divide the dataset. In this cross-validation method, the dataset was first divided into K mutually exclusive subsets of similar sizes, and $K-1$ subsets were trained each time, and the remaining subsets were used for testing. The classification performance of the CNN on the training set is shown in Table 10. The CNN model has the best classification performance for lack of penetration, reaching 100%, with an overall classification accuracy of 97.2%. The classification results demonstrate that the model can effectively identify cracks, pits, incomplete fusion, gas pores, and no defects in the weldments. Therefore, it is feasible to use a multilayer CNN for the classification of weld defect MO images. Figure 17 shows the confusion matrix of the classification results. The classification effect of pores was the worst, and five pores were identified as cracks, indicating that the similarity between MO images of defects may cause confusion.

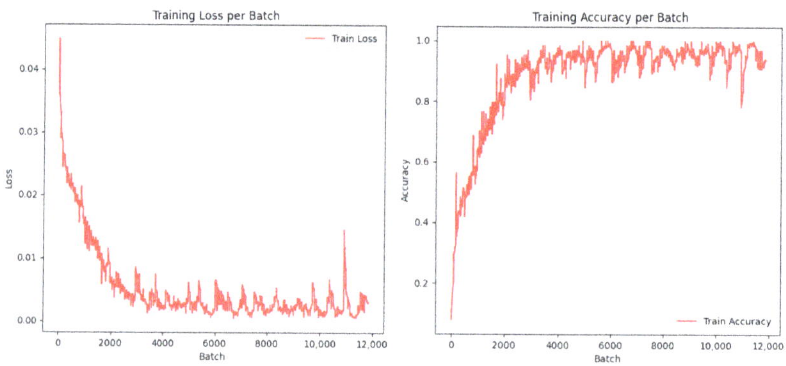

Figure 16. Training accuracy and loss value of CNN model.

Table 10. Classification results of the CNN model for MO images of weld defects.

Defect Types	Number of Images	Train Samples	Valid Samples	Test Samples	Recognition Result	Classification Accuracy/%
Crack	1000	800	100	100	97	97
Pit	1000	800	100	100	97	97
Lack of penetration	1000	800	100	100	100	100
Gas pore	1000	800	100	100	94	94
No defects	1000	800	100	100	98	98
Total	5000	4000	500	500	486	97.2

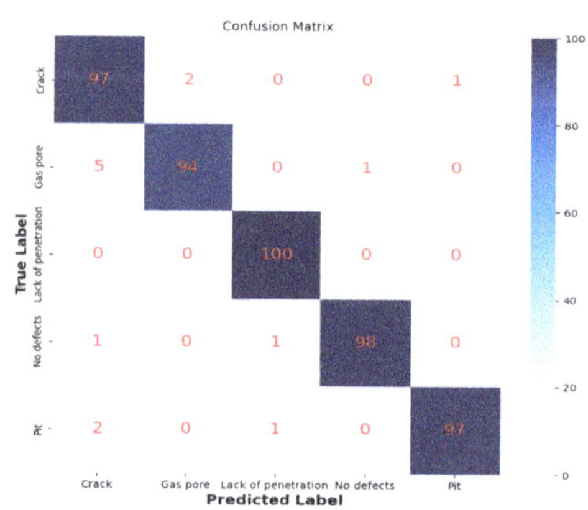

Figure 17. Confusion Matrix for CNN Classification.

4.4. Classification by ResNet50 Model

ResNet is an important improvement in comparison with traditional convolutional neural networks. Introducing the residual unit makes it easier for deep networks to learn the identity mapping, thus solving the degradation problem in the training process of deep neural networks. Compared with ordinary stacked deep neural networks, it is easier to optimize and obtain higher accuracy, thus showing superior performance in image classification. The residual module is the key of the ResNet network. Its structure is shown in Figure 18. After input x is obtained by convolution operation of the main line, the module performs addition operation with the eigenmatrix in the identity mapping. If the original base mapping that the network needs to fit is $F(x)$, then there is $Y = F(x) + x$; that is, the residual module makes it so that the network does not need to fit the real base mapping, but only needs to fit the offset $F(x)$ based on the original input identity mapping. So, there is no need to introduce additional parameters, reduce the computing burden of the network, or make the network performance at least not worse than the original. The ResNet network model has evolved into five structures, such as ResNet18, ResNet34, ResNet50, ResNet101, and ResNet152. With the increase in the number of network layers, the calculation accuracy of ResNet network is gradually improved. The number of computation loads and parameters also increases with the depth of the network. Considering accuracy, parameter counting, and calculation load, the ResNet50 network was used in this study [53]. During the training process, RTX2060 is used to classify and predict MO images of weld defects.

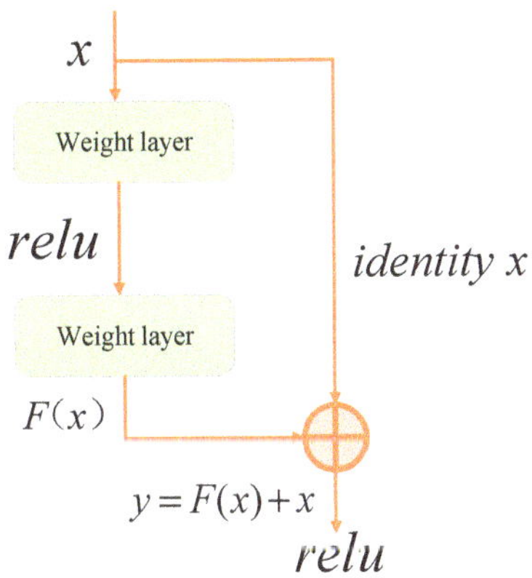

Figure 18. Simplified diagram of the residual module structure.

The experimental dataset is consistent with that in Section 4.3. A total of 4000 images were used as the training set, 500 as the validation set, and 500 as the test set. After noise reduction and downsampling to 50 × 50 pixels, these images were used as the input for training ResNet50. Model weights were updated using optimization techniques, and the model was gradually adapted to a specific classification task through repeated iterative training, and the performance of the model was evaluated on the test set. Figure 19 shows the loss value and accuracy of the ResNet50 training process. According to the data in Figure 19, the accuracy remains at a high level, which indicates that the ResNet50 network model has strong classification ability. The optimal recognition rate is obtained by constantly debugging the training times. The higher the training times, the better. The confusion matrix of the ResNet50 model is shown in Figure 20. The classification results of the ResNet50 model for MO images of weld defects are shown in Table 11. The overall classification performance is excellent, and the classification effect of incomplete fusion and gas pore is the best, both reaching 100%, and the overall classification accuracy is 99%.

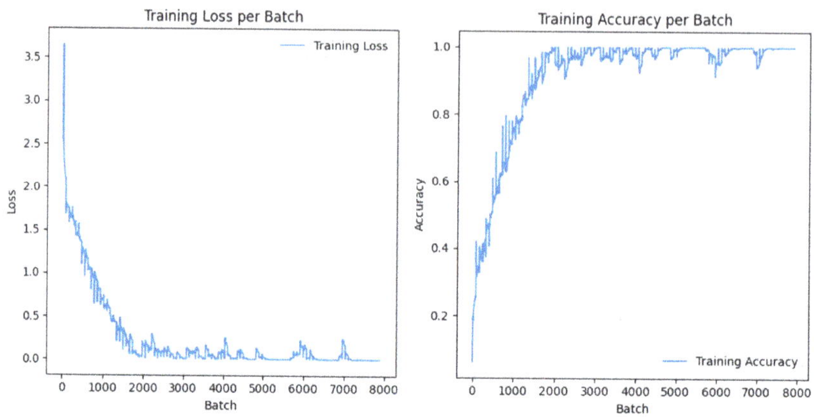

Figure 19. Training accuracy and loss value of ResNet50 model.

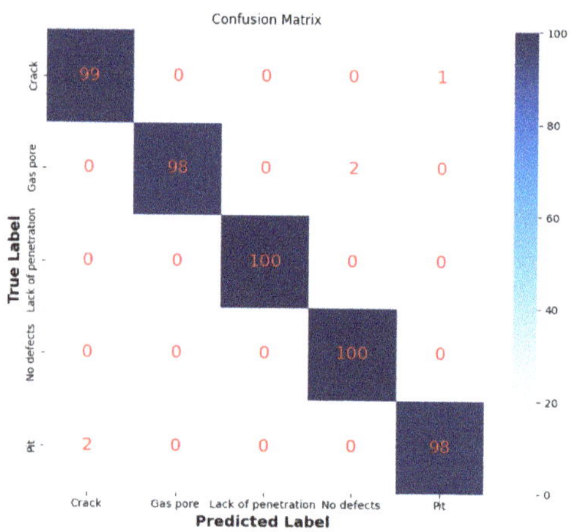

Figure 20. Confusion matrix for ResNet50 classification.

Table 11. Classification results of the ResNet50 model for MO images of weld defects.

Defect Types	Number of Images	Train Samples	Valid Samples	Test Samples	Recognition Result	Classification Accuracy/%
Crack	1000	800	100	100	99	99
Pit	1000	800	100	100	98	98
Lack of penetration	1000	800	100	100	100	100
Gas pore	1000	800	100	100	98	98
No defects	1000	800	100	100	100	100
Total	5000	4000	500	500	495	99

4.5. Experimental Analysis

The classification methods proposed in this article include image downsampling, feature extraction, and PCA-BP, PCA-SVM, CNN, and ResNet50 classification models. The classification results for MO images of weld defects are shown in Tables 4, 5, 10 and 11. The classification accuracy of the PCA-BP model is 90.8%. Since there are significant differences between cracks and no defects compared to other defects, the recognition rates can reach 96% and 98%, respectively. PCA-SVM shows a recognition rate of 94% for incomplete fusion and 98% for no defects. The recognition rates of cracks and gas pores are relatively low, at 86% and 88%, respectively. The classification accuracy of the PCA-SVM model can reach 91.6%, which is 0.8% higher than the PCA-BP model. However, the classification accuracy of gas pores in BP and SVM models is relatively low, which affects the overall classification accuracy. In Table 10, CNN achieves a recognition rate of 100% for MO images of incomplete fusion, demonstrating the model's advantage in processing specific categories. However, the recognition rate for gas pores in MO images is the lowest at just 94%. Compared with the PCA-BP and PCA-SVM models, the classification accuracy of gas pores in the CNN model increased by 16% and 6%, respectively. The overall classification accuracy of the CNN model reaches 97.2%. Compared with PCA-BP and PCA-SVM models, the classification accuracy has been improved by 6.4% and 5.6%, respectively. Experimental results show that compared with the BP neural network model, the overall classification accuracy of the SVM classification model under alternating magnetic field reaches 91.6%, indicating that the SVM model has higher classification accuracy in small samples and high-dimensional pattern recognition. In large samples, the CNN and ResNet50 classification models are less prone to overfitting, and their classification accuracy has been greatly improved. These results highlight the effectiveness of different neural network models in classifying MO images of weld defects and emphasize the importance of deep learning methods in classification accuracy.

As shown in Figure 11, the ResNet50 model achieved classification accuracies of 99% for cracks and 98.0% for pores, surpassing the CNN model by 2% and 4%, respectively. The overall classification accuracy is 1.8% higher than the CNN model. This performance improvement may stem from the advantages of the ResNet50 model in feature extraction, particularly its effectiveness in capturing high-level features when processing complex images. In a deeper analysis of the experimental results, we observed that the classification accuracy for the four types of MO images (cracks, pits, lack of penetration, and no defects) exceeded 95% in both CNN and ResNet50 models, demonstrating good classification performance. The experimental results demonstrate the effectiveness of different neural network models in the classification of weld defect MO images, indicating that ResNet50 has certain advantages in this respect. The comparison of the four proposed algorithms is shown in Table 12. The parameters and floating-point operations per second (FLOPS) of the PCA-BP and PCA-SVM classification models are much lower than those of the CNN and ResNet50 models, which is due to their relatively simple network architectures. Despite the high computational complexity of CNN and ResNet50 deep learning models, they have performance advantages. Especially, the significant increase in accuracy is sufficient to compensate for the burden of computational complexity, which makes the ResNet50 model the best choice. These findings provide an important basis for further research on the identification and processing of MO images of weld defects. The forthcoming study will focus on improving the accuracy of target classification, especially in defect MO images with complex backgrounds, to further enhance the overall robustness of the system.

Table 12. Comparison of the four proposed algorithms.

Methods	Params	FLOPS	Classification Accuracy/%	Complexity
PCA-BP	6.12×10^3	6.17×10^3	90.8	Low
PCA-SVM	3.01×10^5	1.5×10^7	91.6	Intermediate
CNN	1.94×10^6	3.76×10^9	97.2	High
ResNet50	2.35×10^7	1.34×10^{10}	99	Highest

5. Conclusions and Outlook

The MO images of natural weld defects are obtained by the nondestructive testing system excited by an alternating magnetic field. The defect MO images featuring crack, pit, incomplete fusion, gas pore, and no defects have been acquired for weld defect diagnosis. After image enhancement and feature extraction of the collected defect MO images, PCA-

BP model, PCA-SVM model, CNN model, and ResNet50 model are used to classify the MO images of natural weld defects.

Different image filtering techniques of weld defect images are compared, and the filtering effect of Gaussian filter, bilateral filter, and median filter is evaluated by mean square error. At the same time, the dimensionality-reduced feature vectors obtained by PCA are used as the input layers of BP neural network classification model and SVM classification model. The experimental results show that the overall recognition rate of the PCA-BP neural network classification model is 90.8%. The classification accuracy of the PCA-SVM model can reach 91.6%, which is 0.8% higher than the PCA-BP model. However, these two models have poor recognition performance for cracks and gas pores.

The CNN and ResNet50 classification models for MO imaging of natural weld defects are designed, and the forward propagation and iterative optimization process of the CNN model are studied. The experimental results show that the classification accuracy of the ResNet50 model is higher than that of the PCA-SVM model and CNN model, with an overall classification accuracy of 99%. Its total classification accuracy has increased by 7.4% and 1.8%, respectively. In particular, the classification accuracy of the gas pores is 10% and 4% higher than the PCA-SVM model and the CNN model, respectively, indicating that ResNet50 model can effectively improve the classification accuracy of MO images for natural welded defect. Therefore, the proposed method in this paper can provide high-precision diagnostic results for natural welded defects.

At present, the MO imaging detection technology of natural weld defects depends on the magnetic field excitation device, and different excitation modes have different imaging effects. Especially in the actual process of weld defect detection, there are many external factors that affect the effect of defect MO imaging, which require more theoretical research, simulation analysis, and practical experimental verification. Next, it is necessary to further develop a magnetic field excitation device and introduce frequency conversion function into the self-made excitation device to realize the MO imaging technology to detect weld defects at different frequencies, so as to meet the needs of automatic online detection. In future work, the author of this paper aims to investigate effective algorithms for automatically optimizing the parameters of the deep learning method and improved methods for reducing background noise to enhance accuracy. In addition, the method proposed in this article lays the foundation for extending the detection of welded defects using MO imaging technology under alternating magnetic field excitation to weld defect detection under rotating magnetic field excitation.

Author Contributions: Data curation, Y.L. (Yanfeng Li), Y.L. (Yongbiao Luo) and G.L.; Funding acquisition, Y.L. (Yanfeng Li), Y.Z., W.X., C.X. and P.G.; Methodology, Y.L. (Yanfeng Li), W.X., Y.Z., C.X. and J.C.; Project administration, Y.L. (Yanfeng Li), P.G., Y.L. (Yongbiao Luo) and Y.Z.; Writing—original draft, Y.L. (Yanfeng Li), J.C. and X.L.; Writing—review and editing, Y.L. (Yanfeng Li), G.L., X.L., Y.L. (Yongbiao Luo), C.X. and J.C.; Conceptualization, Y.L. (Yanfeng Li), W.X. and Y.Z.; Supervision, Y.L. (Yanfeng Li), G.L., C.X. and W.X. All authors have read and agreed to the published version of the manuscript.

Funding: This research was funded by the Opening Project of Key Laboratory of Ministry of Education (Cultivation), Southwest University of Science and Technology, grant number 24kfsk04; Guangdong Provincial Natural Science Foundation of China, grant number 2023A1515012172; Science and Technology Program of Guangzhou, grant number 202201011405; Guangzhou Municipal Special Fund Project for Scientific and Technological Innovation and Development, grant number 2023B03J1326, 202002030147; Special Talents for Scientific Research Projects of Guangdong Polytechnic Normal University, grant number 2021SDKYA018, 991641218; Guangdong Province Key Construction Discipline Research Ability Improvement Project, grant number 2021ZDJS027.

Institutional Review Board Statement: Not applicable.

Informed Consent Statement: Not applicable.

Data Availability Statement: The original contributions presented in the study are included in the article, further inquiries can be directed to the corresponding authors.

Conflicts of Interest: The authors declare no conflicts of interest.

References

1. Gao, X.; Wang, Y.; Chen, Z.; Ma, B.; Zhang, Y. Analysis of welding process stability and weld quality by droplet transfer and explosion in MAG-laser hybrid welding process. *J. Manuf. Process.* **2018**, *32*, 522–529. [CrossRef]
2. Gao, X.; Wen, Q.; Katayama, S. Analysis of high-power disk laser welding stability based on classification of plume and spatter characteristics. *Trans. Nonferrous Met. Soc. China* **2013**, *23*, 3748–3757. [CrossRef]
3. Zhang, Y.; Gao, X.; You, D.; Zhang, N. Data-driven detection of laser welding defects based on real-time spectrometer signals. *IEEE Sens. J.* **2019**, *19*, 9364–9373. [CrossRef]
4. Wang, L.; Mohammadpour, M.; Yang, B.; Gao, X.; Lavoie, J.P.; Kleine, K.; Kovacevic, R. Monitoring of keyhole entrance and molten pool with quality analysis during adjustable ring mode laser welding. *Appl. Opt.* **2020**, *59*, 1576–1584. [CrossRef] [PubMed]
5. Hawwat, S.; Shah, J.; Wang, H. Machine learning supported ultrasonic testing for characterization of cracks in polyethylene pipes. *Measurement* **2025**, *240*, 115609. [CrossRef]
6. Chen, L.; Li, B.; Zhang, L.; Shang, Z. 3D positioning of defects for gas turbine blades based on digital radiographic projective imaging. *NDT E Int.* **2023**, *133*, 102751. [CrossRef]
7. Gao, X.; Mo, L.; Xiao, Z.; Chen, X.; Katayama, S. Seam tracking based on Kalman filtering of micro-gap weld using magneto-optical image. *Int. J. Adv. Manuf. Technol.* **2016**, *83*, 21–32. [CrossRef]
8. Gao, X.; Chen, Y.; You, D.; Xiao, Z.; Chen, X. Detection of micro gap weld joint by using magneto-optical imaging and Kalman filtering compensated with RBF neural network. *Mech. Syst. Signal Process.* **2017**, *84*, 570–583. [CrossRef]
9. Gao, X.; Du, L.; Ma, N.; Zhou, X. Magneto-optical imaging characteristics of weld defects under alternating and rotating magnetic field excitation. *Opt. Laser Technol.* **2019**, *112*, 188–197. [CrossRef]
10. Arakelyan, S.; Galstyan, O.; Lee, H.; Babajanyan, A.; Lee, J.H.; Friedman, B.; Lee, K. Direct current imaging using a magneto-optical sensor. *Sens. Actuators A Phys.* **2016**, *238*, 397–401. [CrossRef]
11. Cacciola, M.; Megali, G.; Pellicano, D. Heuristic enhancement of magneto-optical images for NDE. *EURASIP J. Adv. Signal Process.* **2010**, *2010*, 44–55. [CrossRef]
12. Li, Y.; Gao, X.; Zhang, Y.; You, D.; Zhang, N.; Wang, C.; Wang, C. Detection model of invisible weld defects by magneto-optical imaging at rotating magnetic field directions. *Opt. Laser Technol.* **2020**, *121*, 105772. [CrossRef]
13. Gao, X.; Ma, N.; Du, L. Magneto-optical imaging characteristics of weld defects under alternating magnetic field excitation. *Opt. Express* **2018**, *26*, 9972–9983. [CrossRef] [PubMed]
14. Su, G.; Qin, Y.; Xu, H.; Liang, J. Automatic real-time crack detection using lightweight deep learning models. *Eng. Appl. Artif. Intell.* **2024**, *138*, 109340. [CrossRef]
15. Zhao, B.; Zhou, X.; Yang, G.; Wen, J.; Zhang, J.; Dou, J.; Li, G.; Chen, X.; Chen, B. High-resolution infrastructure defect detection dataset sourced by unmanned systems and validated with deep learning. *Autom. Constr.* **2024**, *163*, 105405. [CrossRef]
16. Yu, H.; Deng, Y.; Guo, F. Real-time pavement surface crack detection based on lightweight semantic segmentation model. *Transp. Geotech.* **2024**, *48*, 101335. [CrossRef]
17. Jin, G.S.; Oh, S.J.; Lee, Y.S.; Shin, S.C. Extracting weld bead shapes from radiographic testing images with U-Net. *Appl. Sci.* **2021**, *11*, 12051. [CrossRef]
18. Golodov, V.A.; Maltseva, A.A. Approach to weld segmentation and defect classification in radiographic images of pipe welds. *NDT E Int.* **2022**, *127*, 102597. [CrossRef]
19. Zapata, J.; Vilar, R.; Ruiz, R. Automatic inspection system of welding radiographic images based on ANN under a regularisation process. *J. Nondestruct. Eval.* **2012**, *31*, 34–45. [CrossRef]
20. Malarvel, M.; Singh, H. An autonomous technique for weld defects detection and classification using multi-class support vector machine in X-radiography image. *Optik* **2021**, *231*, 166342. [CrossRef]
21. Shen, Q.; Gao, J.; Li, C. Automatic classification of weld defects in radiographic image. *Insight Non-Destr. Test. Cond. Monit.* **2010**, *52*, 134–139. [CrossRef]
22. Kumaresan, S.; Jai Aultrin, K.S.; Kumar, S.S.; Anand, M.D. Transfer learning with CNN for classification of weld defect. *IEEE Access* **2021**, *9*, 95097–95108. [CrossRef]
23. Li, Z.; Li, Y.; Liu, Y.; Wang, P.; Lu, R.; Gooi, H.B. Deep learning based densely connected network for load Forecasting. *IEEE Trans. Power Syst.* **2021**, *36*, 2829–2840. [CrossRef]
24. Singh, A.; Bruzzone, L. Mono- and Dual-Regulated Contractive-Expansive-Contractive deep convolutional networks for classification of Multispectral Remote sensing images. *IEEE Geosci. Remote Sens. Lett.* **2022**, *19*, 5513605. [CrossRef]
25. Mohanasundari, L. Performance analysis of weld image classification using modified Resnet CNN architecture. *Turk. J. Comput. Math. Educ. (TURCOMAT)* **2021**, *12*, 2260–2266.
26. Chinta, B.; Moorthi, M. EEG-dependent automatic speech recognition using deep residual encoder based VGG net CNN. *Comput. Speech Lang.* **2023**, *79*, 101477. [CrossRef]

27. Slama, A.; Sahli, H.; Amri, Y.; Trabelsi, H. Res-Net-VGG19: Improved tumor segmentation using MR images based on Res-Net architecture and efficient VGG gliomas grading. *Appl. Eng. Sci.* **2023**, *16*, 100153. [CrossRef]
28. Palma-Ramírez, D.; Ross-Veitía, B.D.; Font-Ariosa, P.; Espinel-Hernández, A.; Sanchez-Roca, A.; Carvajal-Fals, H.; Hernández-Herrera, H. Deep convolutional neural network for weld defect classification in radiographic images. *Heliyon* **2024**, *10*, e30590. [CrossRef]
29. Yang, D.; Jiang, H.; Ai, S.; Yang, T.; Zhi, Z.; Jing, D.; Gao, J.; Yue, K.; Cheng, H.; Xu, Y. Detection method for weld defects in time-of-flight diffraction images based on multi-image fusion and feature hybrid enhancement. *Eng. Appl. Artif. Intell.* **2024**, *138*, 109442. [CrossRef]
30. Zhang, Q.; Zhang, K.; Pan, K.; Huang, W. Image defect classification of surface mount technology welding based on the improved ResNet model. *J. Eng. Res.* **2024**, *12*, 154–162. [CrossRef]
31. Zhao, D.; Cai, W.; Cui, L. Adaptive thresholding and coordinate attention-based tree-inspired network for aero-engine bearing health monitoring under strong noise. *Adv. Eng. Inform.* **2024**, *61*, 102559. [CrossRef]
32. Zhao, D.; Shao, D.; Cui, L. CTNet: A data-driven time-frequency technique for wind turbines fault diagnosis under time-varying speeds. *ISA Trans.* **2024**, *154*, 335–351. [CrossRef] [PubMed]
33. Chandni; Sachdeva, M.; Kushwaha, A.K.S. Effective Brain Tumor Image Classification using Deep Learning. *Natl. Acad. Sci. Lett.* **2024**, *47*, 257–260. [CrossRef]
34. He, X.; Wang, T.; Wu, K.; Liu, H. Automatic defects detection and classification of low carbon steel WAAM products using improved remanence/magneto-optical imaging and cost-sensitive convolutional neural network. *Measurement* **2020**, *173*, 108633. [CrossRef]
35. Li, Y.; Gao, X.; Liu, J.; Zhang, Y.; Qu, M. Detection and classification of invisible weld defects by magneto-optical imaging under alternating magnetic field excitation. *Sens. Actuators A Phys.* **2024**, *374*, 115507. [CrossRef]
36. Liu, Q.; Ye, G.; Gao, X.; Zhang, Y.; Gao, P.P. Magneto-optical imaging nondestructive testing of welding defects based on image fusion. *NDT E Int. Indep. Nondestruct. Test. Eval.* **2023**, *138*, 102887. [CrossRef]
37. Shen, Y.; Wang, Y.; Wu, B.; Li, P.; Han, Z.; Zhang, C.; Liu, X. A novel sensor based on the composite mechanism of magnetic flux leakage and magnetic field disturbance for comprehensive inspection of defects with varying angles and width. *NDT E Int.* **2024**, *145*, 103131. [CrossRef]
38. Vandendriessche, S.; Valev, V.K.; Verbiest, T. Faraday rotation and its dispersion in the visible region for saturated organic liquids. *Phys. Chem. Chem. Phys.* **2012**, *14*, 1860–1864. [CrossRef]
39. Cheng, Y.; Deng, Y.; Bai, L.; Chen, K. Enhanced laser-based magneto-optic imaging system for nondestructive evaluation applications. *IEEE Trans. Instrum. Meas.* **2013**, *62*, 1192–1198. [CrossRef]
40. Pradhan, K.; Patra, S. A semantic edge-aware parameter efficient image filtering technique. *Comput. Graph.* **2024**, *124*, 104068. [CrossRef]
41. Zhang, C.; Gai, K.; Zhang, S. Matrix normal PCA for interpretable dimension reduction and graphical noise modeling. *Pattern Recognit.* **2024**, *154*, 110591. [CrossRef]
42. Bisheh, H.; Amiri, G. Structural damage detection based on variational mode decomposition and kernel PCA-based support vector machine. *Eng. Struct.* **2023**, *278*, 115565. [CrossRef]
43. Hu, D.; Liu, H.; Zhu, Y.; Sun, J.; Zhang, Z.; Yang, L.; Liu, Q.; Yang, B. Demand response-oriented virtual power plant evaluation based on AdaBoost and BP neural network. *Energy Rep.* **2023**, *9*, 922–931. [CrossRef]
44. Liu, Z.; Yuan, J.; Shen, J.; Hu, Y.; Chen, S. A new DEM calibration method for wet and stick materials based on the BP neural network. *Powder Technol.* **2024**, *448*, 120228. [CrossRef]
45. Hayati, R.; Munawar, A.A.; Lukitaningsih, E.; Earlia, N.; Karma, T. Combination of PCA with LDA and SVM classifiers: A model for determining the geographical origin of coconut in the coastal plantation. *Aceh Prov. Indones. Case Stud. Chem. Environ. Eng.* **2024**, *9*, 100552. [CrossRef]
46. Zhang, G.; Carrasco, C.; Winsler, K.; Bahle, B.; Cong, F.; Luck, S. Assessing the effectiveness of spatial PCA on SVM-based decoding of EEG data. *Neuro Image* **2024**, *293*, 120625. [CrossRef] [PubMed]
47. Zhang, S.; Wu, J.; Shi, E.; Yu, S.; Gao, Y.; Li, L.C.; Kuo, L.R.; Pomeroy, M.J.; Liang, Z.J. MM-GLCM-CNN: A multi-scale and multi-level based GLCM-CNN for polyp classification. *Comput. Med. Imaging Graph.* **2023**, *108*, 102257. [CrossRef]
48. Wang, J.J.; Sharma, A.K.; Liu, S.H.; Zhang, H.; Chen, W.; Lee, T.L. Prediction of vascular access stenosis by lightweight convolutional neural network using blood flow sound signals. *Sensors* **2024**, *24*, 5922. [CrossRef]
49. Zheng, X.; Zhang, L.; Xu, C.; Chen, X.; Cui, Z. An attribution graph-based interpretable method for CNNs. *Neural Netw.* **2024**, *179*, 106597. [CrossRef]
50. Mohajelin, F.; Sheykhivand, S.; Shabani, A.; Danishvar, M.; Danishvar, S.; Lahijan, L.Z. Automatic recognition of multiple emotional classes from EEG signals through the use of graph theory and convolutional neural networks. *Sensors* **2024**, *24*, 5883. [CrossRef]
51. Xie, G.; Wang, L.; Williams, R.A.; Li, Y.; Zhang, P.; Gu, S. Segmentation of wood CT images for internal defects detection based on CNN: A comparative study. *Comput. Electron. Agric.* **2024**, *224*, 109244. [CrossRef]

52. Vrbancic, G.; Podgorelec, V. Efficient ensemble for image-based identification of Pneumonia utilizing deep CNN and SGD with warm restarts. *Expert Syst. Appl.* **2022**, *187*, 115834. [CrossRef]
53. Yu, Q.; Zhang, Y.; Xu, J.; Zhao, Y.; Zhou, Y. Intelligent damage classification for tensile membrane structure based on continuous wavelet transform and improved ResNet50. *Measurement* **2024**, *227*, 114260. [CrossRef]

MDPI AG
Grosspeteranlage 5
4052 Basel
Switzerland
Tel.: +41 61 683 77 34

Sensors Editorial Office
E-mail: sensors@mdpi.com
www.mdpi.com/journal/sensors

www.ingramcontent.com/pod-product-compliance
Lightning Source LLC
Chambersburg PA
CBHW040814250926
51850CB00007BA/1167
* 9 7 8 3 7 2 5 8 6 7 8 8 2 *